Current Research in Circulating Tumor Cells

Current Research in Circulating Tumor Cells

Editor: Jonah Randall

FOSTER ACADEMICS

www.fosteracademics.com

www.fosteracademics.com

FOSTER ACADEMICS

Cataloging-in-Publication Data

Current research in circulating tumor cells / edited by Jonah Randall.
p. cm.
Includes bibliographical references and index.
ISBN 978-1-64646-561-3
1. Cancer cells. 2. Cancer cells--Research. 3. Cancer--Research. 4. Tumor markers.
5. Metastasis. I. Randall, Jonah.
RC269.7 .C87 2023
616.994 07--dc23

Foster Academics,
118-35 Queens Blvd., Suite 400,
Forest Hills, NY 11375, USA

ISBN 978-1-64646-561-3 (Hardback)

Contents

Preface

Circulating tumor cells (CTCs) are a rare subset of cells identified in the blood of patients having solid tumor. It serves as a seed for metastases. Cancer cells generally metastasize in two ways in the bloodstream, through single migratory CTCs or multicellular groupings. They imitate tumor properties and sustain tumor heterogeneity. They are also potentially useful as preclinical models, clinical biomarkers, and therapeutic targets. Furthermore, they have a potential clinical use as a component of liquid biopsy, where they can be used for early diagnosis, prognosis prediction, disease monitoring, molecular phenotyping and assessing therapy response. CTCs are also suitable for developing preclinical models, particularly 3D organoid cultures that might be used in tumor immunity, disease modelling, organoid biobanks, drug screening and genome editing. This book explores the status of current research on circulating tumor cells. Those in search of information to further their knowledge will be greatly assisted by it.

This book is a result of research of several months to collate the most relevant data in the field.

When I was approached with the idea of this book and the proposal to edit it, I was overwhelmed. It gave me an opportunity to reach out to all those who share a common interest with me in this field. I had 3 main parameters for editing this text:

1. Accuracy – The data and information provided in this book should be up-to-date and valuable to the readers.

2. Structure – The data must be presented in a structured format for easy understanding and better grasping of the readers.

3. Universal Approach – This book not only targets students but also experts and innovators in the field, thus my aim was to present topics which are of use to all.

Thus, it took me a couple of months to finish the editing of this book.

I would like to make a special mention of my publisher who considered me worthy of this opportunity and also supported me throughout the editing process. I would also like to thank the editing team at the back-end who extended their help whenever required.

Editor

Single-Cell Analysis of Circulating Tumor Cells: Why Heterogeneity Matters

Su Bin Lim [1,2], Chwee Teck Lim [1,2,3,4] and Wan-Teck Lim [5,6,7,*]

1 NUS Graduate School for Integrative Sciences & Engineering, National University of Singapore, Singapore 117456, Singapore; sblim@u.nus.edu (S.B.L.); ctlim@nus.edu.sg (C.T.L.)
2 Department of Biomedical Engineering, National University of Singapore, Singapore 117583, Singapore
3 Mechanobiology Institute, National University of Singapore, Singapore 117411, Singapore
4 Institute for Health Innovation and Technology (iHealthtech), National University of Singapore, Singapore 117599, Singapore
5 Division of Medical Oncology, National Cancer Centre Singapore, Singapore 169610, Singapore
6 Office of Academic and Clinical Development, Duke-NUS Medical School, Singapore 169857, Singapore
7 IMCB NCC MPI Singapore Oncogenome Laboratory, Institute of Molecular and Cell Biology (IMCB), Agency for Science, Technology and Research (A*STAR), Singapore 138673, Singapore
* Correspondence: darren.lim.w.t@singhealth.com.sg.

Abstract: Unlike bulk-cell analysis, single-cell approaches have the advantage of assessing cellular heterogeneity that governs key aspects of tumor biology. Yet, their applications to circulating tumor cells (CTCs) are relatively limited, due mainly to the technical challenges resulting from extreme rarity of CTCs. Nevertheless, recent advances in microfluidics and immunoaffinity enrichment technologies along with sequencing platforms have fueled studies aiming to enrich, isolate, and sequence whole genomes of CTCs with high fidelity across various malignancies. Here, we review recent single-cell CTC (scCTC) sequencing efforts, and the integrated workflows, that have successfully characterized patient-derived CTCs. We examine how these studies uncover DNA alterations occurring at multiple molecular levels ranging from point mutations to chromosomal rearrangements from a single CTC, and discuss their cellular heterogeneity and clinical consequences. Finally, we highlight emerging strategies to address key challenges currently limiting the translation of these findings to clinical practice.

Keywords: single-cell analysis; cellular heterogeneity; circulating tumor cells

1. Introduction

The concept of intratumoral heterogeneity (ITH), first described in 1982 by Fidler and Hart [1], has been expanded to include genetic, phenotypic and functional heterogeneity within tumors comprising diverse malignant and non-malignant subpopulations. With accumulations of mutations in DNA damage checkpoint control genes and DNA repair genes, divergent cancer clones may evolve and propagate over time through selection processes driven by constantly changing microenvironment and by the use of therapy [2,3]. In addition to genetic heterogeneity in clonal mutations and subclonal de novo mutations, functional heterogeneity related to developmental pathways and epigenetic programs and spatial variability in tumor microenvironment contribute to ITH, which governs key aspects of tumor biology, including tumor invasion, metastasis, and drug resistance [2].

Recent years have also seen the contribution of non-malignant cells, such as stromal fibroblasts, immune cells, bone-marrow-derived cells, mesenchymal stem cells, and endothelial cells, to ITH within a tumor. Malignant and benign cells interact locally through a complex network of extracellular matrix

(ECM) and the related components, collectively termed as matrisome, which has been linked to tumor progression, response to adjuvant therapy, and immune system [4–7], driving tumor phenotypes supporting their metastatic competence. The tumor cells can be shed passively (through corrupted blood vessel) and/or actively (through epithelial–mesenchymal transition (EMT)) from tumors into the circulation, referred to as circulating tumor cells (CTCs), in which only a minority of cell populations survive under the physiological blood flow and succeed in early colonization phases [8]. Despite the low frequency of occurrence, CTCs allow for repeated sampling, which is not clinically practical for tissue biopsy, and may thus be an excellent tool for assessing tumor heterogeneity and for revealing clonal diversity underlying resistance to treatment [9].

The prevalence of phenotypic plasticity involving programmed death-ligand 1 (PD-L1) [10], stemness [11], drug resistance [12], and EMT [13], within CTC populations has fueled investigation of cellular heterogeneity using single-cell high-throughput enrichment and sequencing platforms over the past five years. Emerging single-cell RNA sequencing (scRNA-seq) data further suggest that CTCs may interact with hematopoietic cells or platelets in blood through direct and/or indirect routes, adding another layer of heterogeneity [14]. Bulk-cell approaches traditionally deployed in existing CTC literature, however, have provided insights into cellular processes averaged throughout the enriched blood sample, which largely comprises leukocytes, or white blood cells (WBCs). The leukocyte contamination is inevitable in any given primarily enriched sample due in part to extremely rare CTCs occurring at a frequency of ~1 in 10^7 WBCs in blood from a cancer patient [15] and the relatively low cell capture efficiency of existing cell sorting technologies, which are further limited to isolating only certain CTC subpopulations (refer to the Section 2.1 for further details). It is therefore essential to reach single-cell resolution to precisely characterize CTCs at the genomic level and further to investigate the clinical impact of cellular heterogeneity present within CTC populations.

2. Methods and Technologies

Despite technical challenges, single-cell CTC (scCTC) analyses have so far revealed genomic variations specific to each CTC, including mutations that are not yet present in the Catalogue of Somatic Mutations in Cancer (COSMIC) database [16–20] or subclonal alterations that are not easily discernible from tissue biopsies [16,17], cells of origin of cancer (e.g., bone marrow-derived multiple myeloma) [21], or with bulk-cell approaches [22], providing comprehensive landscape of evolving tumor cells. Such private genomic variations shared by CTCs may represent "CTC phenotypes", including intravasation competency, increased migration/motility, enhanced cell–cell interactions, variation in energy metabolism, interaction with platelet and blood immune cells, resistance to anoikis, and resistance to therapy [19,23]. Single nucleotide variations (SNVs) and/or copy number variations (CNVs) in these putative precursors of metastasis present early in clonal evolution or in tumor progression may be excellent targets for therapeutic intervention. Examined below are scCTC DNA sequencing studies that have successfully assessed DNA alterations in patient-derived CTCs across various cancer types (Table 1).

Table 1. Summary of single-cell circulating tumor cells (scCTC) sequencing studies that analyzed DNA alterations in patient-derived CTCs.

CTC Enrichment	Single-Cell Isolation	CTC Criteria	WGA	Profiling	Investigated Genes	Genomic Analysis	Number of Single CTCs (Patients) [1]	Ref.
Prostate cancer								
MagSweeper	CellCelector	DAPI− CD45− EpCAM+	MDA	NGS	All	SNVs	42 (5)	[24]
Epic Sciences CTC Platform	Micromanipulation	CD45− CK+/−	DOP-PCR	NGS	All	CNVs, LSTs	67 (7)	[25]
NanoVelcro CTC Chip	LCM	CD45− CK+	MDA	NGS, Sanger, aCGH	All	SNVs, SVs, CNVs	12 (1)	[26]
HD-CTC Assay	Micromanipulation	DAPI+ CK+ CD45−	LA-PCR	NGS	All	CNVs	41 (1)	[27]
CellSearch, Spectra Optia Apheresis System	FACS	DAPI+ CK+ CD45−	LA-PCR	aCGH	All	CNVs	205 (14)	[17]
Breast cancer								
MagSweeper	Micromanipulation	DAPI+ CK+ CD45−	No WGA	Sanger	PIK3CA	SNVs	185 (11)	[28]
CellSearch	DEPArray	DAPI+ CK+ CD45−	LA-PCR	Sanger	TP53	SNVs	11 (2)	[29]
CellSearch	DEPArray	DAPI+ CK+ CD45−	LA-PCR	Sanger	PIK3CA	SNVs	241 (43)	[30]
				qPCR	HER2	CNVs	192 (42)	
				aCGH	All	CNVs	37 (15)	
CellSearch	DEPArray	DAPI+ CK+ CD45− CD34−	LA-PCR	Sanger	PIK3CA	SNVs	115 (18)	[18]
CellSearch	DEPArray	DAPI+ CK+ CD45−	LA-PCR	Targeted NGS	50 cancer-related genes	SNVs	14 (4)	[31]
Leukapheresis, CellSearch	Micromanipulation	CK+ CD45−	DOP-PCR	CGH	All	CNVs	65 (19)	[32]
				aCGH	All	CNVs		
CellSearch	MoFlo XDP flow-sorting	DAPI+ CK+ CD45−	LA-PCR	qPCR	CCND1 locus	CNVs	26 (12)	[33]
				Sanger	PIK3CA, TP53	SNVs		
FACS	DEPArray	DAPI− CD45− EpCAM− CD44+ CD24− uPAR+/− intβ1+/−	LA-PCR	MassARRAY	>200 hallmark cancer genes	SNVs	7 (-)	[34]
CellSearch	Micromanipulation	DAPI+ CK+ CD45−	MDA, LA-PCR	Sanger	PIK3CA	SNVs	114 (33)	[35]
CellSearch	DEPArray	DAPI+ CK+ CD45−	LA-PCR	Targeted NGS, ddPCR	50 COSMIC genes	SNVs	40 (5)	[36]
CellSearch	CellCelector	DAPI+ CK+ CD45−	LA-PCR	Targeted NGS	50 COSMIC genes	SNVs	7 (2)	[37]
CellSearch	DEPArray	DAPI+ CK+ CD45−	LA-PCR	Sanger	TP53, HER2, PIK3CA, RB1	SNVs	24 (6)	[38]
Ficoll Separation	Micromanipulation	DAPI+ CK+ CD45−	LA-PCR	Sanger	ESR1	SNVs	8 (4)	[39]
AutoMACS Classic Separator	LCM	CK+ CD45−	LA-PCR	SNP Array	All	CNVs	17 (17)	[23]
CellSearch, Oncoquick	CellCelector	EpCAM+ CD45−	MDA	aCGH, targeted NGS	All	SNVs, CNVs	31 (1)	[40]
oHSV-hTERT-GFP method	FACS	CD45− hTERT+	MALBAC	NGS	All	SNVs, CNVs	11 (8)	[19]
ScreenCell	DEPArray	DAPI+ CK+ CD45−	LA-PCR	Sanger	TP53, ESR1	SNVs	7 (1)	[41]
ClearCell FX System	Manipulation	DAPI+ CK+ CD45−	MALBAC	NGS	All	SNVs	3 (1)	[20]

Table 1. *Cont.*

CTC Enrichment	Single-Cell Isolation	CTC Criteria	WGA	Profiling	Investigated Genes	Genomic Analysis	Number of Single CTCs (Patients) [1]	Ref.
Lung cancer								
CellSearch	DEPArray	DAPI+ CK+ CD45−	LA-PCR	NGS	All	CNVs	72 (13)	[42]
CellSearch	Micromanipulation	DAPI+ CK+ CD45−	MALBAC	NGS, digital PCR, Sanger	All	SNVs, INDELs CNVs	24 (4) 61 (11)	[43]
ClearCell FX System	Microfluidic chip	DAPI+ CK+ CD45−	No WGA	Sanger	EGFR	SNVs	26 (7)	[44]
CellSearch	Micromanipulation	DAPI+ CK+ CD45−	MALBAC	NGS	All	SNVs, INDELs, CNVs, SVs	97 (23)	[45]
CellSearch	Micromanipulation	DAPI+ CK+ CD45−	MALBAC	NGS	All	SNVs, INDELs, CNVs, SVs	91 (10)	[46]
MagSifter	Single-Nanowell Assay	DAPI+ CK+ CD45− TERT+ MET+	No WGA	Multiplex PCR	EGFR	SNVs	202 (7)	[22]
Colorectal cancer								
CellSearch	Micromanipulation	EpCAM+ CD45− CK+	LA-PCR	aCGH Sanger	All KRAS, BRAF, TP53	CNVs SNVs	8 (8) 126 (31)	[47]
CellSearch	Micromanipulation	DAPI+ CK+ CD45−	LA-PCR, MDA	Multiplex PCR aCGH Targeted NGS	NCI/ICG-HNPCC marker panel All 68 colorectal cancer-associated genes	MSI CNVs SNVs	122 (30) 37 (6) 8 (2)	[16]
CellSearch	Micromanipulation	DAPI+ CK+ CD45−	LA-PCR, MDA	qPCR Sanger	EGFR KRAS, BRAF, PIK3CA	CNVs SNVs	26 (3) 69 (5)	[48]
Oncoquick	DEPArray	HOECHST+ CK+ CD45−	LA-PCR	Sanger, pyrosequencing	KRAS	SNVs	- (16)	[49]
Melanoma								
Dynabeads	LCM	HMW− MAA+ CD45− MART-1/gp100+	No WGA	Sanger	BRAF KIT	SNVs SNVs	14 (9) 4 (4)	[50]
Dielectrophoretic microwell array	Micromanipulation	CD45− MART-1/gp100+	No WGA	Sanger	BRAF	SNVs	33 (1)	[51]
Multiple myeloma								
MACS beads	Micromanipulation	CD45−CD138+	MDA	Targeted NGS	35 most commonly mutated loci	SNVs	203 (10)	[21]
Epic Sciences CTC Platform	Micromanipulation	CK+/− CD45−	DOP-PCR	NGS	All	CNVs	9 (1)	[52]
Pancreatic cancer								
NanoVelcro Chip	LCM	HOECHST+ CD45− CK/CEA+	MDA	Sanger	KRAS	SNVs	60 (12)	[53]

[1] QC-passed CTCs that have been included in the final analysis.

2.1. CTC Enrichment

Once collected, blood samples are subjected to CTC enrichment using density gradient centrifugation, 2D/3D microfiltration, microfluidic devices, or immunoaffinity-based technologies (Figure 1; refer to Ref. [14] for detailed summary and comparison of existing CTC enrichment technologies). While CellSearch® remains the choice of primary enrichment tool in scCTC sequencing studies, such immunoaffinity-based enrichment technology relying on epithelial cell surface markers (e.g., EpCAM or CKs) have varying capture efficiency depending on the degree of EMT, stemness, and the resulting differentiation cell state. $EpCAM^{+}CK^{+}$ cells have also been detected by CellSearch® in patients with benign diseases, but in lower frequency compared to the cancer group [54]. Clinical data supporting the metastatic competence of $EpCAM^{-}$ CTCs in numerous studies [34,55] have further fueled the transition in the field to the development of label-free approaches leveraging on biophysical properties of CTCs (e.g., size, density, stiffness).

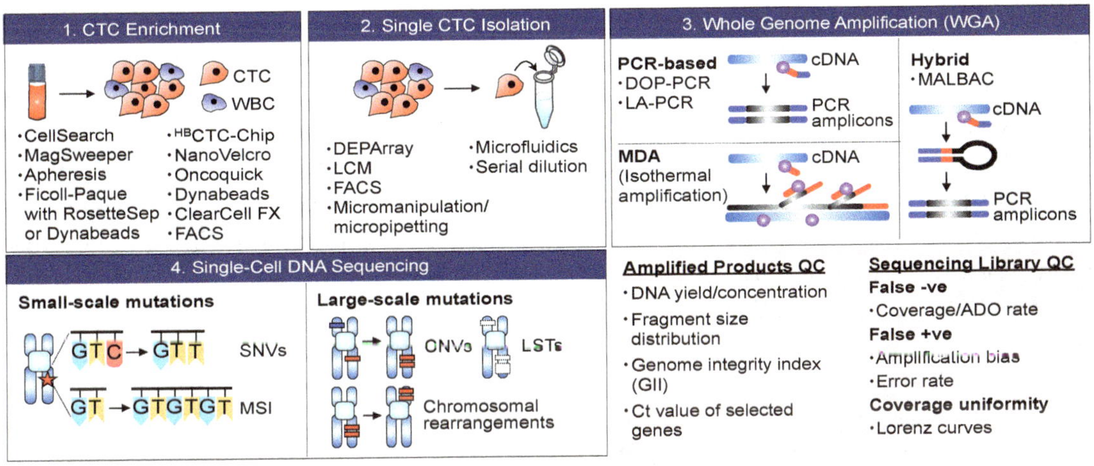

Figure 1. The standard workflow and existing technologies for scCTC sequencing. CTCs are primarily enriched from a whole blood sample, and are isolated as single cells for subsequent downstream molecular analyses. For genomic analysis of whole genome/exome, whole-genome amplification (WGA) is performed and amplified DNA products are QC-checked prior to sequencing.

One of our pioneering efforts in developing such marker-independent technology was the application of inertial microfluidics using spiral microchannels, in which depth and width of each channel can be designed to control the positioning of cells or microparticles in suspension via Dean flow fractionation (DFF). Our technology, so-called ClearCell® FX, enriches intact and viable CTCs from the peripheral blood of cancer patients in a fully automated and high-throughput fashion, with a reported >80% sensitivity and specificity in detecting CTCs from clinical samples (refer to [56] for further details on the system). Mechanobiologically inspired enrichment platforms achieve high sensitivity and are not limited to certain CTC subpopulations, yet the purity of enriched samples may be compromised due to their size distribution overlapping with leukocytes, as observed in breast, colorectal, and prostate cancer [57]. To best interrogate CTCs and their cellular heterogeneity, cell enrichment technology should thus be carefully selected for unbiased capture and recovery of different CTC subpopulations.

2.2. Single-Cell Isolation

CTC enrichment technologies are used in conjunction with microscopic micromanipulators (e.g., Eppendorf Transfer Man NK2/4 micromanipulator, CellCelector™), laser capture microdissection (LCM) [26], microfluidic devices [44], or DEPArray™ to recover putative CTCs as single cells. Enriched cells are often fixed and stained with the nucleic acid dye DAPI and monoclonal antibodies specific to epithelial cell surface marker cytokeratin (CK) and leukocyte marker CD45, and manually selected by the trained and skilled operator based on DAPI and CK positivity and CD45 negativity. Morphology of the captured cells (e.g., cell shape, size) are concurrently assessed to identify cells having 4 to 40 μm diameter, round or oval shape, and/or high nuclear/cytoplasm ratio [37,39,49].

Even though dozens of CTCs might have been captured in CTC enrichment step, only very few single CTCs (<10 CTCs) have been isolated and transferred successfully to a PCR tube per sample [4,19,49]. Such low recovery rate could be attributed to apoptotic CTCs, which are often excluded from the isolation in most scCTC sequencing studies. Consistent with the classic definition of apoptosis, these cells are defined as CK^+CD45^- CTCs with non-intact nuclei having DAPI pattern of chromosomal condensation and/or nuclear fragmentation and blebbing [52]. Given that low burden of apoptotic CTCs has been associated with poor prognosis and aggressive phenotypes across several cancer types [52,58], characterization of CTC apoptosis in situ may facilitate the development of new platform for real-time monitoring of antitumor drug efficacy.

2.3. *Whole-Genome Amplification (WGA)*

WGA is an active area of development with wide applications to study rare tumor cells or single-celled organisms, such as bacteria and archaea. There exist different WGA approaches based on specific, degenerate, and/or hybrid primers: Linker-adapter PCR (LA-PCR), interspersed repetitive sequence PCR (IRS-PCR), primer extension preamplification (PEP-PCR), degenerate oligonucleotide primed PCR (DOP-PCR), displacement degenerate oligonucleotide primed PCR (D-DOP-PCR), multiple displacement amplification (MDA), single primer isothermal amplification (SPIA), and multiple annealing and looping-based amplification cycles (MALBAC; refer to ref. [59] for a comprehensive summary on working principles and characteristics of each WGA method).

Briefly, LA-PCR and IRS-PCR methods utilize specific primers, each amplifying digested DNA litigated to adapter fragments and repeating sequence elements, respectively, while the rest of the methods, such as PEP-PCR, DOP-PCR, and D-DOP-PCR, are based on the use of degenerated primers. MDA has been commonly applied to single-cell sequencing of microorganisms as well as patient-derived CTCs. It employs a unique polymerase with strong strand displacement activity (e.g., phi29 DNA polymerase), which can amplify fragments of up to 100 kb with high replication fidelity compared to purely PCR-based (e.g., Taq polymerase) methods [59]. MALBAC has been proposed as a hybrid PCR/MDA method, relying on two relatively error-prone DNA polymerases, *Bst* DNA polymerase and Taq DNA polymerase, for isothermal strand displacement and PCR, respectively. Each WGA technique has its own advantages and limitations in terms of sensitivity, specificity, uniformity, and amplification bias. For example, while LA-PCR, DOP-PCR, and MALBAC may be the choice of method for detection of CNVs but not SNVs, MDA (REPLI-g™) has proven to be most sensitive in detecting mutations at a single-base resolution compared to LA-PCR methods (GenomePlex™, Ampli1™) [60].

The challenge is that the yield of amplified DNA varies significantly across CTCs, where the success rate of amplification ranges from 11% to 100% [24,61], and WGA step itself is subjected to coverage biases and errors, such as preferential allelic amplification, GC bias, dropout events, and nucleotide copy errors [60]. To account for such variability, studies have established an additional QC step prior to in-depth sequencing to probe only CTCs with yields of DNA greater than negative controls [24] or a fixed concentration level [27] or those showing specific bands corresponding to targets of interest on the Agilent 2100 Bioanalyzer [19,29]. The author-defined QC assays have also been developed to identify CTCs suited for single-cell targeted sequencing and analysis. For example, genome integrity index (GII), which is determined from detectable PCR bands corresponding to three Mse fragments and KRAS fragment, has been proven to be predictive of successful analysis of sequence-based molecular changes, including point mutations, gene amplifications, and CNVs [30,36,42].

2.4. *Sequencing and Profiling*

Amplified DNA samples are subjected to library preparation and quantification. To date, scCTC studies have most commonly employed next-generation sequencing (NGS), Sanger sequencing, single nucleotide polymorphism (SNP), and array comparative genomic hybridization (aCGH) platforms, and conventional PCR technologies to analyze somatic SNVs, structural variations, (SVs), CNVs, and chromosomal breakpoints and rearrangements for whole exome/genome or selected cancer-associated genes, often comparatively with matched primary tumors and/or metastatic tissues or disseminated tumor cells (DTCs).

In the library QC step, the sequencing depth, percentage of area covered, homogeneity of coverage, and/or SNP densities are assessed to only select high-quality CTC libraries based on author-defined assessment techniques, such as autocorrelation analysis [24] and Lorenz curves [26]. Fluorimetric assays (e.g., Fluorometer) and analytical tool provided by the sequencing platform (e.g., Torrent Suite) may also be used to quantify DNA samples and to assess the performance of sequencing runs and the quality of generated data, respectively [19,31,37]. In some cases, the variants identified by NGS were specifically selected and further validated by Sanger sequencing [31,45] or digital droplet PCR (ddPCR) [36] using the same samples.

The sequence queried in single CTCs in prior studies vary from small-scale mutations (<1 kb) to large-scale mutations (1 kb–100 Mb). Targeting larger regions may come with the trade-off of increased number of false variant calls and sequencing costs and reduced number of individual cells to be sequenced [62]. Nevertheless, whole-genome sequencing (WGS) allows new discoveries of genomic variations occurring even in non-coding regions that may add significant values to the analysis of rare tumor cells.

3. CTC Heterogeneity and Clinical Impact

While resolving cellular heterogeneity, single-cell approaches may link specific CTC subpopulation programs to cancer cell phenotypes, metastasis, patient outcomes, and drug resistance, as demonstrated by recent studies. Examined below are genomic aberrations commonly analyzed in CTCs and their clinical impact (Figure 2). Clinical data derived from scCTC transcriptomic analyses are discussed elsewhere [14].

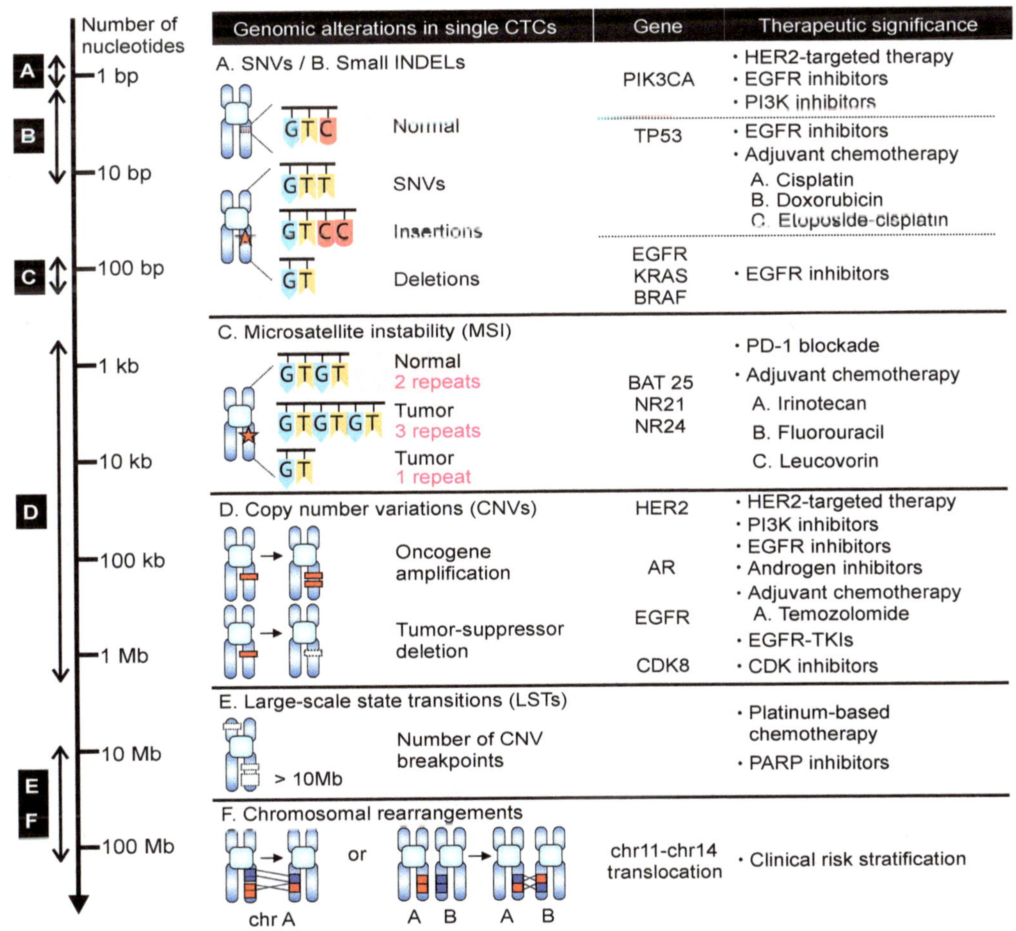

Figure 2. Summary of genomic alterations found in scCTC sequencing studies.

3.1. *Single Nucelotide Variation (SNV)*

3.1.1. PIK3CA

PIK3CA is a gene harboring major driver mutations in many cancer types [63,64]. Its mutational status has increasingly been recognized as a promising predictor of resistance to targeted therapies [65]. In breast cancer, tumors harboring PIK3CA mutations are often resistant to HER2-based therapy [66–68], and are less likely to achieve pathologic complete response to anti-HER2 treatments [69,70]. Though limited to the analysis of EpCAM-expressing CTCs, scCTC studies have applied targeted sequencing approaches to examine mutational hotspots, most commonly in exon 9 and 20 [16,18,28,30,35,37,48,71]. The assessment of pre-existing resistant clones through scCTC analysis prior to the administration of HER2-based therapies has been suggested to be of clinical significance for patients harboring CTCs with HER2 amplification and double-mutant PIK3CA/HER2 [30]. Longitudinal monitoring of therapy response through HER2 mutational analysis of CTCs in this subset of patients will be of particular clinical interest, given the known drug efficacy of PIK3CA pathway inhibitors in patients with HER2$^+$ primary tumors [72],

PIK3CA mutational status in CTCs indicative of resistance against HER2-targeted therapy has been demonstrated for HER2$^-$ metastatic breast cancer patients screened for German multicentric phase III trial (i.e., DETECT III study) harboring HER2$^+$ CTCs [35]. Further, studies have noted a high degree of intrapatient cellular heterogeneity and discordant PIK3CA status between CTCs and matched primary tumors [18,28,35,36], of which PIK3CA was one of two genes (among >2200 COSMIC mutations analyzed) frequently mutated in CTCs, cfDNA, and matched primary tumor in HER2$^-$ breast cancer [36]. PIK3CA mutation has also been implicated in drug resistance of EGFR tyrosine kinase inhibitor (TKI) treatment. Notably, a lung cancer patient harboring PIK3CA mutation in almost all CTCs (7/8 CTCs) but not in primary tumor (low abundance) had progressive disease and presented distant metastasis after one month of treatment with erlotinib. Early detection of such resistant cells in such a less invasive way may thus be tremendously useful in drug selection.

3.1.2. TP53

TP53 is a tumor suppressor gene frequently mutated in most human cancers [73]. TP53 mutations have functional implications in key molecular events in tumor progression, such as EMT [74], stemness [75], cancer prognosis, and survival outcomes [76]. Highly heterogeneous TP53 mutational status was observed across individual CTCs in prostate [24], lung [43], colorectal [16], and breast cancer [29,31,36,38,41]. In metastatic prostate cancer, ubiquitous TP53 mutations were found among multiple foci of the primary tumor and metastases, suggesting divergent cancer evolution from a single ancestor cancer [24].

In breast cancer, TP53 harbored the highest number of mutations across CTCs [31]. Mutant TP53 p.R273C has been associated with cisplatin chemotherapy resistance [77]. Concurrent mutations of RB1 and TP53 genes were also found in the majority of CTCs from a lung cancer patient who experienced a phenotypic transition from adenocarcinoma to small-cell lung cancer (SCLC) [43]. Notably, dramatic clinical response was observed in this patient upon etoposide-cisplatin treatment, which is a standard chemotherapy for SCLC patients [78,79]. These studies altogether highlight how scCTC profiling may provide an early signal of phenotypic transition in tumor to guide new therapeutic regimen.

3.1.3. EGFR

Despite promising efficacy of EGFR in multiple cancer types [80], prediction of response against EGFR inhibition still remains ambiguous. The mere assessment of EGFR expression at the DNA or protein level using bulk primary tumor samples has not been an ideal indicator for predicting the response to anti-EGFR drugs [81,82]. Whole-exome sequencing of lung CTCs revealed specific INDEL in the EGFR gene (p.Lys746_Ala750del) shared by primary tumors and metastases, which could be targeted with TKIs [43]. In our earlier work, we reported highly sensitive detection of EGFR mutations

(T790M and L868R) in microfluidically enriched CTCs, which showed a complete concordance of mutation status with matched primary tumors in non-small-cell lung cancer (NSCLC) patients [44]. Similarly, an integrated method using a magnetic sifter (MagSifter) and nanowell system has been described for accurate detection of EGFR (del19, T790M, and L868R) mutations in CTCs from lung cancer patients [22]. Notably, RT-qPCR readings of bulk blood samples did not reach detectable level to be analyzed for same mutations in this study [22].

3.1.4. KRAS

KRAS, one of the genes involved in the EGFR signaling pathway, may have predictive value of clinical response to anti-EGFR therapies, such as cetuximab [83], panitumumab [84], and gefitinib [85]. Patients exhibited mutational disparity between CTCs and matched primary tumors and/or across individual CTCs in breast [36], colorectal [47–49], and multiple myeloma [24], and breast cancer [36]. This may explain the variable response to anti-EGFR treatment in these cancer patients. Similarly, a highly varying degree of concordance in KRAS mutational status between CTCs and primary/metastatic lesions was observed, reflecting intratumoral heterogeneity of point mutations in KRAS occurring in 48–76% of various cancers [86].

3.1.5. BRAF

Another mutation predictive of response to EGFR-inhibiting therapy is BRAF, which is associated with a very poor prognosis particularly in colorectal cancer and melanoma [87]. Further, the predictive values of V600E and V601E mutations have been demonstrated for the use of RAF kinase inhibitor (vemurafenib) and MEK inhibitor (trametinib), respectively, in BRAF-mutated melanomas [50,88]. Somatic missense mutations in BRAF have been found in approximately 10% and 60% of colorectal tumors and melanoma lesions, respectively [50,89]. Single-cell genomic characterization of CTCs across these cancer types have revealed highly heterogeneous BRAF status across CTCs [47,48,50,51], with disparities in BRAF mutations to the corresponding primary tumor [47,50]. While such considerable heterogeneity observed in these selected genes might be the result of newly acquired mutations in CTCs, it is likely that these mutations were missed by single sector-based tissue biopsies. Through the additional deep sequencing of tissue samples, other groups indeed found mutations that were initially unique to CTCs in the primary tumors and metastases at subclonal level [16].

3.2. Microsatellite Instability (MSI)

Defective DNA mismatch repair (MMR) machinery leads to hypermutation and instability in nucleotide repeat sequences [90]. MSI is an established prognostic [91–93] and predictive marker [94,95] in many cancer types. In colorectal cancer (CRC), tumors with high-level MSI, or MSI-H phenotype account for ~15% of metastatic disease [96], and have distinct pathologic and clinical features [97]. MSI typing may serve as a predictor of benefit from adjuvant 5-fluorouacil chemotherapy for non-MSI-H CRC patients [95,98]. The standard MSI assessment recommended by the National Cancer Institute/International Collaborative Group/HNPCC (NCI/ICG-HNPCC) involves the examination of two mononucleotide repeats and three dinucleotide repeats in tumor and non-tumor adjacent normal tissues [99].

A comprehensive genomic study of CRC identified microsatellite stable (MSS) tumor harboring MSI CTCs, through aCGH, mutational profiling, and MSI analyses at the single-cell level [47]. Of note, immunohistochemical (IHC) expression analysis of DNA MMR proteins (i.e., MLH1, MSH2, MSH6, and PMS2) in multiple sectors obtained from matched primary tumor and metastatic lesions did not indicate MSI-H status in this patient. Such discordance was also found in mutational profiles, where mutations in key genes such as KRAS and TP53 were detectable only in CTCs but not in the tumor. Given that contaminating stromal cells or surrounding non-tumor cells in the tissue may obscure MSI, single-cell approaches will be essential for MSI typing.

3.3. Copy-Number Variation (CNV)

With the advent of SNP/aCGH arrays and NGS technologies, identification and characterization of CNVs in primary and metastatic tumors have provided insight into the role of CNVs in cellular functions and cancer pathogenesis [100–102]. Gain of oncogenes and loss of tumor suppressors are frequent drivers of tumor progression, and are closely associated with therapeutic responses [103]. In CTCs, genome-wide CNVs were found to be highly reproducible from cell to cell within the same individual and across different lung cancer patients with same pathological subtypes [43,45]. This is consistent with homogeneity found in CTCs from SCLC [42,46] and colorectal cancer patients [45].

While harboring a substantial number of genomic aberrations, CTCs exhibit concordant changes in chromosomes observed in matched primary and/or metastatic tumors, supporting their malignant origin, but to different extents across individual CTC, as demonstrated in colorectal [45,47], breast [40], and bladder [52] cancers. For example, phylogenetic analysis of CNV profiles identified homozygous deletion of a chromosomal region containing the tumor suppressor gene, PTEN, in CTCs and lymph node metastases, but not in primary tumor, suggesting that such CTC subsets might have metastasized in colorectal cancer [45].

A unique signature of recurrent CNVs specific to CTCs was also found in breast cancer, consisting of genes and miRNAs related to CTC phenotypes, such as resistance to anoikis, TGFβ signaling, and metastasis [23]. This copy number signature clustered patients into two groups, independent of subtype, revealing distinct functional or metastatic features in different populations. Gain of a chromosomal region harboring the HER2 gene was further consistently observed across CTCs regardless of HER2 status of matched primary tumors in this study, suggesting a potential role of HER2 amplification in CTC biology. Notably, HER2 amplification associated with the degree of chromosomal changes was identified in another multi-scale scCTC study, where a significantly higher number of genomic rearrangements was observed in breast CTCs with HER2 amplification than those without amplification, whereas no such change was seen for PIK3CA mutations [30].

CNV profiling of CTCs could potentially be used as a tool for risk stratification. CNV-based classifiers that can assign SCLC patients as chemosensitive or chemorefractory have been developed [42,46], one of which have been validated in an independent patient cohort [42]. The two patient groups stratified by these classifiers were found to have significant different progression-free survival (PFS) and/or overall survival (OS), demonstrating prognostic and predictive value of CTC CNV profiles [42,46]. Further, apheresis-acquired, CK^+ CTCs harboring >10 chromosomal alterations were associated with risk of early metastasis [32]. Better therapeutic strategies may also be inferred from sequential single-cell characterization of CNV changes in CTCs over the course of treatment; MYC amplification occurred along with AR protein expression and AR amplification in a prostate cancer patient progressing through targeted therapy [27]. While direct targeting of MYC proves to be difficult, scCTC data suggest that co-targeting of c-Myc in conjunction with AR may serve as an alternative path to prevent the emergence of drug-resistant subclones.

3.4. Chromosomal Breakpoints

In addition to focal CNV analyses, whole genome-wide CNV profiles have been extensively analyzed in bulk primary tumors. They are often associated with therapeutic resistance against platinum-based chemotherapy and PARP inhibitors [104]. Large-scale state transition (LST), which is defined as the number of chromosomal breaks between adjacent regions of at least 10Mb, is one of surrogates of such large-scale genomic instability. Significantly higher and heterogeneous LST scores were observed in CTCs from metastatic castration resistant prostate cancer (mCRPC) patients compared to cancer cell lines and WBCs from healthy donors, implying unstable CTC genomes at the single-cell level [25]. The ability to assess genomic instability with LST scoring could potentially improve risk stratification for therapeutic strategies.

3.5. Chromosomal Rearrangement

Two SVs common in all CTCs and both primary and metastatic tumors were found in prostate cancer [26]: TMEM207 in chr3, which facilitates tumor invasion, migration, and metastasis [105,106], and chr13-chr15 translocation. An extensive assessment further revealed heterogeneous status of SVs involving tumor suppressor genes, such as BRCA2, RB1, and PTEN, which are common in prostate cancer [107]. Heterogeneous status of PTEN SVs in single CTCs may suggest the acquirement of variations at different time-points, as PTEN point mutations emerges as a late event during cancer evolution [108]. Parallel transcriptomic profiling for detecting such SVs encoding oncogenes or tumor suppressors that carry founder mutation will further allow discovery of novel fusion products.

4. Longitudinal Studies

While it has long been recognized that CTC count could serve as a robust predictor of patient outcomes in breast, prostate, and colorectal cancers [36,39,109–112], CTC enumeration at a single time point alone may not provide sufficient information in terms of treatment regimen. Alternatively, phenotyping of CTCs with morphometric parameters and protein expression has been shown to be better correlated with therapy response [27,113]. Another major advantage of CTC profiling is the feasibility to analyze longitudinal samples to study mechanisms related to acquired resistance to therapy (Figure 3). Yet, such analyses may be challenging in the absence of evident de novo global chromosomal changes upon relapse and progression, requiring larger patient cohorts [42].

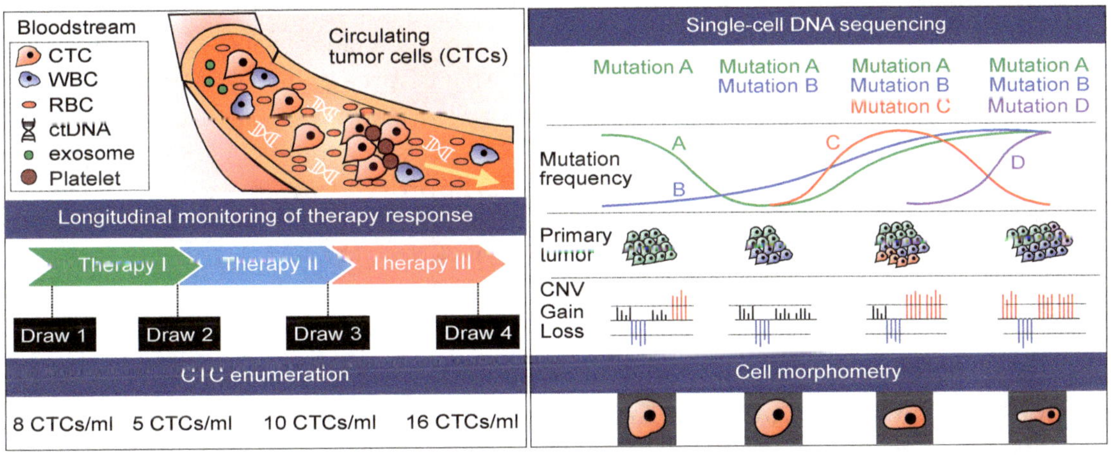

Figure 3. scCTC DNA sequencing for longitudinal monitoring of therapy response.

Selection of patients for targeted therapy has been based on the IHC detection of protein of interest and/or fluorescence in situ hybridization-based analysis of gene-level alterations, typically in known oncogenes, using primary tumor tissues. An underlying assumption is that only marker-positive patients will respond to mutant oncogene-targeted therapy. These treatments may include vemurafenib (BRAF inhibitor) for melanoma patients with V600E/V600K mutation in the BRAF gene [114], Neratinib (HER2 TKI) [115], trastuzumab (HER2 antibody) [116] and pertuzumab (HER2 antibody) [117] for breast cancer patients with amplification and/or overexpression of the HER2 gene, and erlotinib (EGFR TKI) for non-small cell lung cancer patients with exon 19 deletion or L858R mutation in the EGFR gene [118].

Such oncogene-targeted therapies, however, should carefully be applied in the clinical settings. IHC detection of EGFR expression in bulk tumor tissue alone, for example, may not be an ideal tool for prediction of response to gefitinib [119]. Discordance in genomic profiles between primary and recurring/metastatic tumors [120–122] attributed to clonal changes, sampling error in clonal selection, and/or technical flaws in the assay has complicated the decision-making process in treatment selection. Repeat biopsy for marker reassessment does not guarantee improved accuracy, nor is it without false-negative readings [123].

Varying genomic status of oncogenes has thus been re-evaluated with CTC pools over the course of therapy. Studies have examined the feasibility of detection of $HER2^+$ CTCs in patients with $HER2^-$ primary tumors [124,125] and that of KRAS-mutated CTCs in patients with nonmutated colorectal primary tumor and mutated metastases [89]. Notably, it was found that treatment with trastuzumab, an anti-HER2 antibody traditionally prescribed for patients with $HER2^+$ primary tumor, was shown to be effective in improving survival outcomes of patients with $HER2^-$ tumor by eliminating CK^+HER2^+ CTCs [126]. This was the first study that clearly demonstrated the potential of CTCs to effectively monitor evolving mutational landscape reflecting time-varying changes in drug susceptibility. Since then, several prospectively conducted studies have been carried out to facilitate the phenotyping and genetic characterization of CTCs as targeted-therapeutic intervention in metastatic cancer [127].

Longitudinal genomic data on SNVs and CNVs have so far been obtained in scCTC studies where CTCs were collected in a sequential manner during the course of treatment in lung [42,43] and prostate cancer [17]. A unique set of somatic CNV variations distinct from those observed in response to standard chemotherapy, was identified in single prostate CTCs at the time of targeted therapy failure, inferring rapidly evolving genomic organization and the emergent putative-resistant clones [27]. In another study where single-cell characterization was done before and during treatment for a breast cancer patient, while CTCs revealed distinct mutational profiles at different time points, all CTCs harbored mutation in the HER2 gene (p.V777L), regardless of sampling time, indicating resistance to HER2-targeted treatment in this patient. Interestingly, the best treatment response to chemotherapy with capecitabine and vinerolbine was observed in a patient who had the highest number of mutated genes and sequence variants in single breast CTCs [31].

Not all longitudinal studies, however, have demonstrated a clear association between CTCs and metastatic status. PIK3CA mutational status in majority of breast CTCs (7 out of 9 blood draws) sequentially sampled over time was not reflective of bone and lung metastases while DTCs achieved 100% concordance in a patient with progressive metastatic breast cancer [28]. Such highly discordant results, however, may be attributed to EpCAM-based approach employed in this study for CTC enrichment, missing out de-differentiated $EpCAM^-$ or mesenchymally shifted CTCs, all of which may be major constituents of putative metastatic founders [4].

5. Challenges and Emerging Technologies

Clearly, recent CTC studies have uncovered new perspectives in tumor biology and cancer management, through the application of rare cell sorting and single-cell sequencing technologies. Practically, however, there are challenges and technical errors associated with the developed workflows presented in this work.

5.1. Fresh-Frozen Versus Formalin-Fixed

Immunoaffinity-based enrichment technologies, including CellSearch®, involve a fixation step, which makes use of fixatives to stabilize whole blood for up to 96 h. Yet, fixed CTCs may be not suitable for RNA-based measurements, ex vivo culture and expansion, drug screening, and xenograft model-based functional studies, for compromised cell viability and degraded RNAs. Alternatively, multiparametric flow cytometry or FACS can be used to keep viable CTCs, which can further be expanded ex vivo to generate patient-derived 3D-spheroids, as demonstrated in prostate and breast cancers [17,34]. DEPArray™ and microfluidic technologies can also be used to isolate viable CTCs for subsequent molecular screening on a cell-per-cell basis.

To preserve cell viability, peripheral blood sample should be delivered on ice immediately to the laboratory once collected from a cancer patient. RNA degradation occurs within 2–4 h, and sample processing >5 h after the blood draw may result in >60% loss in CTC yield [128]. Similar to tissue acquisition of solid lesions, these requirements impose practical challenges in hospitals and labs, particularly for longitudinal cohort studies. Nevertheless, emerging technologies may be applied to CTCs to circumvent such issues, through the use of nuclear RNA or preservation protocol that retains cell viability and RNA quality for up to 72 h for single-cell transcriptomic profiling. Studies have demonstrated high concordance between nuclear RNA and whole cell RNA in the expression of cell-type specific and metabolic modeling genes [129], and that between fresh and preserved blood in detecting cancer-specific transcripts [128]. Alternatively, frozen-optimized scRNA-seq protocols (e.g., Nuc-seq) may be applied to CTC profiling, which will be particularly useful for serial monitoring of previously inaccessible tissues [2].

5.2. Increasing Number of CTC Libraries

Among all the steps, WGA was found to be the most error-prone step in performing scCTC sequencing, with allele dropout (ADO) being a significant source of failure for scCTC sequencing [53]. Such prevalent limitations may be overcome if the amount of DNA template is increased. For example, it has been suggested that at least 10 CTCs are required to reliably detect point mutations in KRAS from pancreatic CTCs [53]. Similarly, sequencing multiple independent libraries of CTCs has been proposed to improve sensitivity in determining variants and to better represent bulk library from a matched primary tumor in prostate cancer [24]. The development of WGA methods providing improved uniformity in genome-wide coverage of the amplified DNA may further facilitate reproducible and accurate sequencing for clinical use.

Alternatively, highly sensitive CTC enrichment technologies may be applied to scCTC sequencing studies to capture, assumedly, all CTCs present in blood so as to generate more DNA templates. Recently, methods of performing single-cell CNV analysis of CTCs acquired by apheresis was described in prostate and breast cancers [17], where CTCs harvested from an apheresis (mean volume = 59.5 mL) achieved approximately 90-fold increased yield (CTC count = 12,546) [17]. Extrapolation analysis further indicated that CTCs might be concentrated along with the mononuclear cell populations during diagnostic leukapheresis (DLA), thereby enhancing CTC detection frequency even in nonmetastatic cancer patients, as compared to standard CTC blood tests processing a volume of 1–10 mL of peripheral blood [32].

In recent years, scRNA-seq technologies coupled with massively parallel microfluidics have enabled high-throughput analysis of mouse retinal cells [130], human macrophages [131], embryonic stem cells [132], and peripheral blood mononuclear cells (PBMCs) [133]. Their application to CTCs, however, has been relatively limited due to the inefficiency of bead-cell pairing and the inevitable contamination of blood cells even in primarily enriched samples [134]. To overcome these limitations, Cheng et al. have recently developed Hydro-seq, which enable high-throughput contamination-free scRNA-seq of CTCs from breast cancer patients, uncovering cellular heterogeneity in metastasis and therapy related genes [134]. Having the scale-up capability while achieving high cell-capture efficiency and high-fidelity single-cell sequencing results, these technologies will play an increasingly important role in future studies aiming to generate more accurate and reproducible data at high throughput with low cost.

5.3. Multidimensional Measurements

While others have attributed the appearance of $CK^{+}CD45^{+}$ cells to false-positive CK^{+} staining of WBCs [32], such "double-positive" cells may have functional roles or clinical implication within circulation given their occurrence at a much lower frequency in healthy blood samples [135]. Similarly, the exact role of apoptotic CTCs (e.g., $CK^{+}CD45^{-}$ cells with abnormal chromosomal patterns and/or nuclear fragmentation), PD-$L1^{+}CD45^{-}$ cells, and $CK^{-}CD45^{-}$ cells in metastasis is not yet fully clear. Technologies enabling joint profiling of multiple modalities from the same individual cell may provide accurate means to understand clinical implication on the occurrence of these specific CTC subpopulations in circulation.

The high definition-CTC (HD-CTC) is an exemplary technology that facilitates real-time single-cell characterization of morphometric (i.e., cell roundness, cell area, AR subcellular localization) and protein expression changes in AR for prostate cancer [27]. Similarly, the functional EPISPOT assay, which is now named EPIDROP, allows simultaneous single-cell analysis of proteome and secretome of viable CTCs or of CTC clusters [136]. Such functional assays or microfluidic technologies, which have been successfully applied to patient-derived CTCs for analysis of genome [44], transcriptome [4], proteome [137], metabolome [138], and secretome [139] at the single-cell level, may further be integrated into the framework for deriving multidimensional data. Along with the increasingly available genomic data derived from tissue biopsies spanning diverse cancer types [140,141], the combined analysis of tissue and liquid biopsies may further uncover new insights into tumor heterogeneity and provide additional clinical information, as recently shown in lung cancer [4].

6. Conclusions

Recent scCTC studies have focused on how best to (1) identify and isolate extremely heterogeneous, fragile, and rare CTCs in a highly specific and unbiased manner, (2) discriminate false negatives and detect actionable mutations, and (3) relate the findings to clinically meaningful outcomes that could not have been caught by a tissue biopsy. Emerging data pointing towards the prevalence of CTC subpopulations and their differing metastatic potential have further stimulated studies aiming to identify and genetically/phenotypically characterize such premetastatic subsets of CTC populations that are favored to be liberated from primary tumors, survive in the bloodstream, and succeed in the early colonization phases.

While past efforts in deconvoluting the complex nature of CTCs have been largely ineffective with bulk-cell analysis, single-cell approaches are beginning to unmask cellular heterogeneity of CTCs and their clinical significance, providing a foundation for liquid biopsy in the clinic. As we continue to develop sensitive CTC enrichment technologies and generate more sequencing data from patient-derived CTCs, clinicians will continue to find a better way to apply liquid biopsies, possibly (1) at initial diagnosis, for prognostication, (2) after tumor resection, for assessment of residual disease, and (3) after adjuvant therapies, for prediction of early recurrence or relapse. With mounting molecular evidence suggesting prognostic value of CTC-derived biomarkers predictive of response to chemotherapy, targeted therapy, and immune checkpoint inhibitors, we anticipate that, in the near future, liquid biopsies will become a routine screening and monitoring of cancer patients.

Author Contributions: Conceptualization, S.B.L., C.T.L., and W.-T.L.; writing—original draft preparation, S.B.L.; writing—review and editing, S.B.L., C.T.L., and W.-T.L.; visualization, S.B.L.; supervision, C.T.L. and W.-T.L.; project administration, W.-T.L.; funding acquisition, W.-T.L.

Acknowledgments: This work was carried out at the MechanoBioengineering Laboratory at the Department of Biomedical Engineering, National University of Singapore (NUS). The authors thank support provided by the Institute for Health Innovation and Technology (iHealthtech) at NUS. S.B.L. acknowledges support provided by NUS Graduate School for Integrative Sciences and Engineering (NGS), Mogam Science Scholarship Foundation, and Daewoong Foundation.

References

1. Fidler, I.J.; Hart, I.R. Biological diversity in metastatic neoplasms: Origins and implications. *Science* **1982**, *217*, 998–1003. [CrossRef] [PubMed]
2. Tirosh, I.; Suvà, M.L. Deciphering Human Tumor Biology by Single-Cell Expression Profiling. *Annu. Rev. Cancer Biol.* **2019**, *3*, 151–166. [CrossRef]
3. Shi, X.; Chakraborty, P.; Chaudhuri, A. Unmasking tumor heterogeneity and clonal evolution by single-cell analysis. *J. Cancer Metastasis Treat.* **2018**, *4*, 47. [CrossRef]
4. Lim, S.B.; Yeo, T.; Lee, W.D.; Bhagat, A.A.S.; Tan, S.J.; Tan, D.S.W.; Lim, W.T.; Lim, C.T. Addressing cellular heterogeneity in tumor and circulation for refined prognostication. *Proc. Natl. Acad. Sci. USA* **2019**, *116*, 17957–17962. [CrossRef]

5. Lim, S.B.; Tan, S.J.; Lim, W.T.; Lim, C.T. An extracellular matrix-related prognostic and predictive indicator for early-stage non-small cell lung cancer. *Nat. Commun.* **2017**, *8*, 1734. [CrossRef]
6. Lim, S.B.; Lim, C.T.; Lim, D.W.-T. MO2-6-3Matrisomal abnormality: A predictive biomarker for cancer immunotherapy. *Ann. Oncol.* **2019**, *30*. [CrossRef]
7. Lim, S.B.; Chua, M.L.K.; Yeong, J.P.S.; Tan, S.J.; Lim, W.-T.; Lim, C.T. Pan-cancer analysis connects tumor matrisome to immune response. *NPJ Precis. Oncol.* **2019**, *3*, 15. [CrossRef]
8. Francart, M.E.; Lambert, J.; Vanwynsberghe, A.M.; Thompson, E.W.; Bourcy, M.; Polette, M.; Gilles, C. Epithelial-mesenchymal plasticity and circulating tumor cells: Travel companions to metastases. *Dev. Dyn.* **2018**, *247*, 432–450. [CrossRef]
9. Menon, N.V.; Lim, S.B.; Lim, C.T. Microfluidics for personalized drug screening of cancer. *Curr. Opin. Pharmacol.* **2019**, *48*, 155–161. [CrossRef]
10. Koh, Y.; Yagi, S.; Akamatsu, H.; Kanai, K.; Hayata, A.; Tokudome, N.; Akamatsu, K.; Higuchi, M.; Kanbara, H.; Nakanishi, M.; et al. Heterogeneous Expression of Programmed Death Receptor-ligand 1 on Circulating Tumor Cells in Patients with Lung Cancer. *Clin. Lung Cancer* **2019**, *20*, 270–277.e271. [CrossRef]
11. Grillet, F.; Bayet, E.; Villeronce, O.; Zappia, L.; Lagerqvist, E.L.; Lunke, S.; Charafe-Jauffret, E.; Pham, K.; Molck, C.; Rolland, N.; et al. Circulating tumour cells from patients with colorectal cancer have cancer stem cell hallmarks in ex vivo culture. *Gut* **2017**, *66*, 1802. [CrossRef] [PubMed]
12. Jordan, N.V.; Bardia, A.; Wittner, B.S.; Benes, C.; Ligorio, M.; Zheng, Y.; Yu, M.; Sundaresan, T.K.; Licausi, J.A.; Desai, R.; et al. HER2 expression identifies dynamic functional states within circulating breast cancer cells. *Nature* **2016**, *537*, 102–106. [CrossRef] [PubMed]
13. Agnoletto, C.; Corra, F.; Minotti, L.; Baldassari, F.; Crudele, F.; Cook, W.J.J.; Di Leva, G.; d'Adamo, A.P.; Gasparini, P.; Volinia, S. Heterogeneity in Circulating Tumor Cells: The Relevance of the Stem-Cell Subset. *Cancers* **2019**, *11*, 483. [CrossRef] [PubMed]
14. Lim, S.B.; Di Lee, W.; Vasudevan, J.; Lim, W.-T.; Lim, C.T. Liquid biopsy: One cell at a time. *NPJ Precis. Oncol.* **2019**, *3*, 23. [CrossRef]
15. Racila, E.; Euhus, D.; Weiss, A.J.; Rao, C.; McConnell, J.; Terstappen, L.W.M.M.; Uhr, J.W. Detection and characterization of carcinoma cells in the blood. *Proc. Natl. Acad. Sci. USA* **1998**, *95*, 4589. [CrossRef]
16. Heitzer, E.; Auer, M.; Gasch, C.; Pichler, M.; Ulz, P.; Hoffmann, E.M.; Lax, S.; Waldispuehl-Geigl, J.; Mauermann, O.; Lackner, C.; et al. Complex tumor genomes inferred from single circulating tumor cells by array-CGH and next-generation sequencing. *Cancer Res.* **2013**, *73*, 2965–2975. [CrossRef]
17. Lambros, M.B.; Seed, G.; Sumanasuriya, S.; Gil, V.; Crespo, M.; Fontes, M.; Chandler, R.; Mehra, N.; Fowler, G.; Ebbs, B.; et al. Single-Cell Analyses of Prostate Cancer Liquid Biopsies Acquired by Apheresis. *Clin. Cancer Res.* **2018**, *24*, 5635–5644. [CrossRef]
18. Pestrin, M.; Salvianti, F.; Galardi, F.; De Luca, F.; Turner, N.; Malorni, L.; Pazzagli, M.; Di Leo, A.; Pinzani, P. Heterogeneity of PIK3CA mutational status at the single cell level in circulating tumor cells from metastatic breast cancer patients. *Mol. Oncol.* **2015**, *9*, 749–757. [CrossRef]
19. Wang, Y.; Guo, L.; Feng, L.; Zhang, W.; Xiao, T.; Di, X.; Chen, G.; Zhang, K. Single nucleotide variant profiles of viable single circulating tumour cells reveal CTC behaviours in breast cancer. *Oncol. Rep.* **2018**, *39*, 2147–2159. [CrossRef]
20. Yin, J.; Wang, Z.; Li, G.; Lin, F.; Shao, K.; Cao, B.; Hou, Y. Characterization of circulating tumor cells in breast cancer patients by spiral microfluidics. *Cell Biol. Toxicol.* **2019**, *35*, 59–66. [CrossRef]
21. Lohr, J.G.; Kim, S.; Gould, J.; Knoechel, B.; Drier, Y.; Cotton, M.J.; Gray, D.; Birrer, N.; Wong, B.; Ha, G.; et al. Genetic interrogation of circulating multiple myeloma cells at single-cell resolution. *Sci. Transl. Med.* **2016**, *8*, 363ra147. [CrossRef] [PubMed]
22. Park, S.M.; Wong, D.J.; Ooi, C.C.; Kurtz, D.M.; Vermesh, O.; Aalipour, A.; Suh, S.; Pian, K.L.; Chabon, J.J.; Lee, S.H.; et al. Molecular profiling of single circulating tumor cells from lung cancer patients. *Proc. Natl. Acad. Sci. USA* **2016**, *113*, E8379–E8386. [CrossRef] [PubMed]
23. Kanwar, N.; Hu, P.; Bedard, P.; Clemons, M.; McCready, D.; Done, S.J. Identification of genomic signatures in circulating tumor cells from breast cancer. *Int. J. Cancer* **2015**, *137*, 332–344. [CrossRef] [PubMed]
24. Lohr, J.G.; Adalsteinsson, V.A.; Cibulskis, K.; Choudhury, A.D.; Rosenberg, M.; Cruz-Gordillo, P.; Francis, J.M.; Zhang, C.-Z.; Shalek, A.K.; Satija, R.; et al. Whole-exome sequencing of circulating tumor cells provides a window into metastatic prostate cancer. *Nat. Biotechnol.* **2014**, *32*, 479–484. [CrossRef]

25. Greene, S.B.; Dago, A.E.; Leitz, L.J.; Wang, Y.; Lee, J.; Werner, S.L.; Gendreau, S.; Patel, P.; Jia, S.; Zhang, L.; et al. Chromosomal Instability Estimation Based on Next Generation Sequencing and Single Cell Genome Wide Copy Number Variation Analysis. *PLoS ONE* **2016**, *11*, e0165089. [CrossRef] [PubMed]
26. Jiang, R.; Lu, Y.-T.; Ho, H.; Li, B.; Chen, J.-F.; Lin, M.; Li, F.; Wu, K.; Wu, H.; Lichterman, J.; et al. A comparison of isolated circulating tumor cells and tissue biopsies using whole-genome sequencing in prostate cancer. *Oncotarget* **2015**, *6*, 44781–44793. [CrossRef]
27. Dago, A.E.; Stepansky, A.; Carlsson, A.; Luttgen, M.; Kendall, J.; Baslan, T.; Kolatkar, A.; Wigler, M.; Bethel, K.; Gross, M.E.; et al. Rapid phenotypic and genomic change in response to therapeutic pressure in prostate cancer inferred by high content analysis of single circulating tumor cells. *PLoS ONE* **2014**, *9*, e101777. [CrossRef]
28. Deng, G.; Krishnakumar, S.; Powell, A.A.; Zhang, H.; Mindrinos, M.N.; Telli, M.L.; Davis, R.W.; Jeffrey, S.S. Single cell mutational analysis of PIK3CA in circulating tumor cells and metastases in breast cancer reveals heterogeneity, discordance, and mutation persistence in cultured disseminated tumor cells from bone marrow. *BMC Cancer* **2014**, *14*, 456. [CrossRef]
29. Fernandez, S.V.; Bingham, C.; Fittipaldi, P.; Austin, L.; Palazzo, J.; Palmer, G.; Alpaugh, K.; Cristofanilli, M. TP53 mutations detected in circulating tumor cells present in the blood of metastatic triple negative breast cancer patients. *Breast Cancer Res.* **2014**, *16*, 445. [CrossRef]
30. Polzer, B.; Medoro, G.; Pasch, S.; Fontana, F.; Zorzino, L.; Pestka, A.; Andergassen, U.; Meier-Stiegen, F.; Czyz, Z.T.; Alberter, B.; et al. Molecular profiling of single circulating tumor cells with diagnostic intention. *EMBO Mol. Med.* **2014**, *6*, 1371–1386. [CrossRef]
31. De Luca, F.; Rotunno, G.; Salvianti, F.; Galardi, F.; Pestrin, M.; Gabellini, S.; Simi, L.; Mancini, I.; Vannucchi, A.M.; Pazzagli, M.; et al. Mutational analysis of single circulating tumor cells by next generation sequencing in metastatic breast cancer. *Oncotarget* **2016**, *7*, 26107–26119. [CrossRef] [PubMed]
32. Fischer, J.C.; Niederacher, D.; Topp, S.A.; Honisch, E.; Schumacher, S.; Schmitz, N.; Zacarias Fohrding, L.; Vay, C.; Hoffmann, I.; Kasprowicz, N.S.; et al. Diagnostic leukapheresis enables reliable detection of circulating tumor cells of nonmetastatic cancer patients. *Proc. Natl. Acad. Sci. USA* **2013**, *110*, 16580–16585. [CrossRef] [PubMed]
33. Neves, R.P.; Raba, K.; Schmidt, O.; Honisch, E.; Meier-Stiegen, F.; Behrens, B.; Mohlendick, B.; Fehm, T.; Neubauer, H.; Klein, C.A.; et al. Genomic high-resolution profiling of single CKpos/CD45neg flow-sorting purified circulating tumor cells from patients with metastatic breast cancer. *Clin. Chem.* **2014**, *60*, 1290–1297. [CrossRef] [PubMed]
34. Vishnoi, M.; Peddibhotla, S.; Yin, W.; Scamardo, A.T.; George, G.C.; Hong, D.S.; Marchetti, D. The isolation and characterization of CTC subsets related to breast cancer dormancy. *Sci. Rep.* **2015**, *5*, 17533. [CrossRef] [PubMed]
35. Gasch, C.; Oldopp, T.; Mauermann, O.; Gorges, T.M.; Andreas, A.; Coith, C.; Muller, V.; Fehm, T.; Janni, W.; Pantel, K.; et al. Frequent detection of PIK3CA mutations in single circulating tumor cells of patients suffering from HER2-negative metastatic breast cancer. *Mol. Oncol.* **2016**, *10*, 1330–1343. [CrossRef] [PubMed]
36. Shaw, J.A.; Guttery, D.S.; Hills, A.; Fernandez-Garcia, D.; Page, K.; Rosales, B.M.; Goddard, K.S.; Hastings, R.K.; Luo, J.; Ogle, O.; et al. Mutation Analysis of Cell-Free DNA and Single Circulating Tumor Cells in Metastatic Breast Cancer Patients with High Circulating Tumor Cell Counts. *Clin. Cancer Res.* **2017**, *23*, 88–96. [CrossRef] [PubMed]
37. Neumann, M.H.D.; Schneck, H.; Decker, Y.; Schömer, S.; Franken, A.; Endris, V.; Pfarr, N.; Weichert, W.; Niederacher, D.; Fehm, T.; et al. Isolation and characterization of circulating tumor cells using a novel workflow combining the CellSearch®system and the CellSelector™. *Biotechnol. Prog.* **2017**, *33*, 125–132. [CrossRef]
38. Bingham, C.; Fernandez, S.V.; Fittipaldi, P.; Dempsey, P.W.; Ruth, K.J.; Cristofanilli, M.; Katherine Alpaugh, R. Mutational studies on single circulating tumor cells isolated from the blood of inflammatory breast cancer patients. *Breast Cancer Res. Treat.* **2017**, *163*, 219–230. [CrossRef]
39. Babayan, A.; Hannemann, J.; Spotter, J.; Muller, V.; Pantel, K.; Joosse, S.A. Heterogeneity of estrogen receptor expression in circulating tumor cells from metastatic breast cancer patients. *PLoS ONE* **2013**, *8*, e75038. [CrossRef]

40. Heidary, M.; Auer, M.; Ulz, P.; Heitzer, E.; Petru, E.; Gasch, C.; Riethdorf, S.; Mauermann, O.; Lafer, I.; Pristauz, G.; et al. The dynamic range of circulating tumor DNA in metastatic breast cancer. *Breast Cancer Res.* **2014**, *16*, 421. [CrossRef]
41. Mu, Z.; Benali-Furet, N.; Uzan, G.; Znaty, A.; Ye, Z.; Paolillo, C.; Wang, C.; Austin, L.; Rossi, G.; Fortina, P.; et al. Detection and Characterization of Circulating Tumor Associated Cells in Metastatic Breast Cancer. *Int. J. Mol. Sci.* **2016**, *17*, 1665. [CrossRef] [PubMed]
42. Carter, L.; Rothwell, D.G.; Mesquita, B.; Smowton, C.; Leong, H.S.; Fernandez-Gutierrez, F.; Li, Y.; Burt, D.J.; Antonello, J.; Morrow, C.J.; et al. Molecular analysis of circulating tumor cells identifies distinct copy-number profiles in patients with chemosensitive and chemorefractory small-cell lung cancer. *Nat. Med.* **2017**, *23*, 114–119. [CrossRef] [PubMed]
43. Ni, X.; Zhuo, M.; Su, Z.; Duan, J.; Gao, Y.; Wang, Z.; Zong, C.; Bai, H.; Chapman, A.R.; Zhao, J.; et al. Reproducible copy number variation patterns among single circulating tumor cells of lung cancer patients. *Proc. Natl. Acad. Sci. USA* **2013**, *110*, 21083–21088. [CrossRef] [PubMed]
44. Yeo, T.; Tan, S.J.; Lim, C.L.; Lau, D.P.; Chua, Y.W.; Krisna, S.S.; Iyer, G.; Tan, G.S.; Lim, T.K.; Tan, D.S.; et al. Microfluidic enrichment for the single cell analysis of circulating tumor cells. *Sci. Rep.* **2016**, *6*, 22076. [CrossRef]
45. Gao, Y.; Ni, X.; Guo, H.; Su, Z.; Ba, Y.; Tong, Z.; Guo, Z.; Yao, X.; Chen, X.; Yin, J.; et al. Single-cell sequencing deciphers a convergent evolution of copy number alterations from primary to circulating tumor cells. *Genome Res.* **2017**, *27*, 1312–1322. [CrossRef]
46. Su, Z.; Wang, Z.; Ni, X.; Duan, J.; Gao, Y.; Zhuo, M.; Li, R.; Zhao, J.; Ma, Q.; Bai, H.; et al. Inferring the Evolution and Progression of Small-Cell Lung Cancer by Single-Cell Sequencing of Circulating Tumor Cells. *Clin. Cancer Res.* **2019**, *25*, 5049–5060. [CrossRef]
47. Steinert, G.; Schölch, S.; Niemietz, T.; Iwata, N.; García, S.A.; Behrens, B.; Voigt, A.; Kloor, M.; Benner, A.; Bork, U.; et al. Immune Escape and Survival Mechanisms in Circulating Tumor Cells of Colorectal Cancer. *Cancer Res.* **2014**, *74*, 1694–1704. [CrossRef]
48. Gasch, C.; Bauernhofer, T.; Pichler, M.; Langer-Freitag, S.; Reeh, M.; Seifert, A.M.; Mauermann, O.; Izbicki, J.R.; Pantel, K.; Riethdorf, S. Heterogeneity of epidermal growth factor receptor status and mutations of KRAS/PIK3CA in circulating tumor cells of patients with colorectal cancer. *Clin. Chem.* **2013**, *59*, 252–260. [CrossRef]
49. Fabbri, F.; Carloni, S.; Zoli, W.; Ulivi, P.; Gallerani, G.; Fici, P.; Chiadini, E.; Passardi, A.; Frassineti, G.L.; Ragazzini, A.; et al. Detection and recovery of circulating colon cancer cells using a dielectrophoresis-based device: KRAS mutation status in pure CTCs. *Cancer Lett.* **2013**, *335*, 225–231. [CrossRef]
50. Sakaizawa, K.; Goto, Y.; Kiniwa, Y.; Uchiyama, A.; Harada, K.; Shimada, S.; Saida, T.; Ferrone, S.; Takata, M.; Uhara, H.; et al. Mutation analysis of BRAF and KIT in circulating melanoma cells at the single cell level. *Br. J. Cancer* **2012**, *106*, 939–946. [CrossRef]
51. Kiniwa, Y.; Nakamura, K.; Mikoshiba, A.; Akiyama, Y.; Morimoto, A.; Okuyama, R. Diversity of circulating tumor cells in peripheral blood: Detection of heterogeneous BRAF mutations in a patient with advanced melanoma by single-cell analysis. *J. Derm. Sci.* **2018**, *90*, 211–213. [CrossRef] [PubMed]
52. Anantharaman, A.; Friedlander, T.; Lu, D.; Krupa, R.; Premasekharan, G.; Hough, J.; Edwards, M.; Paz, R.; Lindquist, K.; Graf, R.; et al. Programmed death-ligand 1 (PD-L1) characterization of circulating tumor cells (CTCs) in muscle invasive and metastatic bladder cancer patients. *BMC Cancer* **2016**, *16*, 744. [CrossRef] [PubMed]
53. Court, C.M.; Ankeny, J.S.; Sho, S.; Hou, S.; Li, Q.; Hsieh, C.; Song, M.; Liao, X.; Rochefort, M.M.; Wainberg, Z.A.; et al. Reality of Single Circulating Tumor Cell Sequencing for Molecular Diagnostics in Pancreatic Cancer. *J. Mol. Diagn.* **2016**, *18*, 688–696. [CrossRef] [PubMed]
54. Pantel, K.; Deneve, E.; Nocca, D.; Coffy, A.; Vendrell, J.P.; Maudelonde, T.; Riethdorf, S.; Alix-Panabieres, C. Circulating epithelial cells in patients with benign colon diseases. *Clin. Chem.* **2012**, *58*, 936–940. [CrossRef]
55. Zhang, L.; Ridgway, L.D.; Wetzel, M.D.; Ngo, J.; Yin, W.; Kumar, D.; Goodman, J.C.; Groves, M.D.; Marchetti, D. The identification and characterization of breast cancer CTCs competent for brain metastasis. *Sci. Transl. Med.* **2013**, *5*, 180ra148. [CrossRef]
56. Lee, Y.; Guan, G.; Bhagat, A.A. ClearCell®FX, a label-free microfluidics technology for enrichment of viable circulating tumor cells. *Cytom. Part. A* **2018**, *93*, 1251–1254. [CrossRef]

57. Ligthart, S.T.; Coumans, F.A.; Bidard, F.C.; Simkens, L.H.; Punt, C.J.; de Groot, M.R.; Attard, G.; de Bono, J.S.; Pierga, J.Y.; Terstappen, L.W. Circulating Tumor Cells Count and Morphological Features in Breast, Colorectal and Prostate Cancer. *PLoS ONE* **2013**, *8*, e67148. [CrossRef]
58. Kallergi, G.; Konstantinidis, G.; Markomanolaki, H.; Papadaki, M.A.; Mavroudis, D.; Stournaras, C.; Georgoulias, V.; Agelaki, S. Apoptotic circulating tumor cells in early and metastatic breast cancer patients. *Mol. Cancer Ther.* **2013**, *12*, 1886–1895. [CrossRef]
59. Blainey, P.C. The future is now: Single-cell genomics of bacteria and archaea. *FEMS Microbiol. Rev.* **2013**, *37*, 407–427. [CrossRef]
60. Salvianti, F.; Pazzagli, M.; Pinzani, P. Single circulating tumor cell sequencing as an advanced tool in cancer management. *Expert Rev. Mol. Diagn.* **2016**, *16*, 51–63. [CrossRef]
61. Swennenhuis, J.F.; Reumers, J.; Thys, K.; Aerssens, J.; Terstappen, L.W. Efficiency of whole genome amplification of single circulating tumor cells enriched by CellSearch and sorted by FACS. *Genome Med.* **2013**, *5*, 106. [CrossRef] [PubMed]
62. Gawad, C.; Koh, W.; Quake, S.R. Single-cell genome sequencing: Current state of the science. *Nat. Rev. Genet.* **2016**, *17*, 175–188. [CrossRef] [PubMed]
63. Pao, W.; Girard, N. New driver mutations in non-small-cell lung cancer. *Lancet Oncol.* **2011**, *12*, 175–180. [CrossRef]
64. Stephens, P.J.; Tarpey, P.S.; Davies, H.; Van Loo, P.; Greenman, C.; Wedge, D.C.; Nik-Zainal, S.; Martin, S.; Varela, I.; Bignell, G.R.; et al. The landscape of cancer genes and mutational processes in breast cancer. *Nature* **2012**, *486*, 400–404. [CrossRef] [PubMed]
65. Jhawer, M.; Goel, S.; Wilson, A.J.; Montagna, C.; Ling, Y.H.; Byun, D.S.; Nasser, S.; Arango, D.; Shin, J.; Klampfer, L.; et al. PIK3CA mutation/PTEN expression status predicts response of colon cancer cells to the epidermal growth factor receptor inhibitor cetuximab. *Cancer Res.* **2008**, *68*, 1953–1961. [CrossRef] [PubMed]
66. Berns, K.; Horlings, H.M.; Hennessy, B.T.; Madiredjo, M.; Hijmans, E.M.; Beelen, K.; Linn, S.C.; Gonzalez-Angulo, A.M.; Stemke-Hale, K.; Hauptmann, M.; et al. A Functional Genetic Approach Identifies the PI3K Pathway as a Major Determinant of Trastuzumab Resistance in Breast Cancer. *Cancer Cell* **2007**, *12*, 395–402. [CrossRef] [PubMed]
67. Chandarlapaty, S.; Sakr, R.A.; Giri, D.; Patil, S.; Heguy, A.; Morrow, M.; Modi, S.; Norton, L.; Rosen, N.; Hudis, C.; et al. Frequent mutational activation of the PI3K-AKT pathway in trastuzumab-resistant breast cancer. *Clin. Cancer Res.* **2012**, *18*, 6784–6791. [CrossRef]
68. Mukohara, T. Mechanisms of resistance to anti-human epidermal growth factor receptor 2 agents in breast cancer. *Cancer Sci.* **2011**, *102*, 1–8. [CrossRef]
69. Ibrahim, E.M.; Kazkaz, G.A.; Al-Mansour, M.M.; Al-Foheidi, M.E. The predictive and prognostic role of phosphatase phosphoinositol-3 (PI3) kinase (PIK3CA) mutation in HER2-positive breast cancer receiving HER2-targeted therapy: A meta-analysis. *Breast Cancer Res. Treat.* **2015**, *152*, 463–476. [CrossRef]
70. Loibl, S.; von Minckwitz, G.; Schneeweiss, A.; Paepke, S.; Lehmann, A.; Rezai, M.; Zahm, D.M.; Sinn, P.; Khandan, F.; Eidtmann, H.; et al. PIK3CA mutations are associated with lower rates of pathologic complete response to anti-human epidermal growth factor receptor 2 (her2) therapy in primary HER2-overexpressing breast cancer. *J. Clin. Oncol.* **2014**, *32*, 3212–3220. [CrossRef]
71. Neumann, M.; Decker, Y.; Franken, A.; Schömer, S.; Schneck, H.; Fehm, T.; Weichert, W.; Endriss, V.; Neubauer, H.; Niederacher, D. Sequential analysis of circulating tumor cells on genome and protein level: Potential regulation of the invasion marker CapG by PIK3CA. *Senologie Zeitschrift für Mammadiagnostik und Herapie* **2016**, *13*, A95. [CrossRef]
72. Janku, F.; Wheler, J.J.; Westin, S.N.; Moulder, S.L.; Naing, A.; Tsimberidou, A.M.; Fu, S.; Falchook, G.S.; Hong, D.S.; Garrido-Laguna, I.; et al. PI3K/AKT/mTOR inhibitors in patients with breast and gynecologic malignancies harboring PIK3CA mutations. *J. Clin. Oncol.* **2012**, *30*, 777–782. [CrossRef] [PubMed]
73. Olivier, M.; Eeles, R.; Hollstein, M.; Khan, M.A.; Harris, C.C.; Hainaut, P. The IARC TP53 database: New online mutation analysis and recommendations to users. *Hum. Mutat.* **2002**, *19*, 607–614. [CrossRef] [PubMed]
74. Kim, T.; Veronese, A.; Pichiorri, F.; Lee, T.J.; Jeon, Y.-J.; Volinia, S.; Pineau, P.; Marchio, A.; Palatini, J.; Suh, S.-S.; et al. p53 regulates epithelial–mesenchymal transition through microRNAs targeting ZEB1 and ZEB2. *J. Exp. Med.* **2011**, *208*, 875. [CrossRef]

75. Chang, C.-J.; Chao, C.-H.; Xia, W.; Yang, J.-Y.; Xiong, Y.; Li, C.-W.; Yu, W.-H.; Rehman, S.K.; Hsu, J.L.; Lee, H.-H.; et al. p53 regulates epithelial-mesenchymal transition and stem cell properties through modulating miRNAs. *Nat. Cell Biol.* **2011**, *13*, 317–323. [CrossRef]
76. Petitjean, A.; Achatz, M.I.; Borresen-Dale, A.L.; Hainaut, P.; Olivier, M. TP53 mutations in human cancers: Functional selection and impact on cancer prognosis and outcomes. *Oncogene* **2007**, *26*, 2157–2165. [CrossRef]
77. Li, J.; Yang, L.; Gaur, S.; Zhang, K.; Wu, X.; Yuan, Y.-C.; Li, H.; Hu, S.; Weng, Y.; Yen, Y. Mutants TP53 p.R273H and p.R273C but not p.R273G Enhance Cancer Cell Malignancy. *Hum. Mutat.* **2014**, *35*, 575–584. [CrossRef]
78. Kalemkerian, G.P. Chemotherapy for small-cell lung cancer. *Lancet Oncol.* **2014**, *15*, 13–14. [CrossRef]
79. Sutherland, K.D.; Proost, N.; Brouns, I.; Adriaensen, D.; Song, J.Y.; Berns, A. Cell of origin of small cell lung cancer: Inactivation of Trp53 and Rb1 in distinct cell types of adult mouse lung. *Cancer Cell* **2011**, *19*, 754–764. [CrossRef]
80. Nicholson, R.I.; Gee, J.M.; Harper, M.E. EGFR and cancer prognosis. *Eur. J. Cancer* **2001**, *37* (Suppl. S4), S9–S15. [CrossRef]
81. Ciardiello, F.; Tortora, G. Epidermal growth factor receptor (EGFR) as a target in cancer therapy: Understanding the role of receptor expression and other molecular determinants that could influence the response to anti-EGFR drugs. *Eur. J. Cancer* **2003**, *39*, 1348–1354. [CrossRef]
82. Moroni, M.; Veronese, S.; Benvenuti, S.; Marrapese, G.; Sartore-Bianchi, A.; Di Nicolantonio, F.; Gambacorta, M.; Siena, S.; Bardelli, A. Gene copy number for epidermal growth factor receptor (EGFR) and clinical response to antiEGFR treatment in colorectal cancer: A cohort study. *Lancet Oncol.* **2005**, *6*, 279–286. [CrossRef]
83. Lièvre, A.; Bachet, J.-B.; Le Corre, D.; Boige, V.; Landi, B.; Emile, J.-F.; Côté, J.-F.; Tomasic, G.; Penna, C.; Ducreux, M.; et al. KRAS Mutation Status Is Predictive of Response to Cetuximab Therapy in Colorectal Cancer. *Cancer Res.* **2006**, *66*, 3992. [CrossRef]
84. Amado, R.G.; Wolf, M.; Peeters, M.; Van Cutsem, E.; Siena, S.; Freeman, D.J.; Juan, T.; Sikorski, R.; Suggs, S.; Radinsky, R.; et al. Wild-Type KRAS Is Required for Panitumumab Efficacy in Patients with Metastatic Colorectal Cancer. *J. Clin. Oncol.* **2008**, *26*, 1626–1634. [CrossRef]
85. Zhu, C.-Q.; da Cunha Santos, G.; Ding, K.; Sakurada, A.; Cutz, J.-C.; Liu, N.; Zhang, T.; Marrano, P.; Whitehead, M.; Squire, J.A.; et al. Role of KRAS and EGFR as Biomarkers of Response to Erlotinib in National Cancer Institute of Canada Clinical Trials Group Study BR.21. *J. Clin. Oncol.* **2008**, *26*, 4268–4275. [CrossRef]
86. Watanabe, T.; Kobunai, T.; Yamamoto, Y.; Matsuda, K.; Ishihara, S.; Nozawa, K.; Iinuma, H.; Shibuya, H.; Eshima, K. Heterogeneity of KRAS Status May Explain the Subset of Discordant KRAS Status Between Primary and Metastatic Colorectal Cancer. *Dis. Colon Rectum* **2011**, *54*, 1170–1178. [CrossRef]
87. Ducreux, M.; Chamseddine, A.; Laurent-Puig, P.; Smolenschi, C.; Hollebecque, A.; Dartigues, P.; Samallin, E.; Boige, V.; Malka, D.; Gelli, M. Molecular targeted therapy of BRAF-mutant colorectal cancer. *Ther. Adv. Med. Oncol.* **2019**, *11*. [CrossRef]
88. Marconcini, R.; Galli, L.; Antonuzzo, A.; Bursi, S.; Roncella, C.; Fontanini, G.; Sensi, E.; Falcone, A. Metastatic BRAF K601E-mutated melanoma reaches complete response to MEK inhibitor trametinib administered for over 36 months. *Exp. Hematol. Oncol.* **2017**, *6*, 6. [CrossRef]
89. Mostert, B.; Jiang, Y.; Sieuwerts, A.M.; Wang, H.; Bolt-de Vries, J.; Biermann, K.; Kraan, J.; Lalmahomed, Z.; van Galen, A.; de Weerd, V.; et al. KRAS and BRAF mutation status in circulating colorectal tumor cells and their correlation with primary and metastatic tumor tissue. *Int. J. Cancer* **2013**, *133*, 130–141. [CrossRef]
90. Wheeler, J.M.; Bodmer, W.F.; Mortensen, N.J. DNA mismatch repair genes and colorectal cancer. *Gut* **2000**, *47*, 148–153. [CrossRef]
91. Mackay, H.J.; Gallinger, S.; Tsao, M.S.; McLachlin, C.M.; Tu, D.; Keiser, K.; Eisenhauer, E.A.; Oza, A.M. Prognostic value of microsatellite instability (MSI) and PTEN expression in women with endometrial cancer: Results from studies of the NCIC Clinical Trials Group (NCIC CTG). *Eur. J. Cancer* **2010**, *46*, 1365–1373. [CrossRef] [PubMed]
92. Popat, S.; Hubner, R.; Houlston, R.S. Systematic Review of Microsatellite Instability and Colorectal Cancer Prognosis. *J. Clin. Oncol.* **2005**, *23*, 609–618. [CrossRef] [PubMed]
93. Thibodeau, S.N.; Bren, G.; Schaid, D. Microsatellite instability in cancer of the proximal colon. *Science* **1993**, *260*, 816–819. [CrossRef] [PubMed]

94. Kim, G.P.; Colangelo, L.H.; Wieand, H.S.; Paik, S.; Kirsch, I.R.; Wolmark, N.; Allegra, C.J.; National Cancer, I. Prognostic and predictive roles of high-degree microsatellite instability in colon cancer: A National Cancer Institute-National Surgical Adjuvant Breast and Bowel Project Collaborative Study. *J. Clin. Oncol.* **2007**, *25*, 767–772. [CrossRef] [PubMed]
95. Ribic, C.M.; Sargent, D.J.; Moore, M.J.; Thibodeau, S.N.; French, A.J.; Goldberg, R.M.; Hamilton, S.R.; Laurent-Puig, P.; Gryfe, R.; Shepherd, L.E.; et al. Tumor Microsatellite-Instability Status as a Predictor of Benefit from Fluorouracil-Based Adjuvant Chemotherapy for Colon Cancer. *N. Engl. J. Med.* **2003**, *349*, 247–257. [CrossRef]
96. Goldstein, J.; Tran, B.; Ensor, J.; Gibbs, P.; Wong, H.L.; Wong, S.F.; Vilar, E.; Tie, J.; Broaddus, R.; Kopetz, S.; et al. Multicenter retrospective analysis of metastatic colorectal cancer (CRC) with high-level microsatellite instability (MSI-H). *Ann. Oncol.* **2014**, *25*, 1032–1038. [CrossRef] [PubMed]
97. Findeisen, P.; Kloor, M.; Merx, S.; Sutter, C.; Woerner, S.M.; Dostmann, N.; Benner, A.; Dondog, B.; Pawlita, M.; Dippold, W.; et al. T25 repeat in the 3′ untranslated region of the CASP2 gene: A sensitive and specific marker for microsatellite instability in colorectal cancer. *Cancer Res.* **2005**, *65*, 8072–8078. [CrossRef]
98. Carethers, J.M.; Smith, E.J.; Behling, C.A.; Nguyen, L.; Tajima, A.; Doctolero, R.T.; Cabrera, B.L.; Goel, A.; Arnold, C.A.; Miyai, K.; et al. Use of 5-fluorouracil and survival in patients with microsatellite-unstable colorectal cancer. *Gastroenterology* **2004**, *126*, 394–401. [CrossRef]
99. Boland, C.R.; Thibodeau, S.N.; Hamilton, S.R.; Sidransky, D.; Eshleman, J.R.; Burt, R.W.; Meltzer, S.J.; Rodriguez-Bigas, M.A.; Fodde, R.; Ranzani, G.N.; et al. A National Cancer Institute Workshop on Microsatellite Instability for cancer detection and familial predisposition: Development of international criteria for the determination of microsatellite instability in colorectal cancer. *Cancer Res.* **1998**, *58*, 5248–5257.
100. Bergamaschi, A.; Kim, Y.H.; Wang, P.; Sorlie, T.; Hernandez-Boussard, T.; Lonning, P.E.; Tibshirani, R.; Borresen-Dale, A.L.; Pollack, J.R. Distinct patterns of DNA copy number alteration are associated with different clinicopathological features and gene-expression subtypes of breast cancer. *Genes Chromosomes Cancer* **2006**, *45*, 1033–1040. [CrossRef]
101. Chin, K.; DeVries, S.; Fridlyand, J.; Spellman, P.T.; Roydasgupta, R.; Kuo, W.L.; Lapuk, A.; Neve, R.M.; Qian, Z.; Ryder, T.; et al. Genomic and transcriptional aberrations linked to breast cancer pathophysiologies. *Cancer Cell* **2006**, *10*, 529–541. [CrossRef] [PubMed]
102. Zhao, X.; Li, C.; Paez, J.G.; Chin, K.; Jänne, P.A.; Chen, T.-H.; Girard, L.; Minna, J.; Christiani, D.; Leo, C.; et al. An Integrated View of Copy Number and Allelic Alterations in the Cancer Genome Using Single Nucleotide Polymorphism Arrays. *Cancer Res.* **2004**, *64*, 3060. [CrossRef] [PubMed]
103. Cappuzzo, F.; Varella-Garcia, M.; Shigematsu, H.; Domenichini, I.; Bartolini, S.; Ceresoli, G.L.; Rossi, E.; Ludovini, V.; Gregorc, V.; Toschi, L.; et al. Increased HER2 Gene Copy Number Is Associated with Response to Gefitinib Therapy in Epidermal Growth Factor Receptor–Positive Non–Small-Cell Lung Cancer Patients. *J. Clin. Oncol.* **2005**, *23*, 5007–5018. [CrossRef]
104. Watkins, J.A.; Irshad, S.; Grigoriadis, A.; Tutt, A.N. Genomic scars as biomarkers of homologous recombination deficiency and drug response in breast and ovarian cancers. *Breast Cancer Res.* **2014**, *16*, 211. [CrossRef]
105. Humar, B.; Blair, V.; Charlton, A.; More, H.; Martin, I.; Guilford, P. E-cadherin deficiency initiates gastric signet-ring cell carcinoma in mice and man. *Cancer Res.* **2009**, *69*, 2050–2056. [CrossRef]
106. Takeuchi, T.; Adachi, Y.; Nagayama, T. A WWOX-binding molecule, transmembrane protein 207, is related to the invasiveness of gastric signet-ring cell carcinoma. *Carcinogenesis* **2012**, *33*, 548–554. [CrossRef]
107. Berger, M.F.; Lawrence, M.S.; Demichelis, F.; Drier, Y.; Cibulskis, K.; Sivachenko, A.Y.; Sboner, A.; Esgueva, R.; Pflueger, D.; Sougnez, C.; et al. The genomic complexity of primary human prostate cancer. *Nature* **2011**, *470*, 214–220. [CrossRef]
108. Baca, S.C.; Prandi, D.; Lawrence, M.S.; Mosquera, J.M.; Romanel, A.; Drier, Y.; Park, K.; Kitabayashi, N.; MacDonald, T.Y.; Ghandi, M.; et al. Punctuated Evolution of Prostate Cancer Genomes. *Cell* **2013**, *153*, 666–677. [CrossRef]
109. Budd, G.T.; Cristofanilli, M.; Ellis, M.J.; Stopeck, A.; Borden, E.; Miller, M.C.; Matera, J.; Repollet, M.; Doyle, G.V.; Terstappen, L.W.M.M.; et al. Circulating Tumor Cells versus Imaging—Predicting Overall Survival in Metastatic Breast Cancer. *Clin. Cancer Res.* **2006**, *12*, 6403. [CrossRef]

110. Cohen, S.J.; Punt, C.J.A.; Iannotti, N.; Saidman, B.H.; Sabbath, K.D.; Gabrail, N.Y.; Picus, J.; Morse, M.; Mitchell, E.; Miller, M.C.; et al. Relationship of Circulating Tumor Cells to Tumor Response, Progression-Free Survival, and Overall Survival in Patients with Metastatic Colorectal Cancer. *J. Clin. Oncol.* **2008**, *26*, 3213–3221. [CrossRef]
111. De Bono, J.S.; Scher, H.I.; Montgomery, R.B.; Parker, C.; Miller, M.C.; Tissing, H.; Doyle, G.V.; Terstappen, L.W.; Pienta, K.J.; Raghavan, D. Circulating tumor cells predict survival benefit from treatment in metastatic castration-resistant prostate cancer. *Clin. Cancer Res.* **2008**, *14*, 6302–6309. [CrossRef] [PubMed]
112. Hayes, D.F.; Cristofanilli, M.; Budd, G.T.; Ellis, M.J.; Stopeck, A.; Miller, M.C.; Matera, J.; Allard, W.J.; Doyle, G.V.; Terstappen, L.W.W.M. Circulating Tumor Cells at Each Follow-up Time Point during Therapy of Metastatic Breast Cancer Patients Predict Progression-Free and Overall Survival. *Clin. Cancer Res.* **2006**, *12*, 4218. [CrossRef] [PubMed]
113. Miyamoto, D.T.; Lee, R.J.; Stott, S.L.; Ting, D.T.; Wittner, B.S.; Ulman, M.; Smas, M.E.; Lord, J.B.; Brannigan, B.W.; Trautwein, J.; et al. Androgen Receptor Signaling in Circulating Tumor Cells as a Marker of Hormonally Responsive Prostate Cancer. *Cancer Discov.* **2012**, *2*, 995. [CrossRef] [PubMed]
114. McArthur, G.A.; Chapman, P.B.; Robert, C.; Larkin, J.; Haanen, J.B.; Dummer, R.; Ribas, A.; Hogg, D.; Hamid, O.; Ascierto, P.A.; et al. Safety and efficacy of vemurafenib in BRAFV600E and BRAFV600K mutation-positive melanoma (BRIM-3): Extended follow-up of a phase 3, randomised, open-label study. *Lancet Oncol.* **2014**, *15*, 323–332. [CrossRef]
115. Cortés-Funes, H.; Mendiola, C.; Manso, L.; Ciruelos, E. Neratinib, an irreversible pan erB receptor tyrosine kinase inhibitor active for advanced HER2+breast cancer. *Breast Cancer Res.* **2009**, *11*, S19. [CrossRef]
116. Slamon, D.J.; Leyland-Jones, B.; Shak, S.; Fuchs, H.; Paton, V.; Bajamonde, A.; Fleming, T.; Eiermann, W.; Wolter, J.; Pegram, M.; et al. Use of Chemotherapy plus a Monoclonal Antibody against HER2 for Metastatic Breast Cancer That Overexpresses HER2. *N. Engl. J. Med.* **2001**, *344*, 783–792. [CrossRef]
117. Baselga, J.; Gelmon, K.A.; Verma, S.; Wardley, A.; Conte, P.; Miles, D.; Bianchi, G.; Cortes, J.; McNally, V.A.; Ross, G.A.; et al. Phase II Trial of Pertuzumab and Trastuzumab in Patients with Human Epidermal Growth Factor Receptor 2–Positive Metastatic Breast Cancer That Progressed During Prior Trastuzumab Therapy. *J. Clin. Oncol.* **2010**, *28*, 1138–1144. [CrossRef]
118. Zhou, C.; Wu, Y.-L.; Chen, G.; Feng, J.; Liu, X.-Q.; Wang, C.; Zhang, S.; Wang, J.; Zhou, S.; Ren, S.; et al. Erlotinib versus chemotherapy as first-line treatment for patients with advanced EGFR mutation-positive non-small-cell lung cancer (OPTIMAL, CTONG-0802): A multicentre, open-label, randomised, phase 3 study. *Lancet Oncol.* **2011**, *12*, 735–742. [CrossRef]
119. Bailey, R.; Kris, M.; Wolf, M.; Kay, A.; Averbuch, S.; Askaa, J.; Janas, M.; Schmidt, K.; Fukuoka, M. O-242 Gefitinib ('Iressa', ZD1839) monotherapy for pretreated advanced non-small-cell lung cancer in IDEAL 1 and 2: Tumor response is not clinically relevantly predictable from tumor EGFR membrane staining alone. *Lung Cancer* **2003**, *41*, S71. [CrossRef]
120. Amir, E.; Miller, N.; Geddie, W.; Freedman, O.; Kassam, F.; Simmons, C.; Oldfield, M.; Dranitsaris, G.; Tomlinson, G.; Laupacis, A.; et al. Prospective study evaluating the impact of tissue confirmation of metastatic disease in patients with breast cancer. *J. Clin. Oncol.* **2012**, *30*, 587–592. [CrossRef]
121. Lindström, L.S.; Karlsson, E.; Wilking, U.M.; Johansson, U.; Hartman, J.; Lidbrink, E.K.; Hatschek, T.; Skoog, L.; Bergh, J. Clinically Used Breast Cancer Markers Such as Estrogen Receptor, Progesterone Receptor, and Human Epidermal Growth Factor Receptor 2 Are Unstable Throughout Tumor Progression. *J. Clin. Oncol.* **2012**, *30*, 2601–2608. [CrossRef] [PubMed]
122. Turner, N.H.; Di Leo, A. HER2 discordance between primary and metastatic breast cancer: Assessing the clinical impact. *Cancer Treat. Rev.* **2013**, *39*, 947–957. [CrossRef]
123. Pusztai, L.; Viale, G.; Kelly, C.M.; Hudis, C.A. Estrogen and HER-2 Receptor Discordance Between Primary Breast Cancer and Metastasis. *Oncologist* **2010**, *15*, 1164–1168. [CrossRef] [PubMed]
124. Fehm, T.; Muller, V.; Aktas, B.; Janni, W.; Schneeweiss, A.; Stickeler, E.; Lattrich, C.; Lohberg, C.R.; Solomayer, E.; Rack, B.; et al. HER2 status of circulating tumor cells in patients with metastatic breast cancer: A prospective, multicenter trial. *Breast Cancer Res. Treat.* **2010**, *124*, 403–412. [CrossRef] [PubMed]
125. Wallwiener, M.; Hartkopf, A.D.; Riethdorf, S.; Nees, J.; Sprick, M.R.; Schonfisch, B.; Taran, F.A.; Heil, J.; Sohn, C.; Pantel, K.; et al. The impact of HER2 phenotype of circulating tumor cells in metastatic breast cancer: A retrospective study in 107 patients. *BMC Cancer* **2015**, *15*, 403. [CrossRef]

126. Georgoulias, V.; Bozionelou, V.; Agelaki, S.; Perraki, M.; Apostolaki, S.; Kallergi, G.; Kalbakis, K.; Xyrafas, A.; Mavroudis, D. Trastuzumab decreases the incidence of clinical relapses in patients with early breast cancer presenting chemotherapy-resistant CK-19mRNA-positive circulating tumor cells: Results of a randomized phase II study. *Ann. Oncol.* **2012**, *23*, 1744–1750. [CrossRef]
127. Schramm, A.; Friedl, T.W.P.; Schochter, F.; Scholz, C.; de Gregorio, N.; Huober, J.; Rack, B.; Trapp, E.; Alunni-Fabbroni, M.; Müller, V.; et al. Therapeutic intervention based on circulating tumor cell phenotype in metastatic breast cancer: Concept of the DETECT study program. *Arch. Gynecol. Obstet.* **2016**, *293*, 271–281. [CrossRef]
128. Wong, K.H.K.; Tessier, S.N.; Miyamoto, D.T.; Miller, K.L.; Bookstaver, L.D.; Carey, T.R.; Stannard, C.J.; Thapar, V.; Tai, E.C.; Vo, K.D.; et al. Whole blood stabilization for the microfluidic isolation and molecular characterization of circulating tumor cells. *Nat. Commun.* **2017**, *8*, 1733. [CrossRef]
129. Lake, B.B.; Codeluppi, S.; Yung, Y.C.; Gao, D.; Chun, J.; Kharchenko, P.V.; Linnarsson, S.; Zhang, K. A comparative strategy for single-nucleus and single-cell transcriptomes confirms accuracy in predicted cell-type expression from nuclear RNA. *Sci. Rep.* **2017**, *7*, 6031. [CrossRef]
130. Macosko, E.Z.; Basu, A.; Satija, R.; Nemesh, J.; Shekhar, K.; Goldman, M.; Tirosh, I.; Bialas, A.R.; Kamitaki, N.; Martersteck, E.M.; et al. Highly Parallel Genome-wide Expression Profiling of Individual Cells Using Nanoliter Droplets. *Cell* **2015**, *161*, 1202–1214. [CrossRef]
131. Gierahn, T.M.; Wadsworth, M.H., 2nd; Hughes, T.K.; Bryson, B.D.; Butler, A.; Satija, R.; Fortune, S.; Love, J.C.; Shalek, A.K. Seq-Well: Portable, low-cost RNA sequencing of single cells at high throughput. *Nat. Methods* **2017**, *14*, 395–398. [CrossRef] [PubMed]
132. Klein, A.M.; Mazutis, L.; Akartuna, I.; Tallapragada, N.; Veres, A.; Li, V.; Peshkin, L.; Weitz, D.A.; Kirschner, M.W. Droplet barcoding for single-cell transcriptomics applied to embryonic stem cells. *Cell* **2015**, *161*, 1187–1201. [CrossRef] [PubMed]
133. Fan, H.C.; Fu, G.K.; Fodor, S.P. Expression profiling. Combinatorial labeling of single cells for gene expression cytometry. *Science* **2015**, *347*, 1258367. [CrossRef]
134. Cheng, Y.H.; Chen, Y.C.; Lin, E.; Brien, R.; Jung, S.; Chen, Y.T.; Lee, W.; Hao, Z.; Sahoo, S.; Min Kang, H.; et al. Hydro-Seq enables contamination-free high-throughput single-cell RNA-sequencing for circulating tumor cells. *Nat. Commun.* **2019**, *10*, 2163. [CrossRef] [PubMed]
135. Stott, S.L.; Hsu, C.H.; Tsukrov, D.I.; Yu, M.; Miyamoto, D.T.; Waltman, B.A.; Rothenberg, S.M.; Shah, A.M.; Smas, M.E.; Korir, G.K.; et al. Isolation of circulating tumor cells using a microvortex-generating herringbone-chip. *Proc. Natl. Acad. Sci. USA* **2010**, *107*, 18392–18397. [CrossRef]
136. Pantel, K.; Alix-Panabières, C. Liquid biopsy and minimal residual disease—Latest advances and implications for cure. *Nat. Rev. Clin. Oncol.* **2019**, *16*, 409–424. [CrossRef] [PubMed]
137. Sinkala, E.; Sollier-Christen, E.; Renier, C.; Rosàs-Canyelles, E.; Che, J.; Heirich, K.; Duncombe, T.A.; Vlassakis, J.; Yamauchi, K.A.; Huang, H.; et al. Profiling protein expression in circulating tumour cells using microfluidic western blotting. *Nat. Commun.* **2017**, *8*, 14622. [CrossRef]
138. Abouleila, Y.; Onidani, K.; Ali, A.; Shoji, H.; Kawai, T.; Lim, C.T.; Kumar, V.; Okaya, S.; Kato, K.; Hiyama, E.; et al. Live single cell mass spectrometry reveals cancer-specific metabolic profiles of circulating tumor cells. *Cancer Sci.* **2019**, *110*, 697–706. [CrossRef]
139. Deng, Y.; Zhang, Y.; Sun, S.; Wang, Z.; Wang, M.; Yu, B.; Czajkowsky, D.M.; Liu, B.; Li, Y.; Wei, W.; et al. An Integrated Microfluidic Chip System for Single-Cell Secretion Profiling of Rare Circulating Tumor Cells. *Sci. Rep.* **2014**, *4*, 7499. [CrossRef]
140. Lim, S.B.; Tan, S.J.; Lim, W.-T.; Lim, C.T. Compendiums of cancer transcriptomes for machine learning applications. *Sci. Data* **2019**, *6*, 194. [CrossRef]
141. Lim, S.B.; Tan, S.J.; Lim, W.-T.; Lim, C.T. A merged lung cancer transcriptome dataset for clinical predictive modeling. *Sci. Data* **2018**, *5*, 180136. [CrossRef] [PubMed]

2

Circulating Tumor Cells in Pancreatic Cancer: Current Perspectives

Verena Martini [1], Sylvia Timme-Bronsert [2], Stefan Fichtner-Feigl [1,3], Jens Hoeppner [1,3] and Birte Kulemann [1,*]

1 Department of General and Visceral Surgery, Medical Center-University of Freiburg, Faculty of Medicine, University of Freiburg, 79106 Freiburg, Germany; verena.martini@uniklinik-freiburg.de (V.M.); stefan.fichtner@uniklinik-freiburg.de (S.F.-F.); jens.hoeppner@uniklinik-freiburg.de (J.H.)
2 Institute for Surgical Pathology, Medical Center-University of Freiburg, Faculty of Medicine, University of Freiburg, 79106 Freiburg, Germany; sylvia.timme@uniklinik-freiburg.de
3 Comprehensive Cancer Center Freiburg-CCCF, Medical Center-University of Freiburg, 79106 Freiburg, Germany
* Correspondence: birte.kulemann@uniklinik-freiburg.de

Abstract: Pancreatic cancer is the fourth leading cause of cancer-related death in the USA and Europe; early symptoms and screenings are lacking, and it is usually diagnosed late with a poor prognosis. Circulating tumor cells (CTCs) have been promising new biomarkers in solid tumors. In the last twenty years (1999–2019), 140 articles have contained the key words "Circulating tumor cells, pancreatic cancer, prognosis and diagnosis." Articles were evaluated for the use of CTCs as prognostic markers and their correlation to survival in pancreatic ductal adenocarcinoma (PDAC). In the final selected 17 articles, the CTC detection rate varied greatly between different enrichment methodologies and ranged from 11% to 92%; the majority of studies used the antigen-dependent CellSearch© system for CTC detection. Fifteen of the reviewed studies showed a correlation between CTC presence and a worse overall survival. The heterogeneity of CTC-detection methods and the lack of uniform results hinder a comparison of the evaluated studies. However, CTCs can be detected in pancreatic cancer and harbor a hope to serve as an early detection tool. Larger studies are needed to corroborate CTCs as valid biomarkers in pancreatic cancer.

Keywords: pancreatic ductal adenocarcinoma; circulating tumor cells; liquid biopsy; biomarker; overall survival

---

1. Introduction

Pancreatic ductal adenocarcinoma (PDAC) is the fourth leading cause of cancer-related death in the United States and Europe [1,2]. Its incidence almost equals mortality with a five-year survival rate of <6% [2], and it will likely be the second leading cause of cancer-related deaths in 2030, exceeding mortality from breast and colorectal cancer [3]. The disease is often detected in metastatic stages, reflecting its late diagnosis and aggressive biology. Early symptoms are rare and most patients initially present with mild symptoms such as abdominal discomfort, back pain, tiredness and weight loss. Distant organ metastases are the final result of hematogenous cancer cell spread, with which about 80% of patients are diagnosed; no sufficient screening regimen is presently used for the early detection of PDAC.

Currently, most treatment decisions are based on the tumor stage estimated by cross-sectional imaging and endoscopic tumor biopsy. Computed tomography (CT) and magnet resonance imaging (MRI) are widely applied, but MRI has the higher diagnostic sensitivity and accuracy [4]. About one-fifth of the patients are "under-staged" since metastatic disease is often only visible upon operative exploration [4].

To date, only serum CA19.9 is routinely used as a non-invasive blood-based biomarker in PDAC. It does not, however, function as diagnostic tool alone—but high CA19.9 levels often indicate advanced tumors [5]. The concern regarding this biomarker is that an elevated expression of CA19.9 can be found in various benign (pancreatitis, acute cholangitis and cirrhosis) and malignant (colorectal cancer, gastric cancer, bladder cancer and uterine squamous cell carcinoma) diseases in addition to PDAC, resulting in its non-specificity. Additionally, about 10% of the Caucasian population do not express CA19.9, and only 65% of patients with resectable PDAC have elevated levels of CA19.9 [6]. CEA also serves as a biomarker, though it has some of the shortcomings of CA19.9 and also lacks specificity.

Other blood-based biomarkers are not routinely used or have not yet entered clinical practice. Liquid biopsies, such as circulating tumor DNA (ctDNA), exosomes, plasma proteomics and circulating tumor cells (CTCs) have now been investigated for over a decade and have presented encouraging results, which are partly discussed in this review; however, clear clinical recommendations for patients with PDAC are still missing.

The routine treatment of resectable PDAC includes complete tumor resection followed by heterogeneous chemotherapy regimens. Gemcitabine is routinely applied as a first-line therapy with a survival benefit of several months [7]. Historically, neoadjuvant treatment regimens have shown outcome benefits. Without randomized trials, however, physicians stick to upfront surgery in resectable cases [8]. Some new adjuvant chemotherapeutic protocols have lately come into clinical practice. Gemcitabine, FOLFIRINOX and paclitaxel have been applied as part of clinical studies, and the combination of nab-paclitaxel and gemcitabine has recently shown an extended overall survival in metastatic pancreatic carcinoma compared to gemcitabine alone [9]. The same results have been documented for gemcitabine and FOLFIRINOX [10] with higher toxicity and more adverse events in the FOLFIRINOX group in the metastatic stage. The current PANACHE01 trial evaluates the efficacy of two neoadjuvant chemotherapeutic regimens with FOLFOX and mFOLFIRINOX compared to standard treatment with surgery followed by adjuvant gemcitabine in resectable PDAC [11].

The limited diagnostic abilities, as well as the complex and narrow treatment options highlight the need for a prognostic and predictive biomarker. A prognostic biomarker can provide insight into a patient's outcome, independent of the treatment received. This biomarker can give an estimation of survival, time to disease progression, or type of progression. Differently, a predictive biomarker is able to identify a subset of patients who would benefit from specific cancer treatment and hence can guide treatment decisions. An ideal biomarker is non-invasive, repeatedly accessible, reasonably-priced, and simple to analyze.

In this review, we give an overview of the current literature on CTCs in pancreatic cancer in a clinical context. We focus on studies that evaluate the clinical applications of CTCs in patients with PDAC. We also discuss the problems of inhomogeneous CTC isolation and analysis, and we highlight potential research perspectives.

2. Diagnosis and Etiology of Pancreatic Cancer

Patients with pancreatic cancer often present with unspecific symptoms such as weight loss, back pain, abdominal pain, and fatigue. Cross-sectional imaging, endoscopy, serum biomarkers, and histological biopsies are currently potential tools used in clinical routines to diagnose pancreatic cancer [12,13]. When imaging is not conclusive, and before oncologic treatment, endoscopic ultrasound techniques are combined with fine-needle aspiration in order to obtain a biopsy and secure the tumor diagnosis. Due to the anatomical position of the pancreas, this procedure is technically challenging and carries a high level of complication. Additionally, false negative results occur in many cases due to the cannulation of stromal cells which represent the majority of the tumor mass [14]. Repeated invasive biopsies are the consequence and harbor a risk for tumor spread and post-interventional complications.

Some cases require laparoscopic- or even open-surgery to obtain a tumor specimen to confirm the diagnosis of pancreatic cancer. To date, no specific screening tests are available for pancreatic cancer.

The causes of pancreatic cancer (PC) are multifactorial and diverse. High-risk groups are patients with a positive family history of PC, hereditary PC, familial atypical multiple mole melanoma (p16 mutation), cystic fibrosis, Peutz–Jeghers syndrome, Lynch syndrome and HBOC (BRCA1/BRCA2 mutation). Risk factors for the development of PC are high age, male gender, obesity, smoking, pancreatitis, and diabetes mellitus. New-onset diabetes mellitus is considered a risk for PC, but whether it is the cause or effect of PC is still unclear [4,7,15].

PDAC is characterized by multiple genetic variants which have been implicated in tumor formation and progression. Point mutations in the *Kirsten rat sarcoma viral oncogene homolog* (*KRAS*) gene are present in over 90% of tumors [7,16–18] and are thought to be an early event in tumorigenesis. Other driver mutations like TP53, p16 and SMAD4 are also commonly found in PDAC specimens [15,16]. More recently, the theory of a "three-step procedure" of carcinogenesis was introduced for PDAC. These steps are: First, the initiation of the tumor by the acquisition of a driver gene mutation in a cell of origin; second, the clonal expansion of the mutation-carrying cell into a multicellular neoplasm; and third, the introduction of the neoplastic cells into both local and distant microenvironments [19,20]. The first step includes the development of pancreatic intraepithelial neoplasms (PanINs) harboring *KRAS* mutations [21] that function as precursor lesions. In other words: Pancreatic tumorigenesis most likely takes years, which offers a long window of time for tumor diagnosis and early intervention.

3. Circulating Tumor Cells in Pancreatic Cancer

CTCs were first pointed out 1896 by the Australian Thomas Ashworth, who described the microscopic observation of CTCs in the blood of a patient with metastatic breast cancer. He postulated: "Cells identical with those of the cancer itself being seen in the blood may tend to throw some light upon the mode of origin of multiple tumors existing in the same person" [22]. It took almost 150 years after this discovery to establish a routine identification of CTCs.

CTCs in the blood stream are thought to represent disseminated tumor cells that have detached from the primary lesion and that are undetectable by clinical imaging and inaccessible to excision. CTCs are thought to be the basis for distant metastasis [23]. Their rarity among billions of blood cells explains the challenge of specific identification and isolation. While CTCs have been extensively studied in other neoplasms, their significance in PDAC at various stages is not completely understood. There is, however, emerging evidence that CTCs may also serve as a valuable tool for the outcome prediction and understanding of tumor biology in PDAC [24–27].

CTCs have been found in all stages of PDAC, starting even with precursor lesions, as described in a subsequent section of this paper [23,24]. Circulating epithelial cells (CECs) have also been found in benign pancreatic lesions such as pancreatitis [23,28,29]; to date, the significance of these finding remains unclear. The detection rate of CTCs as well as the methods to detect them vary greatly. Detection rates have been described from 11% [30] up to 92% with isolation by size [31] or a NanoVelcro assay [15]. To your knowledge, however, it has not been reported that CTCs and CECs can be found in healthy individuals.

Its central location surrounded by important structures makes biopsies challenging and bears the risk of complications. Even the routine and relatively safe endoscopic ultrasound (EUS)-guided fine-needle aspiration (FNA) sampling, the "gold standard" for PDAC diagnosis, is frustrating in 15%–20% of the cases [32,33]. Additionally, the limited number of cells does not always allow for the complete phenotypic and genetic profiling of the retrieved cells, leading to diagnostic limitations. Due to an extensive proportion of stromal cells in tumors, false negative sampling is also possible. Peripheral blood samples can be easily taken at one or multiple time points over the course of treatment without any harm for the patient, and there has been emerging evidence that liquid biopsy may serve as a surrogate for tumor tissues [34]. However, future studies are needed since tumor heterogeneity is a phenomenon that has only been included as a relevant factor in CTC analysis within the last decade [35–37].

One important point of CTC analysis in PDAC is the concept of epithelial–mesenchymal transition (EMT). This is a biological process in which polarized cells that are usually in contact with the basement membrane undergo multiple biochemical changes and gain mesenchymal properties. This leads to an enhanced migratory capacity, invasiveness and elevated resistance to apoptosis. Cells can detach from the primary lesion and enter into the bloodstream (Figure 1A). At a distant site, they can undergo the reverse process, the mesenchymal–epithelial transition (MET) to induce a new metastasis [38]. EMT is involved in embryo formation and organogenesis (type 1), in wound healing (type 2), and in the formation of neoplastic cells and metastases (type 3) [39].

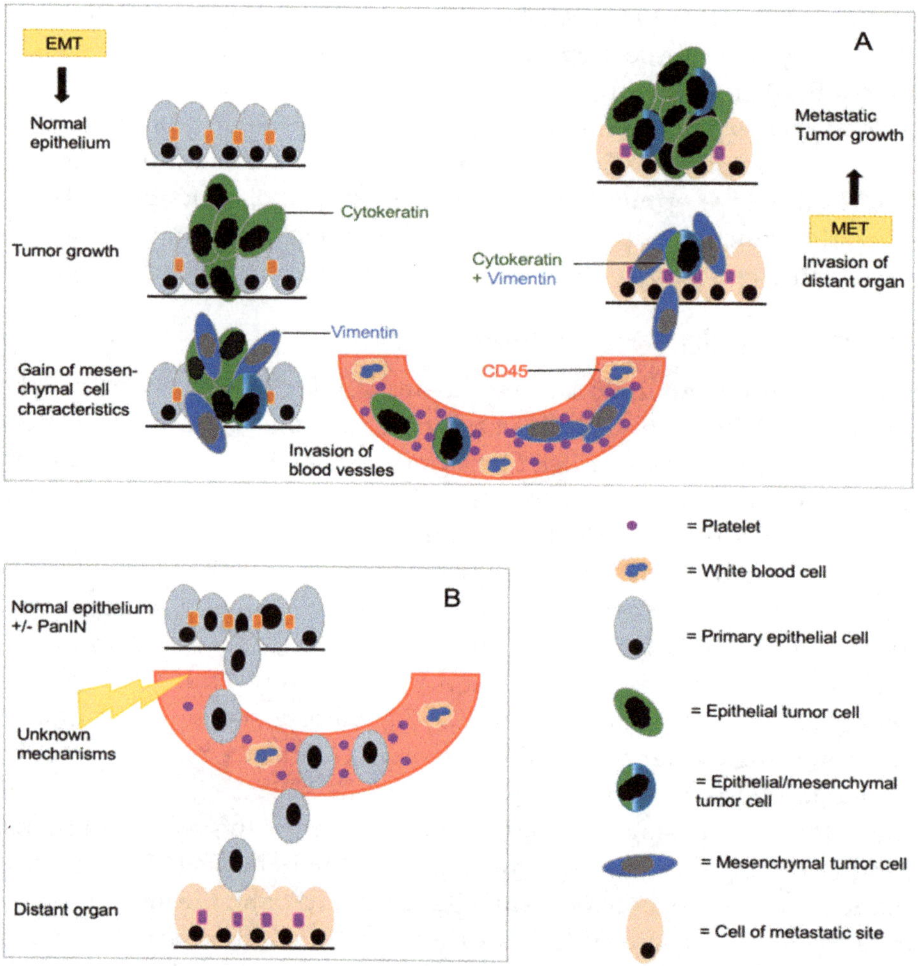

Figure 1. Schematic display of epithelial–mesenchymal transition (EMT) and mesenchymal–epithelial transition (MET) in pancreatic cancer development. Circulating epithelial, circulating mesenchymal and circulating epithelial/mesenchymal cells are displayed to show the heterogeneity of circulating tumor cells (CTCs) (**A**). (**B**) shows the concept of early dissemination before malignancy. Mechanisms involved in this process are widely unknown and part of current research.

EMC in pancreatic cancer seems to be associated with portal vein invasion and lymph node metastasis. Furthermore, premalignant pancreatic lesions (IPMN—intraductal papillary mucinous neoplasm—borderline and carcinoma in situ) also undergo EMT. CTCs are thought to be heterogeneous groups of cells with varying phenotypic and genotypic properties. In lung cancer [37] and breast cancer [40], new studies on single CTC have shown substantial inter- and intra-patient heterogeneity: Different types of CTCs have been found among different patients but also within one patient. Our own studies have also shown a heterogeneous group of cells in the blood with different staining and *KRAS* mutational properties [16,31]. CTCs with both epithelial and mesenchymal markers have been found. The presence of the mesenchymal EMT marker ZEB1 (Zinc finger homebox 1) and the expression of epithelial CK (cytokeratin) have shown no statistically significant impact on survival in patients with PDAC [16]. In contrast, Rhim et al. were able to show that EMT-transformed cells have tumor initiating properties [23].

3.1. Methods of CTC Isolation and Detection

CTCs are very rare. With one cell in 10^6–10^8 leukocytes in 1 mL of whole blood, their detection is a challenge, especially in PDAC due to their low rate of occurrence in comparison to other tumor entities. Thus, CTC isolation is generally a two-step procedure. Step one is CTC enrichment followed by step two, CTC detection.

CTC enrichment techniques are as follows:

- Physical/biological systems isolate CTCs on the basis of electric charge or cell size. The enrichment relies on the fact that CTCs have a higher density, a different electric charge, a different motility, and a larger size than normal blood components. They can thus be isolated with dedicated devices (microfilters, microfluidic chips, electric separation, etc.).
- Surface antibody–based enrichment can be used as "positive" or "negative" enrichment. Positive selection relies on antibodies directed against the surface markers of the CTCs. Common CTC isolation techniques use the epithelial cell adhesion molecule (EpCAM) for positive selection. To eliminate normal blood components, the negative selection systems deplete leukocytes from the specimens. They use CD45 antibodies to bind the blood components and to isolate the CD45-negative CTCs.

The second step—the detection of the enriched CTCs—can also be performed with various approaches.

- Morphological examination supported by immunocytological staining for cancer-specific antibodies is the gold standard for CTC detection and definition. This can be performed after physical enrichment but can also be performed in part with surface-antibody based methods, even though the cytological evaluation is not as good as for regular cytological specimens.
- mRNA analysis for epithelial or pancreatic markers without morphologic control has been performed [41–43] and has shown interesting results in PDAC.
- Mutational analysis of the DNA of enriched cells for tumor-specific mutations in PDAC—typically *KRAS*. This also allows for comparison of the mutational landscape of the CTC with the primary tumor with a high specificity. Our own results showed a heterogeneous picture of the CTC in comparison with the primary tumor; we had about 40% discordant "CTC-primary tumor pairs" [31]. Ankeny et al. found a 100% concordance in five pairs tested [25].

Technologies based on physical properties such as ScreenCell, ISET® (isolation by size of epithelial tumor cells), ApoStream™, ScreenCell®, ClearCell® FX System, and density or gradient centrifugation are used to enrich and detect CTCs. The size-based or fluidic-based selection strategies use the fact that CTCs are physiologically different and notably larger in size than normal blood components [44]. Small CTCs that are cytokeratin-positive, CD45-negative, and morphologically resemble white blood cells are, however, not captured by this technique.

The surface antibody-based technique by CellSearch© (Menarini Biosystems, IT) has been approved by the FDA for CTC detection in breast, colorectal, and prostate cancer, and the authors refer to this technique as the "gold standard" for CTC isolation [45]. The CellSearch© system captures CTCs with the use of the epithelial cell adhesion molecule (EpCAM) followed by the characterization of the captured CTC with cytokeratin positivity and CD45 negativity to exclude white blood cells. Recent studies, however, have shown that the EpCAM-based strategy fails to detect CTCs with low EpCAM expression, since CTCs may lose their epithelial antigens during the EMT process. Moreover, these EpCAM-negative CTCs seem to be an aggressive and invasive subtype [36,46,47]. Cells that express the leukocyte antigen CD45 are not classified as CTCs but as regular blood components by the CellSearch© technique. However, this ignores the CTC–blood cell interaction that leads to "CD45-positive" clusters, some of which may contain CTCs. A recent study also investigated the cell–cell interaction of CTCs and neutrophil granulocytes within the blood stream. The authors found cell cycle genes upregulated in CTC-granulocyte clusters and concluded that the "association between neutrophils and CTCs drives cell cycle progression within the bloodstream and expands the metastatic potential of CTCs, providing a rationale for targeting this interaction in the treatment of breast cancer." [48]. In this manner, EpCAM-based methods may underestimate those CTCs that are highly aggressive and invasive [14].

In summary, there is a broad heterogeneity in the detection methods of CTCs, none of which are perfect. The wide variety of results reported in the literature makes a comparison barely possible. Future studies should combine different isolation techniques to reach a better comparability until an innovative strategy has been developed that combines immunological, physiological and genetic analyses in one device. A single-cell CTC analysis will help us to identify the biologically-relevant fraction of CTCs that cause metastases.

3.2. Clinical Utility of CTCs

3.2.1. Differential Diagnosis and Early Detection

No standardized screening tool exists for PDAC beyond imaging; thus, diagnosis and differential diagnosis rely on cross-sectional imaging and EUS accompanied/confirmed by cytologic specimens. Other preneoplastic lesions (such as IPMN) or benign lesions (such as chronic pancreatitis) have to be distinguished from cancer by imaging and EUS-FNA. An over-interpretation of benign lesions is as much a reality as the false "tumor-negative" classification of a pancreatic mass. A diagnostic tool to screen and to differentiate pancreatic cancer from other pancreatic lesions would be extremely helpful. CTCs or CECs have been evaluated in benign diseases as well as in precursor lesions in pilot studies. A study by Rhim et al. found CECs in 33% of patients with IPMN but without a malignant tumor. In PDAC, they found these cells in 73% of the cases [49]. The same group found pancreatic CECs in premalignant stages in a mouse model before frank malignancy occurred. The cells entered the blood stream much earlier than expected (Figure 1B). Additionally, treatment regimens with dexamethasone led to the almost complete disappearance of CECs in these mice, suggesting that inflammatory stroma facilitate EMT and allow early dissemination [23]. These findings in mice and humans are novel and open a variety of therapeutic opportunities for high-risk patients. The authors stated that "...if these cells represent the earliest forms of cancer, we predict that they would contain a complement of somatic mutations associated with PDAC." Studies evaluating the genetic landscape of these CECs and long-term studies to evaluate the development of PDAC in these patients are currently being performed [49].

Cauley et al. published the results of the CEC-cytology of 179 patients with a variety of pancreatic lesions such as chronic pancreatitis, IPMN, PDAC and neuroendocrine tumors [29]. Interestingly, rates of CEC identification were similar in patients with benign, premalignant, and malignant lesions, and CEC findings in PDAC patients were not associated with a poor prognosis in a follow-up of one year. A more detailed morphologic analysis of these cells revealed a similar cytological appearance in patients with benign and malignant disease [28]. Immunophenotypic and genetic information is lacking in these studies; neither mutational analysis, proliferation analysis (ki67) nor phenotypic analysis have been performed. In summary, there are circulating epithelial cells in transit in patients with premalignant and benign diseases. This leads to the potential for tumor detection in premalignant stages. However, clinical utility remains obscure and needs further investigation.

To our knowledge, there have thus far no publications regarding the utility of circulating tumor cells for early detection and screening in PDAC patients. Circulating tumor DNA (ctDNA), not particularly a part of this review, has, however, gained recent focus in order to improve this field. A study by Cohen et al. found a combination of well-thresholded plasma proteins and a ctDNA analysis for *KRAS* mutations very useful for the earlier detection of PDAC. The sensitivity of this combination test was 64%, with 99.5% specificity [50].

Nevertheless, CTCs harbor the potential for earlier tumor detection and future, large-scale trials are needed. Two relevant studies are registered in the study registry clinicaltrials.gov: One from Rouen, France (NCT02072616) explores CTCs for improvements in diagnostic testing; the other study from Missouri, Columbia (NCT03551951) explores liquid biopsy markers including CTCs in various solid tumors. To our knowledge, there is no large-scale (>1000 patients) CTC screening study for PDAC in preparation.

3.2.2. CTC as Prognostic Marker

CTCs have been shown to serve as prognostic markers in several pilot studies (Table 1). Conventional prognostic factors such as tumor size, nodal status, and perineural invasion can be evaluated only after tumor resection and mostly confirm a poor prognosis. This is different with CTCs, because they offer a window into cancer development and progression in a blood sample. A meta-analysis from 2014 included nine cohort studies with over 600 PDAC patients (UICC stage I–IV). It showed that the patients with positive CTCs (43%) had a poorer progression-free survival (PFS) ($p < 0.001$) and overall survival (OS; $p < 0.001$) than CTC-negative patients, suggesting that CTCs may be promising biomarkers for the prognosis of PDAC [51]. Still, results have greatly varied and are dependent on the isolation technique used and the patient population. Bidard et al. investigated patients with locally advanced PDAC and found 11% of CTC-positive patients with CellSearch®. This finding was, however, clinically relevant; patients with CTCs had a worse OS [30]. Others using the CellSearch® technique also reported a relatively low CTC positivity [52]. Studies which have used antibody-independent CTC isolation have usually had more CTC-positive patients (86%–92%) [24,27,53], but some studies have reported only 34% of positive patients after density gradient separation.

In a recent study of 50 patients with PDAC, Poruk et al. evaluated CTCs isolated with the label-free ISET® device and stained for cytokeratin and the mesenchymal-marker vimentin. The presence of cytokeratin-positive CTCs was shown to be a significant independent predictor of survival by univariate and multivariable analyses after accounting for other prognostic factors. The detection of CTC co-expressing markers, vimentin and cytokeratin, was shown to be predictive for tumor relapse. Patients with vimentin-positive and cytokeratin-positive CTCs had an earlier recurrence than patients without CTCs [54]. The same group evaluated the markers for tumor initiating cells (TIC) and stemness in CTCs found in patients with PDAC (cytokeratin, CD133, CD44, and ALDH). These markers were explored as prognostic factors in a group of 60 patients. They found that ALDH-positive CTCs and triple-positive CTCs were significantly associated with worse survival by univariate analysis, and ALDH-positive CTCs, triple-positive CTCs, and dual cytokeratin- and CD133-positive CTCs were not only independent predictors of tumor recurrence but were also associated with shorter disease-free survival [55].

A group from Baltimore provided the thus far largest study on CTC dynamics over treatment. In the CLUSTER study, multiple blood samples from 200 patients with presumed PDAC were prospectively collected and evaluated for CTCs with the label-free ISET® device over the course of treatment. Out of all the CTCs (tCTC), the authors identified two major groups of CTCs: eCTC, with epithelial phenotype (cytokeratin+ and vimentin−) and mCTC with a mixed epithelial/mesenchymal phenotype (cytokeratin+ and vimentin+). Additionally, they performed a "cocktail" stain to separate different white blood cell (WBC) populations (anti-CD45, CD11b, CD14, and CD34). Of the 136 resected patients, 56 had to undergo neoadjuvant chemotherapy. These patients showed significantly fewer CTCs of all subtypes; surgery let the CTC counts drop severely. It was striking that preoperative numbers of all CTC subpopulations were the only predictors of early recurrence within 12 months after surgery in the pretreated and the primarily resected patients. OS was significantly longer in the pretreated patients without CTCs before therapy.

Table 1. Studies evaluating CTCs in pancreatic ductal adenocarcinoma (PDAC) including survival analysis.

Study	Country	*n*	Tumor Stage	Methodology	CTC Detection Rate *	Outcome	Finding
Soeth et al., 2005 [43]	Germany	172	All stages, mostly IV	Density gradient separation	34%	OS	CTC+: poor OS
Hoffmann et al., 2007 [41]	Germany	37	All stages	Density gradient separation	40%	OS	CTC+: Trend for worse OS
Kurihara et al., 2008 [56]	Japan	26	Stage III and IV	CellSearch®	42%	OS	CTC+: sig. worse OS
Sergeant et al., 2011 [42]	Belgium	48	All stages, 40 resectable	RT-PCR (EpCAM)	25% preoperative 65% postoperative	OS	No correlation between CTC and survival
Khoja et al., 2012 [26]	United Kingdom	54	54% non-resectable	ISET®; CellSearch®	46%/40%	OS; PFS	ISET® detects more CTCs, trend towards decreased OS
De Albuquerque et al., 2012 [57]	Germany	34	All stages (II–IV)	Immunomagnetic (EPCAM)	47%	PFS	CTC + worse PFS
Bidard et al., 2013 [30]	France	79	Non resectable	CellSearch®	11%	OS	CTC+ worse OS
Bissolati et al., 2015 [58]	Italy	20	Resectable Stage IIa and b	CellSearch®	45%	OS; PFS	No correlation to OS or PFS but slightly higher liver metastasis rate in CTC+
Zhang et al., 2015 [59]	China	22	Stage I–IV, all resectable	Immunomagnetic (CEP 8/ CD45)	15%	OS	CTC+: worse OS
Earl et al., 2015 [60]	Spain	35	Stage II–IV	CellSearch®	20%	OS	CTC*: worse OS, almost only metastatic disease
Poruk et al., 2016 [54]	USA	60	All stages	ISET®	90%	OS, PFS	Epithelial CTC+: worse OS Epithelial CTC+: earlier recurrence
Gao et al., 2016 [53]	China	25	All stages	CD45 depletion and SE-FISH	I+II: 92.3% III + IV: 83.3%	OS	Patients with lower CTC count better OS than patients with high numbers of CTC
Kulemann et al., 2017 [31]	Germany	58	All stages	ScreenCell®	68%	OS	CTC+: Trend to worse OS
Okubo et al., 2017 [52]	Japan	65	III–IV	CellSearch®	21%	OS	CTC+ worse OS, more CTC+ in pat. Liver metastases, CTC+ after treatment neg. prong. Factor.
Poruk et al., 2017 [55]	USA	60	All stages mainly I and II	ISET®	78%	OS, PFS	CTC labeled with TIC (tumor initiating cell) are predictive of decreased OS and PFS
Gemenetzis et al., 2018 [61]	USA	165	All stages	ISET®	95% of resectable patients (n = 136)	Identification of mesenchymal-mal and epithelial CTC OS; PFS	Higher CTC counts correlate with earlier recurrence Increase of CTC numbers after neoadjuvant treatment CTC+ correlates with early recurrence and OS in the pretreated group.
Court et al., 2018 [27]	USA	100	All stages 71 localized, 29 metastatic	Nano Velcro Chip enumeration	78%	Identification of occult metastasis; OS	CTC counts correlated with stage and worse OS

* A cut off for CTCs in order to predict prognosis has not been established yet. In different studies, varying amounts of blood and cut offs were used. Maestro et al. detected >2 CTCs/7.5 mL as a prognostically-usable cut off, and others defined this cut off as >1/7.5 mL [26,30]. SeFish: immunostaining-fluorescence in situ hybridization.

In the chemo-naïve group, CTCs were not associated with a worse overall survival. The authors also found an increase of CTCs before recurrence occurred and developed a risk assessment score based on the difference of tCTC increase. This score accurately identified disease recurrence within the next two months, with an accuracy 75% [61]. In summary, there is emerging evidence for (subsets of) CTCs as prognostic markers and clinically relevant tools. All studies mentioned above are displayed in Table 1.

3.2.3. CTC as Predictive Marker

A recent meta-analysis on intensified chemotherapy in PDAC revealed that only about 20–30% of the patients show a response to this treatment according to the RECIST criteria [62]. Thus, a substantial number of the patients received this toxic combination without benefit. The toxicity of FOLFIRINOX compared to gemcitabine is a serious problem. Identifying non-responders is essential to avoid unnecessary costs and, more importantly, to prevent exposing patients to the adverse events of ineffective and toxic treatment. CTCs could function as biomarkers to address this problem, but studies evaluating this problem have thus been lacking.

A study by Yu et al. investigated a pharmacogenomic (PGx) model to predict treatment response of a PDAC patient to chemotherapy regimens based on the genetic mutations in his CTCs. The authors found that PGx profiling of CTCs can predict treatment response. Additionally, they were able to stratify patients as "sensitive" and "insensitive" for specific chemotherapy regimens, showing the benefits of progression-free and overall survival [63]. It remains unclear, however, whether these models mimic reality in PDAC patients.

3.2.4. Advanced Clinical Utility in Other Entities

Prostate, lung and breast cancer are the pioneer fields in the area of solid cancers in which CTCs detection are used as prognostic and predictive parameters and to evaluate their biologic potential [36,48,64]. These tumors shed higher numbers of CTCs into the blood stream, which makes it more feasible to study their potential [37,65,66]. The number of CTCs before treatment has been described as an independent marker of overall survival in metastatic breast cancer [67] and has also entered the UICC classification as micro-metastases called cM0(i+) [68]. Recently-published results from the SUCCESS trial on over 1000 women with breast cancer showed a CTC positivity of 18.2% two years after chemotherapy. CTCs at baseline and in follow-up were identified as independent markers of an inferior OS [69]. Specific subtypes of CTCs with different expressions of HER2 have been described as clinically relevant [70] and as changing over the course of treatment [71]. Patients with HER2-negative breast cancer but HER2-positive CTCs were investigated in a recent study. In the study, the HER2-targeting trastuzumab did not decrease the detection of HER2-positive CTCs and did not significantly influence survival [72]. In another study by Jakabova et al., the HER2 and ESR status of CTCs differed from the status of the primary tumor. Over the course of treatment, even switches of the HER2 status of the CTCs occurred in some patients [71]. Even larger studies are also needed in this regard. One view for future applications of CTCs in lung cancer was recently given by Palliers et al., who investigated different resistance mutations in the single CTCs of patients with lung cancer resistant to crizotinib, an ALK-rearrangement targeting drug. They found multiple mutations in various genes in ALK-independent pathways and showed the genetic heterogeneity and clinical utility of CTCs to identify therapeutic resistance mutations in ALK-rearranged patients [37]. Single-CTC sequencing may be a unique tool to assess heterogeneous resistance mechanisms and help clinicians to personalize treatment and resistance options in patients with lung cancer but also with other malignancies such as pancreatic cancer.

4. Materials and Methods

In September 2019, PubMed, Embase, Medline, Web of Science, Cochrane Library, and Google Scholar were searched using the search terms "Circulating tumor cells, pancreatic cancer, prognosis and diagnosis," which revealed 145 studies over the last twenty years (1999–2019). This review addresses the role of circulating tumor cells in pancreatic cancer as prognostic and diagnostic tools. The literature search identified articles and reviews addressing this topic. One-hundred-and-two studies were excluded due to their titles and abstracts, wrong tumor entity, or irrelevance in other details. The remaining 43 full text articles were screened for eligibility. 26 were excluded due to non-sufficient data or no relevance to PDAC or CTCs. The remaining 17 articles were analyzed for this review. The criteria identified were the number of patients, country of origin, tumor stage, methods used for the enrichment and detection of CTCs, the detection rate of CTCs in the cohorts, and the outcome parameters (OS, and PFS). Letters, non-English language studies, editorials, and expert opinions were excluded. The bibliographies of included studies were searched for other relevant publications.

5. Conclusions and Future Perspectives

PDAC is hardly accessible to tissue biopsy due to its retroperitoneal position. Pre- and intra-therapeutic CTC enrichment and detection can be performed multiple times over a treatment course. The clinical utility of CTCs as prognostic markers and also as risk factors for disease recurrence have been found in patients with PDAC in pilot and prospective midsize studies. However, large-scale studies in PDAC patients with consistent CTC isolation techniques are greatly needed to validate the diagnostic and prognostic potential of CTCs. The predictive value of CTCs has as yet to be sufficiently investigated.

The development from the initiation of mutation to metastatic cancer takes years, giving rise to the potential to interfere before the manifestation of locally advanced or even metastatic tumors. Thus, screening for and the early detection of PDAC in the "windows of opportunity" should be the focus of future trials. Additionally, agreement is needed on a "gold standard" for CTC isolation in PDAC. Due to the EMT features of pancreatic CTCs, this should most likely be a physical enrichment technique followed by a yet-to-be-determined detection.

A single cell analysis of CTCs will prospectively give us more insight into tumor heterogeneity. may reveal particularly aggressive subtypes of CTCs, and may help us to identify treatment targets to fight this devastating disease. A true clinical advantage for patients would be the prevention of invasive carcinoma. Potentially, we will perform more preventive surgery in the future.

Author Contributions: V.M. did the bibliographic search; V.M. and B.K. wrote the initial manuscript. S.T.-B., S.F.-F., and J.H. critically revised the manuscript and actively contributed to it.

Abbreviations

CTC	Circulating tumor cell	OS	Overall survival
CEC	Circulating epithelial cell	PFS	Progression-free survival
tCTC	Total CTC	EUS	Endoscopic ultrasound
mCTC	Mesenchymal/epithelial CTC	FNA	Fine-needle aspiration
eCTC	Epithelial CTC	CK	Cytokeratin
PDAC	Pancreatic ductal adenocarcinoma	ZEB1	Zinc finger homebox 1
IPMN	Intraductal papillary mucinous neoplasm	SeFish	Immunostaining-fluorescence in situ hybridization
EMT	Epithelial-mesenchymal transition	HER2	Human epidermal growth factor receptor 2
MET	Mesenchymal-epithelial transition	TIC	Tumor initiating cell

References

1. Jemal, A.; Siegel, R.; Xu, J.; Ward, E. Cancer statistics. *CA Cancer J. Clin.* **2010**, *60*, 277–300.
2. Siegel, R.L.; Miller, K.D.; Jemal, A. Cancer statistics. *CA Cancer J. Clin.* **2015**, *65*, 5–29.

3. Rahib, L.; Smith, B.D.; Aizenberg, R.; Rosenzweig, A.B.; Fleshman, J.M.; Matrisian, L.M. Projecting cancer incidence and deaths to 2030: The unexpected burden of thyroid, liver, and pancreas cancers in the united states. *Cancer Res.* **2014**, *74*, 2913–2921. [CrossRef] [PubMed]
4. Becker, A.E.; Hernandez, Y.G.; Frucht, H.; Lucas, A.L. Pancreatic ductal adenocarcinoma: Risk factors, screening, and early detection. *World J. Gastroenterol.* **2014**, *20*, 11182–11198. [CrossRef]
5. Ferrone, C.R.; Finkelstein, D.M.; Thayer, S.P.; Muzikansky, A.; Fernandez-delCastillo, C.; Warshaw, A.L. Perioperative CA19-9 levels can predict stage and survival in patients with resectable pancreatic adenocarcinoma. *J. Clin. Oncol.* **2006**, *24*, 2897–2902. [CrossRef]
6. Goggins, M. Molecular markers of early pancreatic cancer. *J. Clin. Oncol.* **2005**, *23*, 4524–4531. [CrossRef]
7. Vincent, A.; Herman, J.; Schulick, R.; Hruban, R.H.; Goggins, M. Pancreatic cancer. *Lancet* **2011**, *378*, 607–620. [CrossRef]
8. Lutz, M.P.; Zalcberg, J.R.; Ducreux, M.; Aust, D.; Bruno, M.J.; Büchler, M.W.; Delpero, J.R.; Gloor, B.; Glynne-Jones, R.; Hartwig, W.; et al. 3rd St. Gallen EORTC Gastrointestinal Cancer Conference: Consensus recommendations on controversial issues in the primary treatment of pancreatic cancer. *Eur. J. Cancer* **2017**, *79*, 41–49. [CrossRef]
9. Von Hoff, D.D.; Ervin, T.; Arena, F.P.; Chiorean, E.G.; Infante, J.; Moore, M.; Seay, T.; Tjulandin, S.A.; Ma, W.W.; Saleh, M.N.; et al. Increased survival in pancreatic cancer with nab-paclitaxel plus gemcitabine. *N. Engl. J. Med.* **2013**, *369*, 1691–1703. [CrossRef]
10. Conroy, T.; Desseigne, F.; Ychou, M.; Bouché, O.; Guimbaud, R.; Bécouarn, Y.; Adenis, A.; Raoul, J.-L.; Gourgou-Bourgade, S.; de la Fouchardière, C.; et al. FOLFIRINOX versus gemcitabine for metastatic pancreatic cancer. *N. Engl. J. Med.* **2011**, *364*, 1817–1825. [CrossRef]
11. Schwarz, L.; Vernerey, D.; Bachet, J.-B.; Tuech, J.-J.; Portales, F.; Michel, P.; Sa Cunha, A. Resectable pancreatic adenocarcinoma neo-adjuvant FOLF(IRIN)OX-based chemotherapy-a multicenter, non-comparative, randomized, phase II trial (PANACHE01-PRODIGE48 study). *BMC Cancer* **2018**, *18*, 762. [CrossRef] [PubMed]
12. Diehl, S.J.; Lehmann, K.J.; Sadick, M.; Lachmann, R.; Georgi, M. Pancreatic cancer: Value of dual-phase helical CT in assessing resectability. *Radiology* **1998**, *206*, 373–378. [CrossRef] [PubMed]
13. Sefrioui, D.; Blanchard, F.; Toure, E.; Basile, P.; Beaussire, L.; Dolfus, C.; Perdrix, A.; Paresy, M.; Antonietti, M.; Iwanicki-Caron, I.; et al. Diagnostic value of CA19.9, circulating tumour DNA and circulating tumour cells in patients with solid pancreatic tumours. *Br. J. Cancer* **2017**, *117*, 1017–1025. [CrossRef] [PubMed]
14. Reimers, N.; Pantel, K. Liquid biopsy: Novel technologies and clinical applications. *Clin. Chem. Lab. Med.* **2019**, *57*, 312–316. [CrossRef]
15. Court, C.M.; Ankeny, J.S.; Hou, S.; Tseng, H.-R.; Tomlinson, J.S. Improving pancreatic cancer diagnosis using circulating tumor cells: Prospects for staging and single-cell analysis. *Expert Rev. Mol. Diagn.* **2015**, *15*, 1491–1504. [CrossRef]
16. Kulemann, B.; Liss, A.S.; Warshaw, A.L.; Seifert, S.; Bronsert, P.; Glatz, T.; Pitman, M.-B.; Hoeppner, J. KRAS mutations in pancreatic circulating tumor cells: A pilot study. *Tumour Biol.* **2016**, *37*, 7547–7554. [CrossRef]
17. Koorstra, J.B.; Hustinx, S.R.; Offerhaus, G.J.; Maitra, A. Pancreatic carcinogenesis. *Pancreatology* **2008**, *8*, 110–125. [CrossRef]
18. Almoguera, C.; Shibata, D.; Forrester, K.; Martin, J.; Arnheim, N.; Perucho, M. Most human carcinomas of the exocrine pancreas contain mutant c-K-ras genes. *Cell* **1988**, *53*, 549–554. [CrossRef]
19. Makohon-Moore, A.; Iacobuzio-Donahue, C.A. Pancreatic cancer biology and genetics from an evolutionary perspective. *Nat. Rev. Cancer* **2016**, *16*, 553–565. [CrossRef]
20. Nagrath, S.; Jack, R.M.; Sahai, V.; Simeone, D.M. Opportunities and Challenges for Pancreatic Circulating Tumor Cells. *Gastroenterology* **2016**, *151*, 412–426. [CrossRef]
21. Kanda, M.; Matthaei, H.; Wu, J.; Hong, S.-M.; Yu, J.; Borges, M.; Hruban, R.H.; Maitra, A.; Kinzler, K.; Vogelstein, B.; et al. Presence of somatic mutations in most early-stage pancreatic intraepithelial neoplasia. *Gastroenterology* **2012**, *142*, 730–733. [CrossRef] [PubMed]
22. Ashworth, T. A case of cancer in which cells similar to those in the tumors were seen in the blood after death. *Aust. Med. J.* **1869**, *14*, 146–149.
23. Rhim, A.D.; Mirek, E.T.; Aiello, N.M.; Maitra, A.; Bailey, J.M.; McAllister, F.; Reichert, M.; Beatty, G.L.; Rustgi, A.K.; Vonderheide, R.; et al. EMT and dissemination precede pancreatic tumor formation. *Cell* **2012**, *148*, 349–361. [CrossRef] [PubMed]

24. Kulemann, B.; Pitman, M.B.; Liss, A.S.; Valsangkar, N.; Fernández-Del Castillo, C.; Lillemoe, K.D.; Hoeppner, J.; Mino-Kenudson, M.; Warshaw, A.L.; Thayer, S. Circulating tumor cells found in patients with localized and advanced pancreatic cancer. *Pancreas* **2015**, *44*, 547–550. [CrossRef]
25. Ankeny, J.S.; Court, C.M.; Hou, S.; Li, Q.; Song, M.; Wu, D.; Lee, T.; Lin, M.; Sho, S.; Rochefort, M.M.; et al. Circulating tumour cells as a biomarker for diagnosis and staging in pancreatic cancer. *Br. J. Cancer* **2016**, *114*, 1367–1375. [CrossRef]
26. Khoja, L.; Backen, A.; Sloane, R.; Menasce, L.; Ryder, D.; Krebs, M.; Board, R.; Clack, G.; Hughes, A.; Blackhall, F.; et al. A pilot study to explore circulating tumour cells in pancreatic cancer as a novel biomarker. *Br. J. Cancer* **2012**, *106*, 508–516. [CrossRef]
27. Court, C.M.; Ankeny, J.S.; Sho, S.; Winograd, P.; Hou, S.; Song, M.; Wainberg, Z.A.; Girgis, M.; Graeber, T.G.; Agopian, V.; et al. Circulating Tumor Cells Predict Occult Metastatic Disease and Prognosis in Pancreatic Cancer. *Ann. Surg. Oncol.* **2018**, *25*, 1000–1008. [CrossRef]
28. Rosenbaum, M.W.; Cauley, C.E.; Kulemann, B.; Liss, A.S.; Castillo, C.F.-D.; Warshaw, A.L.; Lillemoe, K.D.; Thayer, S.P.; Pitman, M.B. Cytologic characteristics of circulating epithelioid cells in pancreatic disease. *Cancer Cytopathol.* **2017**, *125*, 332–340. [CrossRef]
29. Cauley, C.E.; Pitman, M.B.; Zhou, J.; Perkins, J.; Kuleman, B.; Liss, A.S.; Fernandez-Del Castillo, C.; Warshaw, A.L.; Lillemoe, K.D.; Thayer, S.P.; et al. Circulating Epithelial Cells in Patients with Pancreatic Lesions: Clinical and Pathologic Findings. *J. Am. Coll. Surg.* **2015**, *221*, 699–707. [CrossRef]
30. Bidard, F.C.; Huguet, F.; Louvet, C.; Mineur, L.; Bouché, O.; Chibaudel, B.; Artru, P.; Desseigne, F.; Bachet, J.B.; Mathiot, C.; et al. Circulating tumor cells in locally advanced pancreatic adenocarcinoma: The ancillary CirCe 07 study to the LAP 07 trial. *Ann. Oncol.* **2013**, *24*, 2057–2061. [CrossRef]
31. Kulemann, B.; Rösch, S.; Seifert, S.; Timme, S.; Bronsert, P.; Seifert, G.; Martini, V.; Kuvendjiska, J.; Glatz, T.; Hussung, S.; et al. Pancreatic cancer: Circulating Tumor Cells and Primary Tumors show Heterogeneous KRAS Mutations. *Sci. Rep.* **2017**, *7*, 4510. [CrossRef] [PubMed]
32. Brugge, W.R. Endoscopic approach to the diagnosis and treatment of pancreatic disease. *Curr. Opin. Gastroenterol.* **2013**, *29*, 559–565. [CrossRef] [PubMed]
33. Kitano, M.; Yoshida, T.; Itonaga, M.; Tamura, T.; Hatamaru, K.; Yamashita, Y. Impact of endoscopic ultrasonography on diagnosis of pancreatic cancer. *J. Gastroenterol.* **2019**, *54*, 19–32. [CrossRef] [PubMed]
34. Luchini, C.; Veronese, N.; Nottegar, A.; Cappelletti, V.; Daidone, M.G.; Smith, L.; Parris, C.; Brosens, L.A.A.; Caruso, M.G.; Cheng, L.; et al. Liquid Biopsy as Surrogate for Tissue for Molecular Profiling in Pancreatic Cancer: A Meta-Analysis Towards Precision Medicine. *Cancers (Basel)* **2019**, *11*, 1152. [CrossRef] [PubMed]
35. Aktas, B.; Tewes, M.; Fehm, T.; Hauch, S.; Kimmig, R.; Kasimir-Bauer, S. Stem cell and epithelial-mesenchymal transition markers are frequently overexpressed in circulating tumor cells of metastatic breast cancer patients. *Breast Cancer Res.* **2009**, *11*, 46. [CrossRef] [PubMed]
36. Königsberg, R.; Obermayr, E.; Bises, G.; Pfeiler, G.; Gneist, M.; Wrba, F.; de Santis, M.; Zeillinger, R.; Hudec, M.; Dittrich, C. Detection of EpCAM positive and negative circulating tumor cells in metastatic breast cancer patients. *Acta Oncol.* **2011**, *50*, 700–710. [CrossRef]
37. Pailler, E.; Faugeroux, V.; Oulhen, M.; Mezquita, L.; Laporte, M.; Honore, A.; Lecluse, Y.; Queffelec, P.; Ngo-Camus, M.; Nicotra, C.; et al. Acquired Resistance Mutations to ALK-Inhibitors Identified by Single Circulating Tumor Cell Sequencing in ALK-Rearranged Non-Small-Cell Lung Cancer. *Clin. Cancer Res.* **2019**. [CrossRef]
38. Kalluri, R.; Neilson, E.G. Epithelial-mesenchymal transition and its implications for fibrosis. *J. Clin. Investig.* **2003**, *112*, 1776–1784. [CrossRef]
39. Kalluri, R.; Weinberg, R.A. The basics of epithelial-mesenchymal transition. *J. Clin. Investig.* **2009**, *119*, 1420–1428. [CrossRef]
40. Cheng, Y.-H.; Chen, Y.-C.; Lin, E.; Brien, R.; Jung, S.; Chen, Y.-T.; Lee, W.; Hao, Z.; Sahoo, S.; Kang, H.M.; et al. Hydro-Seq enables contamination-free high-throughput single-cell RNA-sequencing for circulating tumor cells. *Nat. Commun.* **2019**, *10*, 2163. [CrossRef]
41. Hoffmann, K.; Kerner, C.; Wilfert, W.; Mueller, M.; Thiery, J.; Hauss, J.; Witzigmann, H. Detection of disseminated pancreatic cells by amplification of cytokeratin-19 with quantitative RT-PCR in blood, bone marrow and peritoneal lavage of pancreatic carcinoma patients. *World J. Gastroenterol.* **2007**, *13*, 257–263. [CrossRef] [PubMed]

42. Sergeant, G.; Roskams, T.; van Pelt, J.; Houtmeyers, F.; Aerts, R.; Topal, B. Perioperative cancer cell dissemination detected with a real-time RT-PCR assay for EpCAM is not associated with worse prognosis in pancreatic ductal adenocarcinoma. *BMC Cancer* **2011**, *11*, 47. [CrossRef] [PubMed]
43. Soeth, E.; Grigoleit, U.; Moellmann, B.; Röder, C.; Schniewind, B.; Kremer, B.; Kalthoff, H.; Vogel, I. Detection of tumor cell dissemination in pancreatic ductal carcinoma patients by CK 20 RT-PCR indicates poor survival. *J. Cancer Res. Clin. Oncol.* **2005**, *131*, 669–676. [CrossRef] [PubMed]
44. Lin, H.K.; Zheng, S.; Williams, A.J.; Balic, M.; Groshen, S.; Scher, H.I.; Fleisher, M.; Stadler, W.; Datar, R.H.; Tai, Y.-C.; et al. Portable filter-based microdevice for detection and characterization of circulating tumor cells. *Clin. Cancer Res.* **2010**, *16*, 5011–5018. [CrossRef] [PubMed]
45. Yadav, D.K.; Bai, X.; Yadav, R.K.; Singh, A.; Li, G.; Ma, T.; Chen, W.; Liang, T. Liquid biopsy in pancreatic cancer: The beginning of a new era. *Oncotarget* **2018**, *9*, 26900–26933. [CrossRef] [PubMed]
46. Gorges, T.M.; Tinhofer, I.; Drosch, M.; Röse, L.; Zollner, T.M.; Krahn, T.; von Ahsen, O. Circulating tumour cells escape from EpCAM-based detection due to epithelial-to-mesenchymal transition. *BMC Cancer* **2012**, *12*, 178. [CrossRef]
47. Punnoose, E.A.; Atwal, S.K.; Spoerke, J.M.; Savage, H.; Pandita, A.; Yeh, R.-F.; Pirzkall, A.; Fine, B.M.; Amler, L.C.; Chen, D.S.; et al. Molecular biomarker analyses using circulating tumor cells. *PLoS ONE* **2010**, *5*, e12517. [CrossRef]
48. Szczerba, B.M.; Castro-Giner, F.; Vetter, M.; Krol, I.; Gkountela, S.; Landin, J.; Scheidmann, M.C.; Donato, C.; Scherrer, R.; Singer, J.; et al. Neutrophils escort circulating tumour cells to enable cell cycle progression. *Nature* **2019**, *566*, 553–557. [CrossRef]
49. Rhim, A.D.; Thege, F.I.; Santana, S.M.; Lannin, T.B.; Saha, T.N.; Tsai, S.; Maggs, L.R.; . Kochman, M.L.; Ginsberg, G.G.; Lieb, J.G.; et al. Detection of circulating pancreas epithelial cells in patients with pancreatic cystic lesions. *Gastroenterology* **2014**, *146*, 647–651. [CrossRef]
50. Cohen, J.D.; Javed, A.A.; Thoburn, C.; Wong, F.; Tie, J.; Gibbs, P.; Schmidt, C.; Yip-Schneider, M.T.; Allen, P.; Schattner, M.; et al. Combined circulating tumor DNA and protein biomarker based liquid biopsy for the earlier detection of pancreatic cancers. *Proc. Natl. Acad. Sci. USA* **2017**, *114*, 10202–10207. [CrossRef]
51. Han, L.; Chen, W.; Zhao, Q. Prognostic value of circulating tumor cells in patients with pancreatic cancer: A meta-analysis. *Tumour Biol.* **2014**, *35*, 2473–2480. [CrossRef] [PubMed]
52. Okubo, K.; Uenosono, Y.; Arigami, T.; Mataki, Y.; Matsushita, D.; Yanagita, S.; Kurahara, H.; Sakoda, M.; Kijima, Y.; Maemura, K.; et al. Clinical impact of circulating tumor cells and therapy response in pancreatic cancer. *Eur. J. Surg. Oncol.* **2017**, *43*, 1050–1055. [CrossRef] [PubMed]
53. Gao, Y.; Zhu, Y.; Zhang, Z.; Zhang, C.; Huang, X.; Yuan, Z. Clinical significance of pancreatic circulating tumor cells using combined negative enrichment and immunostaining-fluorescence in situ hybridization. *J. Exp. Clin. Cancer Res.* **2016**, *35*, 66. [CrossRef] [PubMed]
54. Poruk, K.E.; Valero, V.; Saunders, T.; Blackford, A.L.; Griffin, J.F.; Poling, J.; Hruban, R.H.; Anders, R.A.; Herman, J.; Zheng, L.; et al. Circulating Tumor Cell Phenotype Predicts Recurrence and Survival in Pancreatic Adenocarcinoma. *Ann. Surg.* **2016**, *264*, 1073–1081. [CrossRef]
55. Poruk, K.E.; Blackford, A.L.; Weiss, M.J.; Cameron, J.L.; He, J.; Goggins, M.; Rasheed, Z.A.; Wolfgang, C.L.; Wood, L.D. Circulating Tumor Cells Expressing Markers of Tumor-Initiating Cells Predict Poor Survival and Cancer Recurrence in Patients with Pancreatic Ductal Adenocarcinoma. *Clin. Cancer Res.* **2017**, *23*, 2681–2690. [CrossRef]
56. Kurihara, T.; Itoi, T.; Sofuni, A.; Itokawa, F.; Tsuchiya, T.; Tsuji, S.; Ishii, K.; Ikeuchi, N.; Tsuchida, A.; Kasuya, K.; et al. Detection of circulating tumor cells in patients with pancreatic cancer: A preliminary result. *J. Hepatobiliary Pancreat Surg.* **2008**, *15*, 189–195. [CrossRef]
57. de Albuquerque, A.; Kubisch, I.; Ernst, D.; Breier, G.; Stamminger, G.; Fersis, N.; Stölzel, U.; Boese-Landgraf, J.; Eichler, A.; Kaul, S. Development of a molecular multimarker assay for the analysis of circulating tumor cells in adenocarcinoma patients. *Clin. Lab.* **2012**, *58*, 373–384.
58. Bissolati, M.; Sandri, M.T.; Burtulo, G.; Zorzino, L.; Balzano, G.; Braga, M. Portal vein-circulating tumor cells predict liver metastases in patients with resectable pancreatic cancer. *Tumour Biol.* **2015**, *36*, 991–996. [CrossRef]
59. Zhang, Y.; Wang, F.; Ning, N.; Chen, Q.; Yang, Z.; Guo, Y.; Xu, D.; Zhang, D.; Zhan, T.; Cui, W. Patterns of circulating tumor cells identified by CEP8, CK and CD45 in pancreatic cancer. *Int. J. Cancer* **2015**, *136*, 1228–1233. [CrossRef]

60. Earl, J.; Garcia-Nieto, S.; Martinez-Avila, J.C.; Montans, J.; Sanjuanbenito, A.; Rodríguez-Garrote, M.; Lisa, E.; Mendía, E.; Lobo, E.; Malats, N.; et al. Circulating tumor cells (Ctc) and kras mutant circulating free Dna (cfdna) detection in peripheral blood as biomarkers in patients diagnosed with exocrine pancreatic cancer. *BMC Cancer* **2015**, *15*, 797. [CrossRef]
61. Gemenetzis, G.; Groot, V.P.; Yu, J.; Ding, D.; Teinor, J.A.; Javed, A.A.; Wood, L.D.; Burkhart, R.A.; Cameron, J.L.; Makary, M.A.; et al. Circulating Tumor Cells Dynamics in Pancreatic Adenocarcinoma Correlate with Disease Status: Results of the Prospective CLUSTER Study. *Ann. Surg.* **2018**, *268*, 408–420. [CrossRef] [PubMed]
62. Thibodeau, S.; Voutsadakis, I.A. Folfirinox Chemotherapy in Metastatic Pancreatic Cancer: A Systematic Review and Meta-Analysis of Retrospective and Phase II Studies. *J. Clin. Med.* **2018**, *7*, 7. [CrossRef] [PubMed]
63. Yu, K.H.; Ricigliano, M.; Hidalgo, M.; Abou-Alfa, G.K.; Lowery, M.A.; Saltz, L.B.; Crotty, J.F.; Gary, K.; Cooper, B.; Lapidus, R.; et al. Pharmacogenomic modeling of circulating tumor and invasive cells for prediction of chemotherapy response and resistance in pancreatic cancer. *Clin. Cancer Res.* **2014**, *20*, 5281–5289. [CrossRef] [PubMed]
64. Miller, M.C.; Doyle, G.V.; Terstappen, L.W.M.M. Significance of Circulating Tumor Cells Detected by the CellSearch System in Patients with Metastatic Breast Colorectal and Prostate Cancer. *J. Oncol.* **2010**, *2010*, 617421. [CrossRef] [PubMed]
65. Fujii, T.; Reuben, J.M.; Huo, L.; Espinosa Fernandez, J.R.; Gong, Y.; Krupa, R.; Suranenin, M.V.; Graf, R.P.; Lee, J.; Greene, S.; et al. Androgen receptor expression on circulating tumor cells in metastatic breast cancer. *PLoS ONE* **2017**, *12*, e0185231. [CrossRef] [PubMed]
66. Broncy, L.; Paterlini-Bréchot, P. Clinical Impact of Circulating Tumor Cells in Patients with Localized Prostate Cancer. *Cells* **2019**, *8*, 676. [CrossRef] [PubMed]
67. Cristofanilli, M.; Budd, G.T.; Ellis, M.J.; Stopeck, A.; Matera, J.; Miller, M.C.; Reuben, J.M.; Doyle, G.V.; Allard, W.J.; Terstappen, L.W.M.M.; et al. Circulating tumor cells, disease progression, and survival in metastatic breast cancer. *N. Engl. J. Med.* **2004**, *351*, 781–791. [CrossRef]
68. Brierley, J.D.; Gospodarowicz, M.K.; Wittekind, C. *TNM Classification of Malignant Tumours*, 8th ed.; John Wiley & Sons, Ltd.: Chichester West Sussex, UK, 2016; ISBN 978-1-119-26357-9.
69. Trapp, E.; Janni, W.; Schindlbeck, C.; Jückstock, J.; Andergassen, U.; de Gregorio, A.; Alunni-Fabbroni, M.; Tzschaschel, M.; Polasik, A.; Koch, J.; et al. Presence of Circulating Tumor Cells in High-Risk Early Breast Cancer During Follow-Up and Prognosis. *J. Natl. Cancer Inst.* **2019**, *111*, 380–387. [CrossRef]
70. Ignatiadis, M.; Rothé, F.; Chaboteaux, C.; Durbecq, V.; Rouas, G.; Criscitiello, C.; Metallo, J.; Kheddoumi, N.; Singhal, S.K.; Michiels, S.; et al. HER2-positive circulating tumor cells in breast cancer. *PLoS ONE* **2011**, *6*, e15624. [CrossRef]
71. Jakabova, A.; Bielcikova, Z.; Pospisilova, E.; Matkowski, R.; Szynglarewicz, B.; Staszek-Szewczyk, U.; Zemanova, M.; Petruzelka, L.; Eliasova, P.; Kolostova, K.; et al. Molecular characterization and heterogeneity of circulating tumor cells in breast cancer. *Breast Cancer Res. Treat.* **2017**, *166*, 695–700. [CrossRef]
72. Ignatiadis, M.; Litière, S.; Rothe, F.; Riethdorf, S.; Proudhon, C.; Fehm, T.; Aalders, K.; Forstbauer, H.; Fasching, P.A.; Brain, E.; et al. Trastuzumab versus observation for HER2 nonamplified early breast cancer with circulating tumor cells (EORTC 90091–10093, BIG 1–12, Treat CTC): A randomized phase II trial. *Ann. Oncol.* **2018**, *29*, 1777–1783. [CrossRef] [PubMed]

3

NF-kappa B Signaling-Related Signatures are Connected with the Mesenchymal Phenotype of Circulating Tumor Cells in Non-Metastatic Breast Cancer

Marta Popeda [1], Tomasz Stokowy [2], Natalia Bednarz-Knoll [1], Anna Jurek [1], Magdalena Niemira [3], Agnieszka Bielska [3], Adam Kretowski [3], Leszek Kalinowski [4,5], Jolanta Szade [6], Aleksandra Markiewicz [1] and Anna J. Zaczek [1,*]

1 Laboratory of Translational Oncology, Intercollegiate Faculty of Biotechnology, Medical University of Gdansk, 80-211 Gdansk, Poland; marta.popeda@gumed.edu.pl (M.P.); nbk@gumed.edu.pl (N.B.-K.); annajurek@gumed.edu.pl (A.J.); aleksandra.markiewicz@gmail.com (A.M.)

2 Department of Clinical Science, University of Bergen, 5021 Bergen, Norway; tomasz.stokowy@k2.uib.no

3 Clinical Research Centre, Medical University of Bialystok, 15-276 Bialystok, Poland; magdalena.niemira@umb.edu.pl (M.N.); agnieszka.bielska@umb.edu.pl (A.B.); adamkretowski@wp.pl (A.K.)

4 Department of Medical Laboratory Diagnostics and Central Bank of Frozen Tissues & Genetic Specimens, Medical University of Gdansk, 80-211 Gdansk, Poland; lekal@gumed.edu.pl

5 Biobanking and Biomolecular Resources Research Infrastructure Poland (BBMRI.PL), 80-211 Gdansk, Poland

6 Department of Pathomorphology, Medical University of Gdansk, 80-211 Gdansk, Poland; jszade@gumed.edu.pl

* Correspondence: azaczek@gumed.edu.pl.

Abstract: The role of circulating tumor cells (CTCs), tumor microenvironment (TME), and the immune system in the formation of metastasis is evident, yet the details of their interactions remain unknown. This study aimed at exploring the immunotranscriptome of primary tumors associated with the status of CTCs in breast cancer (BCa) patients. The expression of 730 immune-related genes in formalin-fixed paraffin-embedded samples was analyzed using the multigenomic NanoString technology and correlated with the presence and the phenotype of CTCs. Upregulation of 37 genes and downregulation of 1 gene were observed in patients characterized by a mesenchymal phenotype of CTCs when compared to patients with epithelial CTCs. The upregulated genes were involved in NF-kappa B signaling and in the production of type I interferons. The clinical significance of the differentially expressed genes was evaluated using The Cancer Genome Atlas (TCGA) data of a breast invasive carcinoma (BRCA) cohort. Five of the upregulated genes—*PSMD7, C2, IFNAR1, CD84*, and *CYLD*—were independent prognostic factors in terms of overall and disease-free survival. To conclude, our data identify a group of genes that are upregulated in BCa patients with mesenchymal CTCs and reveal their prognostic potential, thus indicating that they merit further investigation.

Keywords: breast cancer; circulating tumor cells; epithelial–mesenchymal transition; NF-kappa B signaling; type I interferons; tumor microenvironment; immune-related transcriptome

1. Introduction

Distant metastases account for most of cancer-related deaths. Yet, fundamental questions regarding mechanisms that promote or inhibit the formation of metastasis still remain unanswered. It is evident that in breast cancer (BCa), tumor cell dissemination occurs already at early stages of the disease [1,2], with circulating tumor cells (CTCs) being direct initiators of metastasis [3]. With current analytic methods, CTCs are detectable in the peripheral blood, and their presence is considered an adverse prognostic factor in numerous solid tumors, including BCa [4–11]. Tumor dissemination is known to be facilitated by the process of epithelial–mesenchymal transition (EMT) [12]. CTCs are shown to possess phenotypic plasticity that relates to their ability to display various EMT states in the circulation [13–15], with either the mesenchymal [16] or the intermediate [17] phenotype associated with increased tumor-initiating ability. The presence of mesenchymal CTCs is associated with disease progression and worse prognosis for metastatic BCa [18–20] and even operable breast cancer patients [21–23]. In fact, in metastatic BCa patients, the expression of mesenchymal markers is higher than in early-stage BCa patients [24,25] and correlates with lymph node involvement [26], suggesting that the EMT phenotype is directly related to the metastatic potential of CTCs.

The ability to evade the immune system is one of the hallmarks of cancer [27,28]. Tumor cells use multiple strategies to escape immune surveillance, mainly avoiding immune recognition and instigating an immunosuppressive microenvironment [29]. The interactions between cancer cells and their microenvironment (tumor microenvironment, TME) are involved in tumor dissemination. TME is assumed to modulate the capability of CTCs to evade the innate immune response and CTCs survival [30,31]. We showed that absence of ALDH1-positive stromal cells correlates with the presence of disseminated tumor cells (DTCs) in the bone marrow of BCa patients [32]. Nevertheless, the precise profile of immune-related factors influencing tumor dissemination, in particular the presence and phenotype of CTCs, is scarcely known. Therefore, the current study aimed at exploring the association between the immunotranscriptome of primary breast tumors and the status of CTCs. We hypothesized that the presence and the EMT phenotype of CTCs are connected with a specific immune-related gene signature of primary tumors. To verify this hypothesis, we evaluated the expression of 730 immune-related genes in primary tumors of BCa patients with well-described molecular and clinicopathological features, including CTCs presence and phenotype.

2. Results

2.1. *Expression of Immune-Related Genes within Primary Tumours Correlated with the Phenotype of CTCs*

To investigate the immune transcriptome associated with each phenotype of CTCs, we applied NanoString multigene expression analysis to samples of primary breast tumors. We observed that the mesenchymal phenotype of CTCs ($n = 9$) was associated with the upregulation of 37 genes and the downregulation of 1 gene in primary tumors (p-value ≤ 0.05, false discovery rate (FDR) ≤ 0.2) when compared to the epithelial phenotype of CTCs ($n = 14$) (Table 1, Figure S1; all results in Table S1). Due to the limited number of patients included in our study, we employed the conservative FDR method of multiple testing correction.

Our aim was also to explore the association between the expression of immune-related genes within the primary tumors and the overall presence of CTCs. We found no statistically significant differences between the primary tumors' immunotranscriptomes in relation to patients' CTC status (positive vs. negative).

Table 1. Genes up- and downregulated in primary tumors of patients with mesenchymal circulating tumor cells (CTCs) (compared to patients with epithelial CTCs).

Gene Symbol	Gene Name	FC	p-Value	FDR
LAIR2	Leukocyte-associated immunoglobulin-like receptor 2	3.19	0.007	0.144
FADD	Fas-associated via death domain	3.14	0.003	0.082
TLR7	Toll-like receptor 7	2.86	0.011	0.172
CCRL2	C–C motif chemokine receptor-like 2	2.80	0.003	0.087
PBK	PDZ-binding kinase	2.26	0.009	0.161
CD3EAP	CD3e molecule-associated protein	2.25	0.001	0.063
CD84	CD84 molecule	2.22	0.003	0.087

Table 1. *Cont.*

Gene Symbol	Gene Name	FC	*p*-Value	FDR
TNFSF13	TNF superfamily member 13	2.21	0.007	0.144
BIRC5	Baculoviral IAP repeat-containing 5	2.08	0.002	0.082
KLRC1	Killer cell lectin-like receptor C1	1.98	0.001	0.063
CD63	CD63 molecule	1.93	0.003	0.082
TNFRSF11A	TNF receptor superfamily member 11a	1.91	0.013	0.199
C2	Complement C2	1.85	0.001	0.069
IL1RAP	Interleukin 1 receptor accessory protein	1.79	0.011	0.172
TAPBP	TAP-binding protein	1.78	0.003	0.082
NUP107	nucleoporin 107	1.78	0.002	0.082
PSMD7	Proteasome 26S subunit, non-ATPase 7	1.72	<0.001	0.063
TICAM1	Toll-like receptor adaptor molecule 1	1.67	0.013	0.199
ICAM3	Intercellular adhesion molecule 3	1.63	0.009	0.161
IRF3	Interferon regulatory factor 3	1.59	<0.001	0.048
BCL10	BCL10 immune signaling adaptor	1.56	0.001	0.069
IKBKE	Inhibitor of nuclear factor kappa B kinase subunit epsilon	1.55	0.011	0.172
ELK1	ETS transcription factor ELK1	1.55	0.001	0.069
TRAF6	TNF receptor-associated factor 6	1.53	0.003	0.087
RELA	RELA proto-oncogene, NF-kB subunit	1.52	0.003	0.082
IKBKG	Inhibitor of nuclear factor kappa B kinase regulatory subunit gamma	1.49	0.003	0.087
TBK1	TANK-binding kinase 1	1.46	0.001	0.063
STAT6	Signal transducer and activator of transcription 6	1.40	0.004	0.106
ATG10	Autophagy-related 10	1.40	0.007	0.144
CD74	CD74 molecule	1.38	0.009	0.161
ICOSLG	Inducible T cell costimulator ligand	1.38	0.002	0.082
FCGR2A	Fc fragment of IgG receptor IIa	1.37	0.007	0.144
PSMB10	Proteasome 20S subunit beta 10	1.36	0.011	0.172
CYLD	CYLD lysine 63 deubiquitinase	1.36	0.011	0.172
IFNAR1	Interferon alpha and beta receptor subunit 1	1.35	0.001	0.069
CCND3	Cyclin D3	1.30	0.002	0.082
MAP2K1	Mitogen-activated protein kinase kinase 1	1.25	<0.001	0.027
MERTK	MER proto-oncogene, tyrosine kinase	0.68	0.003	0.082

Fold change (FC) based on median normalized counts of the probe in each group; differences in median normalized counts between groups analyzed with Mann–Whitney U test (*p*-value) with Benjamini–Hochberg correction for multiple comparisons (false discovery rate, FDR); only statistically significant results are presented; gene names according to HUGO Gene Nomenclature.

2.2. The Mesenchymal Phenotype of CTCs Is Associated with the Upregulation of Genes Involved in NF-kappa B Signalling and Type I Interferons Production in Matched Primary Tumours

Here, we observed that multiple genes differentially expressed in patients with epithelial and mesenchymal CTC phenotypes (Table 1) play a role in the NF-kappa B signaling pathway. Consequently, we decided to interrogate this link more carefully. We applied the Functional Annotation Tool by DAVID 6.8 [33,34] to associate the selected genes with specific functional annotations. Genes upregulated in tumors with mesenchymal CTCs were generally involved in the activation and regulation of immune response (Table S2). Interestingly, 15 out of 37 upregulated genes (*FADD, TLR7, TNFRSF11A, IL1RAP, PSMD7, TICAM1, IRF3, BCL10, IKBKE, TRAF6, RELA, IKBKG, TBK1, PSMB10,* and *CYLD*) were implicated in the regulation of NF-kappa B signaling and activity (GO:0043122, GO:0051092, and GO:0038061). A literature search provided a number of links between other 11 differentially expressed genes (*CCRL2, PBK, TNFSF13, BIRC5, TAPBP, ELK1, STAT6, ATG10, IFNAR1, CCND3,* and *MAP2K1*) and the NF-kappa B pathway (top 12 genes depicted in Figure 1).

Analysis of Gene Ontology also revealed that nine of the upregulated genes regulate the production of type I interferons (GO:0032479; *TLR7, TICAM1, IRF3, IKBKE, RELA, TBK1, STAT6, CYLD,* and *IFNAR1*), with a particular role in the stimulation of interferon beta (GO:0032728; *TLR7, TICAM1, IRF3, TBK1,* and *IFNAR1*).

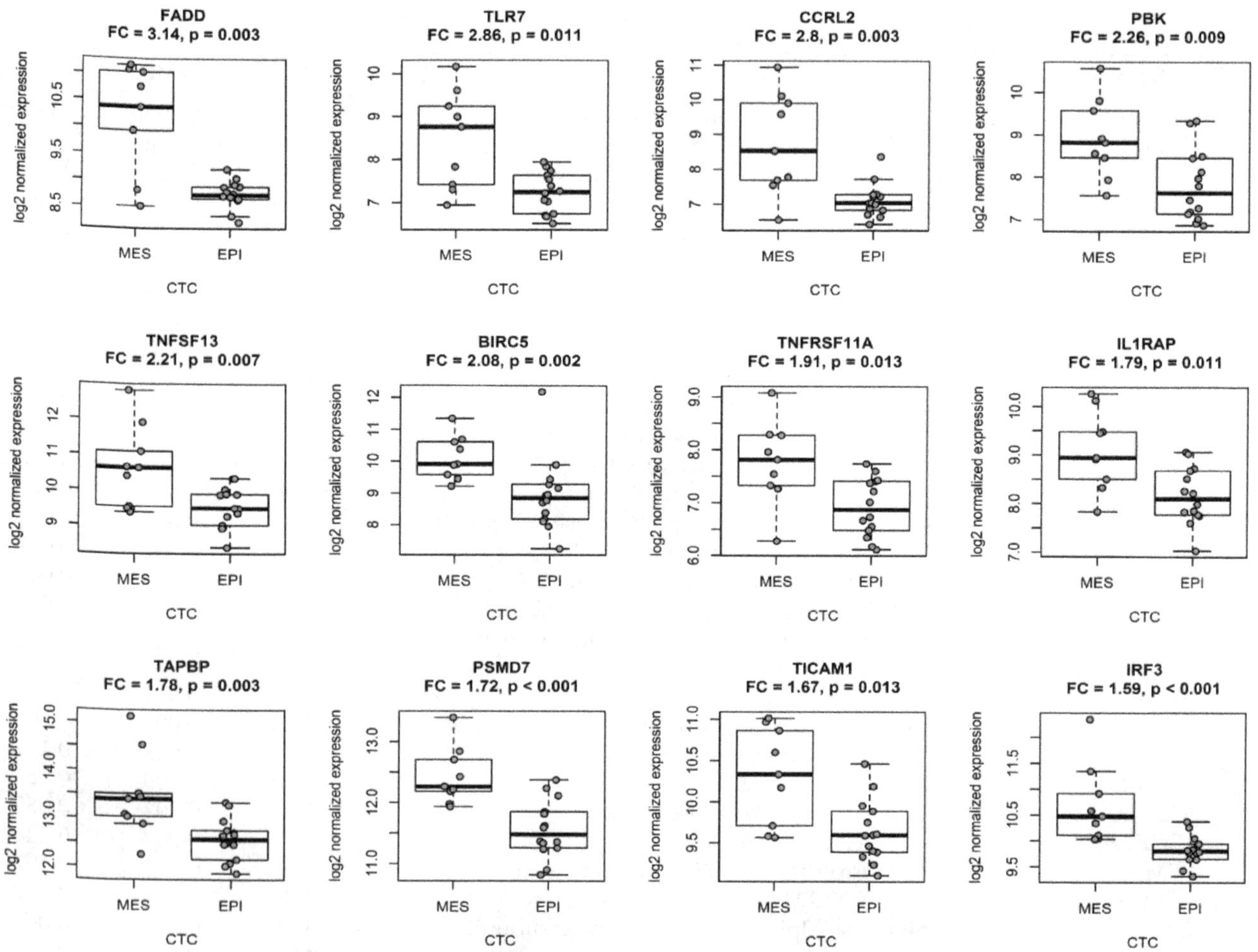

Figure 1. Genes implicated in NF-kappa B signaling were upregulated in primary tumors of breast cancer patients with mesenchymal CTCs (MES, $n = 9$) when compared to patients with epithelial CTCs (EPI, $n = 14$); the top 12 upregulated genes are presented. Gene expression depicted as number of counts of each probe and normalized to the four most stable reference genes (*ABCF1*, *EDC3*, *HDAC3*, and *CNOT4*); FC calculated on the basis of the median normalized counts of the probe in each group; differences in median normalized counts between groups analyzed with the Mann–Whitney U test (*p*); the bars correspond to the interquartile range (IQR), the whiskers cover 1.5 IQR from the median.

2.3. *Immune-Related Genes Connected with the Mesenchymal Phenotype of CTCs Are Potent Negative Prognostic Factors in Breast Cancer*

We have previously demonstrated that the mesenchymal phenotype of CTCs correlates with a poor prognosis in breast cancer patients [21]. Consequently, we decided to evaluate the prognostic significance of the immune-related genes that we found significantly up- or downregulated in primary tumors of BCa patients with mesenchymal CTCs. To this end, we turned to The Cancer Genome Atlas (TCGA) database and analyzed the available RNA-seq data on gene expression in a breast invasive carcinoma (BRCA) cohort ($n = 877$) [35,36]. Five out of 38 genes (*PSMD7*, *C2*, *IFNAR1*, *CD84*, and *CYLD*) associated with the mesenchymal phenotype of CTCs demonstrated a negative prognostic impact in the TCGA cohort. Moderate (higher than the first quartile, Q1 in Table S3) expression of *PSMD7* correlated with shorter overall survival (OS) in comparison to low expression of *PSMD7* in primary tumors (HR = 1.75, 95% CI: 1.08–2.82, $p = 0.022$; Figure 2A). A higher risk of recurrence was observed for tumors with moderate (higher than the first quartile, Q1 in Table S3) expression of *C2* (HR = 4.51, 95% CI: 1.36–14.96, $p = 0.014$; Figure 2B) and *IFNAR1* (HR = 2.68, 95% CI: 1.03–6.97, $p = 0.043$; Figure 2C) in comparison to tumors with low expression of these genes; high (higher than the third quartile, Q3 in Table S3) expression of *CD84* (HR = 2.48, 95% CI: 1.22–5.03, $p = 0.012$; Figure 2D) and *CYLD*

(HR = 2.20, 95% CI: 1.09–4.43, p = 0.028; Figure 2E) was also linked to shorter disease-free survival (DFS) in comparison to low expression of these genes in the primary tumors. Multivariate analysis including the clinical stage confirmed the significance of the aforementioned genes as independent prognostic factors (Table S3).

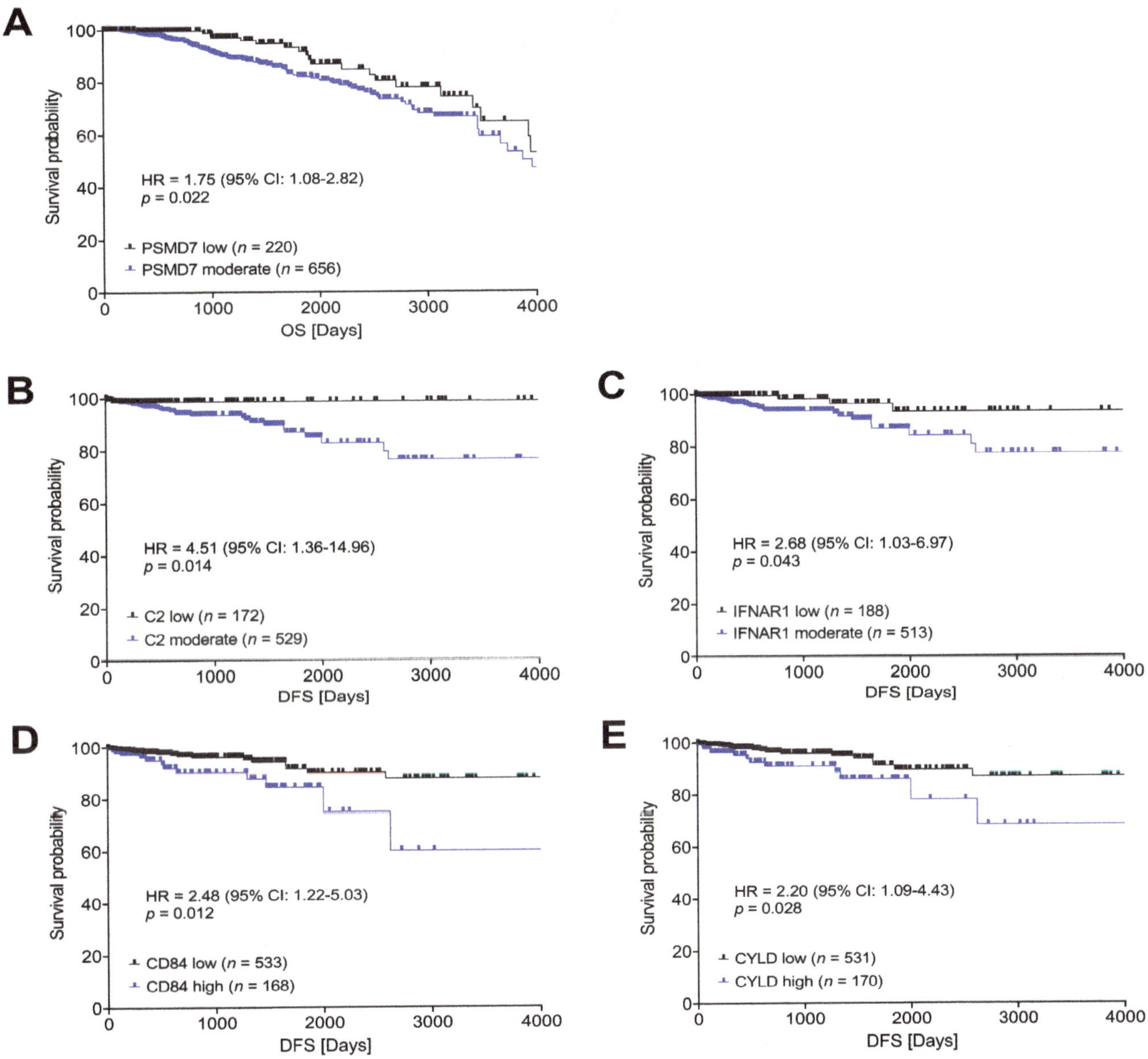

Figure 2. Genes associated with the mesenchymal phenotype of CTCs had a negative impact on survival (**A**) and recurrence (**B–E**) in patients in a BRCA cohort of The Cancer Genome Atlas (TCGA). Low/moderate status of gene expression relative to the first quartile (Q1); low/high status of gene expression relative to the third quartile (Q3); hazard ratios (HR) with 95% confidence intervals (95% CI) computed using Cox proportional hazards regression; OS: overall survival, DFS: disease-free survival.

3. Discussion

The knowledge about immune signatures related to tumor dissemination is still limited. Our current study aimed to identify the immunotranscriptomic profiles of primary tumors associated with the presence of CTCs and the CTCs phenotype in non-metastatic BCa patients.

Our data revealed that 38 genes were differentially expressed in the primary tumors of patients with mesenchymal CTCs when compared to patients with epithelial CTCs. Intriguingly, we did not observe any statistically significant difference between primary tumor transcriptomes when comparing patients according to the presence or absence of CTCs in the circulation. The compared groups (CTC-positive and CTC-negative) were not biased in terms of any clinicopathological parameter (Table S4); hence, we believe that the observed difference in the activation of dissemination at BCa tumors may be cell context-dependent and definitely requires a more thorough analysis. On the other hand, our results demonstrate a substantial connection between the mesenchymal phenotype of CTCs and the NF-kappa B pathway. According to the NanoString gene expression assay and a literature search, 26 out of the 37 genes upregulated in mesenchymal-CTC patients in comparison to epithelial-CTC patients are implicated in NF-kappa B signaling at various levels of the transduction pathway (Figure 3A) and demonstrate a complex network of interactions at the protein level (Figure 3B). Importantly, the enrichment of NF-kappa B-related transcripts was consistently observed when we applied a stricter gene inclusion criteria and limited the analysis to a set of 330 genes with the highest expression (log2 mean count of a gene in all samples >9; Table S5).

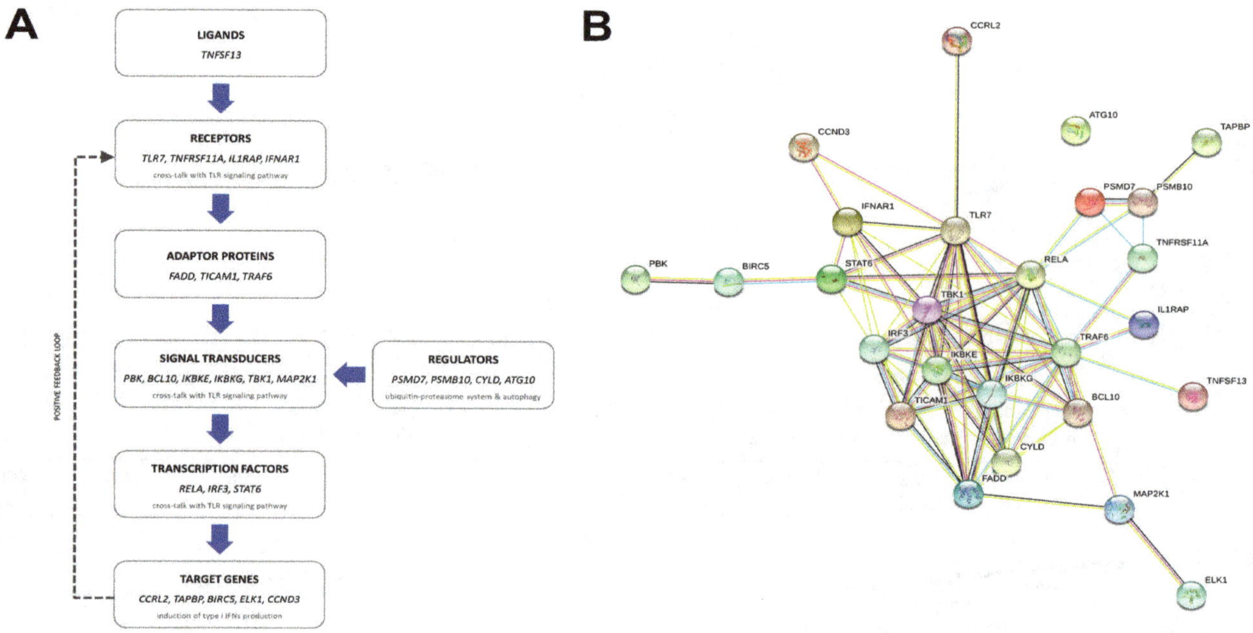

Figure 3. Genes upregulated in the primary tumors of patients with mesenchymal CTCs function at various levels of the NF-kappa B signaling pathway (**A**), within a complex network of interactions (**B**); image depicting a protein–protein association network, generated using the STRING tool; edge (line) coloring defines the type of interaction: blue—from curated databases, pink—experimentally determined, green—gene neighborhood, red—gene fusions, dark blue—gene co-occurrence, yellow—text mining, black—co-expression, violet—protein homology.

NF-kappa B signaling is a potent regulator of numerous vital physiological processes, including survival, inflammation, and immune responses [37]. The activation of the pathway is mediated by numeros receptors. Our enriched set (Table 1) includes genes encoding both specific ligands (APRIL (*TNFSF13*)) and receptors (*TLR7*, *TNFRSF11A*, *IL1RAP*, and *IFNAR1*), as well as universal adaptor proteins (*FADD*, *TICAM1*, and *TRAF6*) that facilitate the transduction of the signal from the receptors in the cell membrane to the effectors in the nucleus. Namely, we observed the upregulation of transducers involved in the canonical cascade (*BCL10* and *IKBKG*) as well as in Toll-like receptor-mediated activation of NF-kappa B signaling (*PBK*, *IKBKE*, and *TBK1*). Moreover, the enhanced expression of *MAP2K1* gene points to the possible role of ERK-mediated stimulation of NF-kappa B signaling in tumors with mesenchymal CTCs [38].

On the other hand, the activity of NF-kappa B is known to be regulated by the proteasome and ubiquitin-mediated proteolysis. Here, we report the upregulation of genes that are implicated in the ubiquitin-proteasome system (*PSMD7*, *PSMB10*, and *CYLD*) [39] and the autophagy cascade (*ATG10*) [40–42]. Eventually, we observed an increased expression of one the subunits of the NF-kappa

B transcription factor—p65 (*RELA*), as well as of two other co-operating transcription factors (*IRF3* and *STAT6*). The enhanced signaling resulted in the upregulation of five target genes—*CCRL2, TAPBP, BIRC5, ELK1*, and *CCND3*.

The NF-kappa B pathway is a well-known driver of EMT during both embryonic and tumor development [37]. In general, a constant stimulation of this pathway in cancer cells results in abnormal proliferation and differentiation, enhanced metastasis, and treatment resistance [43]. In breast cancer, NF-kappa B directly regulates the transcription of genes encoding EMT-inducing transcription factors [44]. In fact, the increased expression of NF-kappa B is a common feature of breast cancer cell lines and tissues, correlating with intensified activation of both the canonical and the non-canonical pathway [45–47]. What is more, several reports point to an interesting association between NF-kappa B and HER2 [47–49], with evidence for predominant NF-kappa B activation in ER–/HER2+ breast tumors [45,49].

Our data revealed another interesting pattern of enrichment, with upregulation of several NF-kappa B-related genes that are particularly involved in the positive regulation of type I interferon production (Table S2). The cross-talk between NF-kappa B and Toll-like receptors (TLR)-mediated signaling results in an increased pro-inflammatory response that is additionally enhanced in an autocrine and paracrine manner by a positive feedback loop (Figure 3A) [50,51].

Noteworthy, among the NF-kappa B-unrelated genes, we found markers of platelet activation (*CD63* and *CD84*) [52,53], which is in line with literature reports on the co-operation between platelets and CTCs in the induction of EMT and metastasis formation [54,55]. Of note, tumor dissemination may also be supported by other populations of cells within the intratumoral stroma. The elevated NF-kappa B activity may result from increased release of pro-inflammatory cytokines by macrophages at the tumor site [56–58]. In fact, NF-kappa B seems to be involved in the polarization of tumor-associated macrophages [57]. We have previously reported the negative prognostic significance of CTCs of mesenchymal phenotype in BCa patients [21]. Due to the low number of patients in this cohort, in the current study we analyzed the impact of genes linked with mesenchymal CTCs in TCGA BCa cohort. In fact, in TCGA data 5 out of the 38 genes of our interest were associated with worse prognosis (overall survival or risk of recurrence), namely, *PSMD7, C2, IFNAR1, CD84*, and *CYLD*. None of the corresponding proteins is currently included in the routine histopathology for breast tumor, thus they need to be validated at the protein level in a large cohort of patients in order to prove their clinical importance and diagnostic applicability.

4. Materials and Methods

4.1. Patients

The study group consisted of 35 breast cancer patients staged I–III, who had undergone surgical treatment at the Medical University Hospital in Gdansk between April 2011 and May 2013. The study was approved by the Ethical Committee of the Medical University of Gdansk (NKBBN 94/2017), and informed consent was collected from all participants. Patients were characterized by different clinicopathological parameters (Table S4), with particular focus on CTC status—negative (n = 12) or positive (n = 23)—and molecular phenotype of CTCs—epithelial (n = 14) or mesenchymal (n = 9)—as described previously [26].

4.2. RNA Extraction

RNA was extracted from formalin-fixed paraffin-embedded (FFPE) primary breast tumor samples (four 10 µm-thick, unstained FFPE sections per patient) using the RNeasy Mini Kit (Qiagen, Hilden, Germany) according to the manufacturer's protocol. RNA concentration and purity were determined using NanoDrop 1000 spectrophotometer (Thermo Scientific, Wilmington, DE, USA). RNA integrity was assessed using Agilent 2100 Bioanalyzer (Agilent Technologies, Santa Clara, CA, USA) with Agilent RNA 6000 Pico Kit (Agilent Technologies).

4.3. *nCounter Gene Expression Assay*

Extracted RNA (4 µl) was pre-amplified using the nCounter Low RNA Input Kit (NanoString Technologies, Seattle, WA, USA) with the dedicated Primer Pool covering the sequences of 730 immune-related genes included in the nCounter PanCancer Immune Profiling Panel (NanoString Technologies). Pre-amplified samples were analyzed using the NanoString nCounter Analysis System (NanoString Technologies) according to the manufacturer's procedures for hybridization, detection, and scanning.

4.4. *Data Analysis*

For each tumor sample analyzed with the NanoString technology, the background level was estimated using the mean plus 2 standard deviations of the counts of the negative control probes included in the assay. Data were normalized using the geometric mean of the positive controls included in the assay and 4 most stably expressed housekeeping genes included in the PanCancer Immune Profiling Panel—*ABCF1*, *EDC3*, *HDAC3*, and *CNOT4*—(expression stability assessed with NormFinder, SD range 173.5–228.4 counts). Background thresholding and normalization were performed using nSolver 4.0 software (NanoString Technologies).

Low-expression genes (log2 mean count of a gene in all samples <6) were excluded, leaving 584 genes for further analysis. Subsequently, the genes differentiating each CTC status were selected on the basis of fold change in comparison to the control; fold change was calculated on basis of the median normalized counts of the probe in each group. The following comparisons were performed: CTC-positive vs. CTC-negative; CTC-epithelial vs. CTC-negative; CTC-mesenchymal vs. CTC-negative; CTC-mesenchymal vs. CTC-epithelial. Genes with FC > 1 were considered upregulated; genes with FC < 1 were considered downregulated.

Data were analyzed using the R statistical computing environment (3.6.1) [59]. Differences in gene expression between groups were analyzed using the Mann–Whitney U test with Benjamini–Hochberg correction for multiple comparisons; p-values ≤ 0.05 and FDR values ≤ 0.2 were considered statistically significant.

For the differing genes, gene ontology was analyzed using the Functional Annotation Tool by DAVID Bioinformatics Resources 6.8 [33,34]. EASE Score, a modified Fisher exact p-value, was used to assess gene enrichment. Multiple testing was corrected using FDR correction.

For the NF-kappa B-related genes, a protein–protein association network was depicted using STRING v11 [60].

4.5. *Survival Analysis in TCGA Cohort*

RNA-seq (RNASeqV2, RSEM_ normalized) and clinical data of BRCA cohort were obtained from TCGA portal [35,36] (data status of 28 January, 2016). The group was limited to T1-3M0 patients, and records with missing clinical or expression values were excluded, leaving 877 out of 1098 BCa patients for the analysis. OS was defined according to the "days_to_death" variable for survival time and the "vital_status" variable for event; DFS was defined according to the "days_to_last_follow-up" variable for survival time and the "person_neoplasm_cancer_status" variable for event. For the genes of interest, low/moderate status of gene expression was determined according to the 1st quartile (Q1) cut-off, while low/high status of gene expression was determined according to the 3rd quartile (Q3) cut-off. For each gene, the expression status (low vs. moderate; low vs. high) was tested in both univariate and multivariate analyses including the clinical stage. Hazard ratios (HR) with 95% confidence intervals (95% CI) were computed using Cox proportional hazards regression using the R statistical computing environment (3.6.1) [59].

5. Conclusions

To summarize, this study points to the potential link between the expression of immune-related genes in cells within the primary tumor and the EMT state of circulating tumor cells. Increased NF-kappa B signaling-related signatures in the tumor mass might possibly promote EMT in CTCs, thus contributing to their more aggressive phenotype and worse patient prognosis. It merits further investigation whether such effect is due to the action of cancer cells or that of normal cells in the surrounding TME. The potential prognostic relevance of selected genes associated with mesenchymal CTCs is promising and deserves further validation.

Supplementary Materials: Table S1: Results of gene expression analysis according to each CTC status or phenotype, Table S2: List of GO terms enriched in the genes upregulated in mesenchymal-CTC patients, Table S3: Results of multivariate survival analysis depending on the expression of genes up- and downregulated in mesenchymal-CTC patients; low/moderate status of gene expression relative to the 1st quartile (Q1); low/high status of gene expression relative to the 3rd quartile (Q3); hazard ratios (HR) with 95% confidence intervals (95% CI) computed using Cox proportional hazards regression; significant results are highlighted in green, Table S4: Clinicopathological characteristics of the patients, Table S5: List of genes upregulated in the primary tumors of patients with mesenchymal CTCs (compared to patients with epithelial CTCs), computed for a set of 330 high-expression genes (log2 mean count of a gene in all samples > 9), Figure S1: Expression of genes up- and downregulated in primary tumors of patients with mesenchymal CTCs (MES, n = 9) and patients with epithelial CTCs (EPI, n = 14); expression depicted as number of counts of each probe and normalized to the 4 most stable reference genes (*ABCF1*, *EDC3*, *HDAC3*, and *CNOT4*); fold change (FC) based on the median normalized counts of the probe in each group; differences in median normalized counts between groups analyzed with the Mann–Whitney U test (p); bars correspond to IQR, whiskers cover 1.5 IQR from the median.

Author Contributions: Conceptualization, T.S., M.P., A.M., N.B.-K., and A.J.Z.; methodology, T.S., A.J., M.P., M.N., A.M., and N.B.-K.; formal analysis, T.S., M.P., and A.J.Z.; investigation, A.M., N.B.-K., M.P., A.J., and A.B.; resources, J.S., A.K., and L.K.; data curation, M.P.; writing—original draft preparation, M.P., T.S., and A.J.Z.; writing—review and editing, all authors; visualization, T.S. and M.P.; supervision, T.S. and A.J.Z.; project administration, A.J.Z.; funding acquisition, A.J.Z.

Acknowledgments: The authors thank Michał Bieńkowski for the preparation of the pathological material.

References

1. Klein, C.A. Parallel progression of primary tumours and metastases. *Nat. Rev. Cancer* **2009**, *9*, 302–312. [CrossRef] [PubMed]
2. Marusyk, A.; Almendro, V.; Polyak, K. Intra-tumour heterogeneity: A looking glass for cancer? *Nat. Rev. Cancer* **2012**, *12*, 323–334. [CrossRef] [PubMed]
3. Aceto, N.; Toner, M.; Maheswaran, S.; Haber, D.A. En Route to Metastasis: Circulating Tumor Cell Clusters and Epithelial-to-Mesenchymal Transition. *Trends Cancer* **2015**, *1*, 44–52. [CrossRef] [PubMed]
4. De Bono, J.S.; Scher, H.I.; Montgomery, R.B.; Parker, C.; Miller, M.C.; Tissing, H.; Doyle, G.V.; Terstappen, L.W.W.M.; Pienta, K.J.; Raghavan, D. Circulating tumor cells predict survival benefit from treatment in metastatic castration-resistant prostate cancer. *Clin. Cancer Res.* **2008**, *14*, 6302–6309. [CrossRef] [PubMed]
5. Armstrong, A.J.; Halabi, S.; Luo, J.; Nanus, D.M.; Giannakakou, P.; Szmulewitz, R.Z.; Danila, D.C.; Healy, P.; Anand, M.; Rothwell, C.J.; et al. Prospective Multicenter Validation of Androgen Receptor Splice Variant 7 and Hormone Therapy Resistance in High-Risk Castration-Resistant Prostate Cancer: The PROPHECY Study. *J. Clin. Oncol.* **2019**, *37*, 1120–1129. [CrossRef]
6. Cristofanilli, M.; Pierga, J.-Y.; Reuben, J.; Rademaker, A.; Davis, A.A.; Peeters, D.J.; Fehm, T.; Nolé, F.; Gisbert-Criado, R.; Mavroudis, D.; et al. The clinical use of circulating tumor cells (CTCs) enumeration for staging of metastatic breast cancer (MBC): International expert consensus paper. *Crit. Rev. Oncol. Hematol.* **2019**, *134*, 39–45. [CrossRef]
7. Bidard, F.-C.; Michiels, S.; Riethdorf, S.; Mueller, V.; Esserman, L.J.; Lucci, A.; Naume, B.; Horiguchi, J.; Gisbert-Criado, R.; Sleijfer, S.; et al. Circulating Tumor Cells in Breast Cancer Patients Treated by Neoadjuvant Chemotherapy: A Meta-analysis. *J. Natl. Cancer Inst.* **2018**, *110*, 560–567. [CrossRef]

8. Lindsay, C.R.; Blackhall, F.H.; Carmel, A.; Fernandez-Gutierrez, F.; Gazzaniga, P.; Groen, H.J.M.; Hiltermann, T.J.N.; Krebs, M.G.; Loges, S.; López-López, R.; et al. EPAC-lung: Pooled analysis of circulating tumour cells in advanced non-small cell lung cancer. *Eur. J. Cancer* **2019**, *117*, 60–68. [CrossRef]
9. Lucci, A.; Hall, C.S.; Lodhi, A.K.; Bhattacharyya, A.; Anderson, A.E.; Xiao, L.; Bedrosian, I.; Kuerer, H.M.; Krishnamurthy, S. Circulating tumour cells in non-metastatic breast cancer: A prospective study. *Lancet Oncol.* **2012**, *13*, 688–695. [CrossRef]
10. Cobain, E.F.; Paoletti, C.; Smerage, J.B.; Hayes, D.F. Clinical Applications of Circulating Tumor Cells in Breast Cancer. *Recent Results Cancer Res.* **2020**, *215*, 147–160.
11. Todenhöfer, T.; Pantel, K.; Stenzl, A.; Werner, S. Pathophysiology of Tumor Cell Release into the Circulation and Characterization of CTC. *Recent Results Cancer Res.* **2020**, *215*, 3–24. [PubMed]
12. Pearson, G.W. Control of Invasion by Epithelial-to-Mesenchymal Transition Programs during Metastasis. *J. Clin. Med.* **2019**, *8*, 646. [CrossRef] [PubMed]
13. McInnes, L.M.; Jacobson, N.; Redfern, A.; Dowling, A.; Thompson, E.W.; Saunders, C.M. Clinical implications of circulating tumor cells of breast cancer patients: Role of epithelial-mesenchymal plasticity. *Front. Oncol.* **2015**, *5*, 42. [CrossRef] [PubMed]
14. Bulfoni, M.; Gerratana, L.; Del Ben, F.; Marzinotto, S.; Sorrentino, M.; Turetta, M.; Scoles, G.; Toffoletto, B.; Isola, M.; Beltrami, C.A.; et al. In patients with metastatic breast cancer the identification of circulating tumor cells in epithelial-to-mesenchymal transition is associated with a poor prognosis. *Breast Cancer Res.* **2016**, *18*, 30. [CrossRef] [PubMed]
15. Markiewicz, A.; Żaczek, A.J. The Landscape of Circulating Tumor Cell Research in the Context of Epithelial-Mesenchymal Transition. *Pathobiology* **2017**, *84*, 264–283. [CrossRef]
16. Pattabiraman, D.R.; Bierie, B.; Kober, K.I.; Thiru, P.; Krall, J.A.; Zill, C.; Reinhardt, F.; Tam, W.L.; Weinberg, R.A. Activation of PKA leads to mesenchymal-to-epithelial transition and loss of tumor-initiating ability. *Science* **2016**, *351*, aad3680. [CrossRef]
17. Jolly, M.K.; Somarelli, J.A.; Sheth, M.; Biddle, A.; Tripathi, S.C.; Armstrong, A.J.; Hanash, S.M.; Bapat, S.A.; Rangarajan, A.; Levine, H. Hybrid epithelial/mesenchymal phenotypes promote metastasis and therapy resistance across carcinomas. *Pharmacol. Ther.* **2019**, *194*, 161–184. [CrossRef]
18. Yu, M.; Bardia, A.; Wittner, B.S.; Stott, S.L.; Smas, M.E.; Ting, D.T.; Isakoff, S.J.; Ciciliano, J.C.; Wells, M.N.; Shah, A.M.; et al. Circulating breast tumor cells exhibit dynamic changes in epithelial and mesenchymal composition. *Science* **2013**, *339*, 580–584. [CrossRef]
19. Satelli, A.; Brownlee, Z.; Mitra, A.; Meng, Q.H.; Li, S. Circulating tumor cell enumeration with a combination of epithelial cell adhesion molecule- and cell-surface vimentin-based methods for monitoring breast cancer therapeutic response. *Clin. Chem.* **2015**, *61*, 259–266. [CrossRef]
20. Guan, X.; Ma, F.; Li, C.; Wu, S.; Hu, S.; Huang, J.; Sun, X.; Wang, J.; Luo, Y.; Cai, R.; et al. The prognostic and therapeutic implications of circulating tumor cell phenotype detection based on epithelial-mesenchymal transition markers in the first-line chemotherapy of HER2-negative metastatic breast cancer. *Cancer Commun.* **2019**, *39*, 1. [CrossRef]
21. Markiewicz, A.; Nagel, A.; Szade, J.; Majewska, H.; Skokowski, J.; Seroczynska, B.; Stokowy, T.; Welnicka-Jaskiewicz, M.; Zaczek, A.J. Aggressive Phenotype of Cells Disseminated via Hematogenous and Lymphatic Route in Breast Cancer Patients. *Transl. Oncol.* **2018**, *11*, 722–731. [CrossRef] [PubMed]
22. Markiewicz, A.; Topa, J.; Nagel, A.; Skokowski, J.; Seroczynska, B.; Stokowy, T.; Welnicka-Jaskiewicz, M.; Zaczek, A.J. Spectrum of Epithelial-Mesenchymal Transition Phenotypes in Circulating Tumour Cells from Early Breast Cancer Patients. *Cancers* **2019**, *11*, 59. [CrossRef] [PubMed]
23. Mego, M.; Karaba, M.; Minarik, G.; Benca, J.; Silvia, J.; Sedlackova, T.; Manasova, D.; Kalavska, K.; Pindak, D.; Cristofanilli, M.; et al. Circulating Tumor Cells with Epithelial-to-mesenchymal Transition Phenotypes Associated With Inferior Outcomes in Primary Breast Cancer. *Anticancer Res.* **2019**, *39*, 1829–1837. [CrossRef] [PubMed]
24. Kallergi, G.; Papadaki, M.A.; Politaki, E.; Mavroudis, D.; Georgoulias, V.; Agelaki, S. Epithelial to mesenchymal transition markers expressed in circulating tumour cells of early and metastatic breast cancer patients. *Breast Cancer Res.* **2011**, *13*, R59. [CrossRef] [PubMed]
25. Papadaki, M.A.; Kallergi, G.; Zafeiriou, Z.; Manouras, L.; Theodoropoulos, P.A.; Mavroudis, D.; Georgoulias, V.; Agelaki, S. Co-expression of putative stemness and epithelial-to-mesenchymal transition markers on single circulating tumour cells from patients with early and metastatic breast cancer. *BMC Cancer* **2014** *14*, 651. [CrossRef]

26. Markiewicz, A.; Książkiewicz, M.; Wełnicka-Jaśkiewicz, M.; Seroczyńska, B.; Skokowski, J.; Szade, J.; Żaczek, A.J. Mesenchymal phenotype of CTC-enriched blood fraction and lymph node metastasis formation potential. *PLoS ONE* **2014**, *9*, e93901. [CrossRef]
27. Hanahan, D.; Weinberg, R.A. Hallmarks of cancer: The next generation. *Cell* **2011**, *144*, 646–674. [CrossRef]
28. Hanahan, D.; Weinberg, R.A. The hallmarks of cancer. *Cell* **2000**, *100*, 57–70. [CrossRef]
29. Gonzalez, H.; Hagerling, C.; Werb, Z. Roles of the immune system in cancer: From tumor initiation to metastatic progression. *Genes Dev.* **2018**, *32*, 1267–1284. [CrossRef]
30. Ingangi, V.; Minopoli, M.; Ragone, C.; Motti, M.L.; Carriero, M.V. Role of Microenvironment on the Fate of Disseminating Cancer Stem Cells. *Front. Oncol.* **2019**, *9*, 82. [CrossRef]
31. Mohme, M.; Riethdorf, S.; Pantel, K. Circulating and disseminated tumour cells—Mechanisms of immune surveillance and escape. *Nat. Rev. Clin. Oncol.* **2017**, *14*, 155–167. [CrossRef] [PubMed]
32. Bednarz-Knoll, N.; Nastały, P.; Żaczek, A.; Stoupiec, M.G.; Riethdorf, S.; Wikman, H.; Müller, V.; Skokowski, J.; Szade, J.; Sejda, A.; et al. Stromal expression of ALDH1 in human breast carcinomas indicates reduced tumor progression. *Oncotarget* **2015**, *6*, 26789–26803. [CrossRef] [PubMed]
33. Huang, D.W.; Sherman, B.T.; Lempicki, R.A. Systematic and integrative analysis of large gene lists using DAVID bioinformatics resources. *Nat. Protoc.* **2009**, *4*, 44–57. [CrossRef] [PubMed]
34. Huang, D.W.; Sherman, B.T.; Lempicki, R.A. Bioinformatics enrichment tools: Paths toward the comprehensive functional analysis of large gene lists. *Nucleic Acids Res.* **2009**, *37*, 1–13. [CrossRef] [PubMed]
35. Cancer Genome Atlas Network. Comprehensive molecular portraits of human breast tumours. *Nature* **2012**, *490*, 61–70. [CrossRef]
36. Ciriello, G.; Gatza, M.L.; Beck, A.H.; Wilkerson, M.D.; Rhie, S.K.; Pastore, A.; Zhang, H.; McLellan, M.; Yau, C.; Kandoth, C.; et al. Comprehensive Molecular Portraits of Invasive Lobular Breast Cancer. *Cell* **2015**, *163*, 506–519. [CrossRef]
37. Pires, B.R.B.; Silva, R.C.M.C.; Ferreira, G.M.; Abdelhay, E. NF-kappaB: Two Sides of the Same Coin. *Genes* **2018**, *9*, 24. [CrossRef]
38. Choi, J.E.; Matthews, A.J.; Michel, G.; Vuong, B.Q. AID Phosphorylation Regulates Mismatch Repair-Dependent Class Switch Recombination and Affinity Maturation. *J. Immunol.* **2019**. [CrossRef]
39. Kravtsova-Ivantsiv, Y.; Ciechanover, A. The ubiquitin-proteasome system and activation of NF-κB: Involvement of the ubiquitin ligase KPC1 in p105 processing and tumor suppression. *Mol. Cell Oncol.* **2015**, *2*, e1054552. [CrossRef]
40. Song, C.; Mitter, S.K.; Qi, X.; Beli, E.; Rao, H.V.; Ding, J.; Ip, C.S.; Gu, H.; Akin, D.; Dunn, W.A.; et al. Oxidative stress-mediated NFκB phosphorylation upregulates p62/SQSTM1 and promotes retinal pigmented epithelial cell survival through increased autophagy. *PLoS ONE* **2017**, *12*, e0171940. [CrossRef]
41. Duran, A.; Linares, J.F.; Galvez, A.S.; Wikenheiser, K.; Flores, J.M.; Diaz-Meco, M.T.; Moscat, J. The signaling adaptor p62 is an important NF-kappaB mediator in tumorigenesis. *Cancer Cell* **2008**, *13*, 343–354. [CrossRef] [PubMed]
42. Trocoli, A.; Djavaheri-Mergny, M. The complex interplay between autophagy and NF-κB signaling pathways in cancer cells. *Am. J. Cancer Res.* **2011**, *1*, 629–649. [PubMed]
43. Xia, L.; Tan, S.; Zhou, Y.; Lin, J.; Wang, H.; Oyang, L.; Tian, Y.; Liu, L.; Su, M.; Wang, H.; et al. Role of the NFκB-signaling pathway in cancer. *Onco Targets Ther.* **2018**, *11*, 2063. [CrossRef] [PubMed]
44. Pires, B.R.B.; Mencalha, A.L.; Ferreira, G.M.; de Souza, W.F.; Morgado-Díaz, J.A.; Maia, A.M.; Corrêa, S.; Abdelhay, E.S.F.W. NF-kappaB Is Involved in the Regulation of EMT Genes in Breast Cancer Cells. *PLoS ONE* **2017**, *12*, e0169622. [CrossRef]
45. Nakshatri, H.; Bhat-Nakshatri, P.; Martin, D.A.; Goulet, R.J.; Sledge, G.W. Constitutive activation of NF-kappaB during progression of breast cancer to hormone-independent growth. *Mol. Cell. Biol.* **1997**, *17*, 3629–3639. [CrossRef]
46. Cogswell, P.C.; Guttridge, D.C.; Funkhouser, W.K.; Baldwin, A.S. Selective activation of NF-kappa B subunits in human breast cancer: Potential roles for NF-kappa B2/p52 and for Bcl-3. *Oncogene* **2000**, *19*, 1123–1131. [CrossRef]
47. Zhou, Y.; Eppenberger-Castori, S.; Eppenberger, U.; Benz, C.C. The NFkappaB pathway and endocrine-resistant breast cancer. *Endocr. Relat. Cancer* **2005**, *12* (Suppl. S1), S37–S46. [CrossRef]
48. Hou, Y.; Liang, H.; Rao, E.; Zheng, W.; Huang, X.; Deng, L.; Zhang, Y.; Yu, X.; Xu, M.; Mauceri, H.; et al.

Non-canonical NF-κB Antagonizes STING Sensor-Mediated DNA Sensing in Radiotherapy. *Immunity* **2018**, *49*, 490–503.e4. [CrossRef]

49. Biswas, D.K.; Shi, Q.; Baily, S.; Strickland, I.; Ghosh, S.; Pardee, A.B.; Iglehart, J.D. NF-kappa B activation in human breast cancer specimens and its role in cell proliferation and apoptosis. *Proc. Natl. Acad. Sci. USA* **2004**, *101*, 10137–10142. [CrossRef]
50. Hervas-Stubbs, S.; Perez-Gracia, J.L.; Rouzaut, A.; Sanmamed, M.F.; Le Bon, A.; Melero, I. Direct effects of type I interferons on cells of the immune system. *Clin. Cancer Res.* **2011**, *17*, 2619–2627. [CrossRef]
51. Jin, J.; Hu, H.; Li, H.S.; Yu, J.; Xiao, Y.; Brittain, G.C.; Zou, Q.; Cheng, X.; Mallette, F.A.; Watowich, S.S.; et al. Noncanonical NF-κB pathway controls the production of type I interferons in antiviral innate immunity. *Immunity* **2014**, *40*, 342–354. [CrossRef] [PubMed]
52. Cuenca, M.; Sintes, J.; Lányi, Á.; Engel, P. CD84 cell surface signaling molecule: An emerging biomarker and target for cancer and autoimmune disorders. *Clin. Immunol.* **2019**, *204*, 43–49. [CrossRef] [PubMed]
53. Pols, M.S.; Klumperman, J. Trafficking and function of the tetraspanin CD63. *Exp. Cell Res.* **2009**, *315*, 1584–1592. [CrossRef] [PubMed]
54. Labelle, M.; Begum, S.; Hynes, R.O. Direct signaling between platelets and cancer cells induces an epithelial-mesenchymal-like transition and promotes metastasis. *Cancer Cell* **2011**, *20*, 576–590. [CrossRef] [PubMed]
55. Labelle, M.; Begum, S.; Hynes, R.O. Platelets guide the formation of early metastatic niches. *Proc. Natl. Acad. Sci. USA* **2014**, *111*, E3053–E3061. [CrossRef] [PubMed]
56. Hoesel, B.; Schmid, J.A. The complexity of NF-κB signaling in inflammation and cancer. *Mol. Cancer* **2013**, *12*, 86. [CrossRef]
57. Ben-Neriah, Y.; Karin, M. Inflammation meets cancer, with NF-κB as the matchmaker. *Nat. Immunol.* **2011**, *12*, 715–723. [CrossRef]
58. Xia, Y.; Shen, S.; Verma, I.M. NF-κB, an active player in human cancers. *Cancer Immunol. Res.* **2014**, *2*, 823–830. [CrossRef]
59. R Core Team. *R: A Language and Environment for Statistical Computing*; R Foundation for Statistical Computing: Vienna, Austria, 2013.
60. Szklarczyk, D.; Gable, A.L.; Lyon, D.; Junge, A.; Wyder, S.; Huerta-Cepas, J.; Simonovic, M.; Doncheva, N.T.; Morris, J.H.; Bork, P.; et al. STRING v11: Protein-protein association networks with increased coverage, supporting functional discovery in genome-wide experimental datasets. *Nucleic Acids Res.* **2019**, *47*, D607–D613. [CrossRef]

4

Gene Expression Alterations during Development of Castration-Resistant Prostate Cancer are Detected in Circulating Tumor Cells

Andreas Josefsson [1,2,3], Karin Larsson [1], Eva Freyhult [4], Jan-Erik Damber [1] and Karin Welén [1,*]

1 Department of Urology, Sahlgrenska Cancer Center, Institute of Clinical Sciences, Sahlgrenska Academy, University of Gothenburg, 405 30 Gothenburg, Sweden; andreas.josefsson@urology.gu.se (A.J.); karin.e.larsson@gu.se (K.L.); jan-erik.damber@urology.gu.se (J.-E.D.)
2 Department of Surgical and Perioperative Sciences, Urology & Andrology, Umeå University, 901 87 Umeå, Sweden
3 Wallenberg Center for Molecular Medicine, Umeå University, 901 87 Umeå, Sweden
4 Department of Medical Sciences, National Bioinformatics Infrastructure Sweden, Science for Life Laboratory, Uppsala University, 751 85 Uppsala, Sweden; Eva.freyhult@medsci.uu.se
* Correspondence: karin.welen@urology.gu.se.

Abstract: Development of castration-resistant prostate cancer (CRPC) is associated with alterations in gene expression involved in steroidogenesis and androgen signaling. This study investigates whether gene expression changes related to CRPC development can be identified in circulating tumor cells (CTCs). Gene expression in paired CTC samples from 29 patients, before androgen deprivation therapy (ADT) and at CRPC relapse, was compared using a panel including 47 genes related to prostate cancer progression on a qPCR platform. Fourteen genes displayed significantly changed gene expression in CTCs at CRPC relapse compared to before start of ADT. The genes with increased expression at CRPC relapse were related to steroidogenesis, AR-signaling, and anti-apoptosis. In contrast, expression of prostate markers was downregulated at CRPC. We also show that midkine (MDK) expression in CTCs from metastatic hormone-sensitive prostate cancer (mHSPC) was associated to short cancer-specific survival (CSS). In conclusion, this study shows that gene expression patterns in CTCs reflect the development of CRPC, and that MDK expression levels in CTCs are prognostic for cancer-specific survival in mHSPC. This study emphasizes the role of CTCs in exploring mechanisms of therapy resistance, as well as a promising biomarker for prognostic and treatment-predictive purposes in advanced mHSPC.

Keywords: CTC; hormone-sensitive prostate cancer; CRPC; liquid biopsy; biomarker; resistance mechanisms

1. Introduction

Metastatic castration-resistant prostate cancer (mCRPC) is responsible for most deaths from prostate cancer. CRPC develops during androgen deprivation therapy (ADT), the standard therapy for metastatic prostate cancer (PC). ADT inhibits the testicular production of testosterone resulting in low circulating levels and inhibited growth of androgen-dependent PC cells. However, androgen receptor signaling is crucial for most PCs, and several mechanisms for its sustained activity exist, enabling the relapse and growth of CRPC. Although CRPC is still lethal, today many life-prolonging therapies are in clinical use, and patients now live longer with CRPC than in the hormone-dependent phase of metastatic PC [1].

It is well established that increased expression of steroidogenic enzymes and constitutively active forms of the androgen receptor are characteristics of CRPC [2–4]. The steroidogenic enzymes confer an intratumoral steroid synthesis resulting in higher testosterone levels within tumors compared to the castrate levels present systemically during ADT [5]. In addition to increased steroid levels, the androgen receptor (AR) presents itself in several differently mutated or spliced variants with increased sensitivity, stability, or activity to maintain sufficient signaling for the PC cells to survive and proliferate [6–9].

Both these strategies can partly be targeted by the new generation of hormone-related drugs. Abiraterone acetate inhibits CYP17A1, which converts pregnenolone and progesterone to DHEA and androstenedione, respectively, both important precursors to testosterone [10]. Enzalutamide is a new-generation AR inhibitor that is more efficient compared to, for example, bicalutamide [11]. However, CRPC patients with expression of the constitutively active AR splice variant AR-V7 often display resistance to both these therapies [12,13]. In addition to AR-targeting drugs, other therapies such as chemotherapy (docetaxel, cabazitaxel), targeted radiation (radium-223), and to some extent immunotherapy (Sipuleucel-T) contribute to the prolonged survival of patients with CRPC [14].

To achieve the most efficient use of different drugs for the benefit for the patients, individualized treatment protocols are needed. In metastatic disease, accurate characterization of tumor biology is challenging. However, circulating tumor cells in the blood stream have been shown to mirror metastatic phenotype [15], and analysis of circulating tumor cells (CTCs) may be a tool to get information for clinical decisions.

In the present study, we isolated CTCs from patients with advanced metastatic PC undergoing ADT to investigate whether the changes occurring in the tumors can be identified in CTCs. The correlation of CTC gene expression before therapy with progression on ADT was also investigated.

2. Results

Forty patients with metastatic hormone-sensitive prostate cancer (mHSPC) were included, and 37 of these displayed signals for at least two of the genes analyzed, not related to the leucocyte contamination, and were defined as CTC positive and included in the study. The median follow-up time for all 40 included patients was 27.5 months (range 0.7–72.8 months). The three CTC negative patients did not have detected metastases at start of ADT, and their follow-up times were 23.5, 43.2, and 62.1 months. For the 37 CTC positive patients included in the survival analysis, the follow-up times, cancer-specific survival (CSS), and time to CRPC can be found in Table 1.

For the 28 of these patients that died of PC within the study, the median time from ADT to PC-death was 17.3 months (Q1: 10.9, Q3: 32.1), and the time from CRPC to PC-death was 10.5 months (Q1: 5.1, Q3: 27.5). For the patients still alive at last follow-up ($n = 7$, two died of other causes), the times from ADT and CRPC to last follow-up were 45.1 months (Q1: 38.7, Q3: 52.9) and 40.6 months (Q1: 33.2, Q3: 53.3), respectively. At CRPC relapse, eight of the 37 patients were either CTC negative ($n = 3$) or were not sampled ($n = 5$), leaving 29 patients for comparisons of CTC gene expression alterations during ADT.

Of the 47 assays included in the PC-panel used for detection of gene expression, seven genes (*ESR1*, *ESR2*, *PTCH1*, *CYP11A1*, *CYP17A1*, *CYP19A1*, and *MET1*) were excluded from analysis, since they could only rarely be detected in the CTCs sampled. Six of the genes (*MYC*, *TP53*, *ANXAR2*, *AKT*, *ALDHA1*, and *RUNX2*) were excluded since they were strongly correlated to *CD45* (statistically significant correlation ($p < 0.05$) with a correlation coefficient >0.5), thus the signal may come from the contaminating population of white blood cells, making interpretations about expression in CTCs difficult. In addition, the four control genes (*GAPDH*, *GUSB*, *CD44*, and *CD45*) were not included in the analysis. The numbers of detected signals for the remaining 30 genes are presented in Table 2.

Table 1. Patient characteristics of 37 circulating tumor cell (CTC)-positive patients).

Clinical Variables	Number or Median (Q1; Q3)
PSA before androgen deprivation therapy (ADT)	590 ng/mL (270; 1900)
Gleason sum	
6–7	4
8	9
9–10	17
x	7
Type of ADT	
GnRH agonist	28
GnRH antagonist	7
Orchidectomy	2
Time from ADT to CRPC relapse	8.4 months (4.9; 12.9)
Castration-resistant prostate cancer CRPC Therapy	
Total androgen blockade (bicalutamide)	25
Abiraterone acetate or enzalutamide	11
Radium-223	9
Docetaxel or cabazitaxel	8
Cyclophosphamide	2
None	5
Time from ADT to prostate cancer death or last follow-up (cancer-specific survival, CSS) (n = 35 *)	25.7 months (13.6; 39.3)
Total follow-up time (n = 37)	24.1 months (11.1; 39.0)

* Two patients died of other causes.

Table 2. Gene expression signals detected and included in paired analysis.

Gene	Before ADT (n = 40) (A)		At CRPC Relapse (n = 32)		Pairs Included in Comparison (A vs. CRPC) **
	Detected Signals	Substituted * Missing Signals	Detected Signals	Substituted * Missing Signals	
AGR2	30	8	14	18	25
AHR	13	24	9	20	13
AKR1C3	23	17	21	11	24
AR	23	18	17	14	19
ARV7	9	26	10	8	12
AURKA	21	19	19	10	22
BCL2	16	23	17	11	21
CDH1	22	14	14	15	17
CDH2	4	30	4	24	4
DDR1	13	23	13	16	14
EGFR	12	9	8	9	9
EMP2	22	16	13	17	16
EPCAM	36	2	23	9	29
FOLH1	30	7	15	14	22
UPA	11	25	10	15	13
HER2	24	16	16	16	21
KLK3	33	5	21	11	27
KRT19	28	9	17	14	25
MDK	29	11	16	15	25
POU5F1	19	20	18	14	22
PSCA	17	20	9	21	14
SNAI1	10	19	4	22	9
SPINK1	20	20	12	20	18
SRD5A1	25	15	27	4	28
TACSTD2	29	9	18	8	24
TOP2A	26	13	18	8	21
TP53	34	6	28	4	28
TUBB3	21	15	14	16	21
TWIST1	21	17	11	21	19
VEGFA	24	16	17	15	23

* Missing signals in samples with too low CTC load for individual genes to be detected are not substituted. ** Pairs need to contain at least one detected signal to be included in analysis.

2.1. *Alterations in Gene Expression during ADT*

Changes in gene expression in CTCs during ADT were investigated by a pair-wise comparison of the ΔCq values of the CTC samples before start of ADT and at CRPC relapse. Missing signals due to undetected gene expression were replaced with a calculated ΔCq value representing low expression when appropriate (see Methods section). A pair consisting of two calculated ΔCq replaced for missing signals was excluded from analysis, resulting in a varying number of patients included in the analysis for different genes (Table 2).

In the comparative analysis, 14 of the 30 genes displayed altered expression levels at CRPC relapse compared to before start of ADT. Increased expression was detected for *AR* ($p < 0.01$), *AR-V7* ($p < 0.05$) and the steroidogenic enzymes *AKR1C3* ($p < 0.05$) and *SRD5A1* ($p < 0.01$) (Figure 1A). In contrast, the expression of the prostate cancer marker genes *KLK3* ($p < 0.01$), *FOLH1* ($p < 0.05$), and *PSCA* ($p < 0.05$) was decreased in CTCs at CRPC relapse (Figure 1B). Expression of genes related to an epithelial phenotype (*EPCAM* ($p < 0.01$), *KRT19* ($p < 0.05$), and *HER2* ($p < 0.05$)) was found to be decreased at CRPC relapse (Figure 1C). Other genes with altered expression levels were the anti-apoptotic *BCL2* (upregulated; $p < 0.05$), the epithelial-to-mesenchymal transition marker *TWIST1* (downregulated; $p < 0.01$), the stem cell marker *TACSTD2* (downregulated; $p < 0.05$), and *AGR2*, an androgen-regulated gene associated with metastasis in PC [16–18] (downregulated; $p < 0.01$) (Figure 1D).

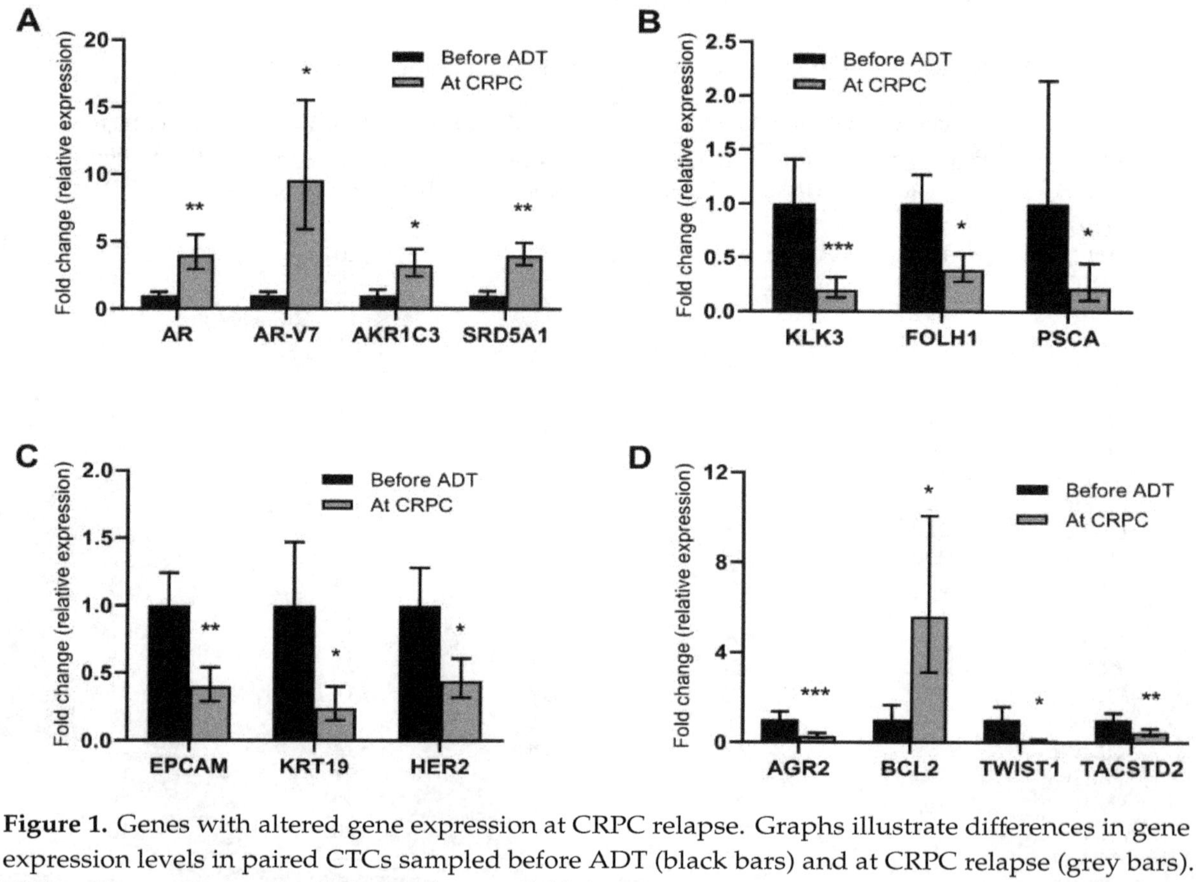

Figure 1. Genes with altered gene expression at CRPC relapse. Graphs illustrate differences in gene expression levels in paired CTCs sampled before ADT (black bars) and at CRPC relapse (grey bars). Differences are displayed as relative changes (fold change) in relation to levels before ADT for (**A**) genes related to androgen signaling; (**B**) prostate markers; (**C**) genes related to epithelial phenotype; and (**D**) other genes with altered expression levels. Bars represent mean fold change ± standard error of the mean (SEM); the black bars for before ADT always shows 1 ± SEM by definition of the method. Statistically significant differences are denoted with * = $p < 0.05$, ** = $p < 0.01$, and *** = $p < 0.001$.

2.2. Gene Expression in CTCs as Prognostic Markers for Survival

The possible prognostic information of gene expression levels in CTCs was assessed by Cox regression analysis relating the gene expression in CTCs before start of ADT to either time to development of CRPC or cancer-specific survival (CSS). To ensure that no false correlations due to an exaggerated estimated low value would be detected, a stricter cut-off level for substituted low values was applied for this analysis. Based on the highest mCq at which a specific gene could be detected, the cut-off was set one Cq lower, i.e., missing signals were only replaced and included in samples with a higher CTC content (one mCq lower) than the one representing the detection limit for the specific gene. Thus, the number of data points included varies among the genes analyzed (Table 2), and a lack of correlation could be a consequence of few data points in the analysis for certain genes.

The only gene that was significantly associated to time to development of CRPC was *EPCAM* ($p = 0.012$). This is most likely due to the fact that EPCAM is the epitope used for capture of the CTCs, and that it therefore, also after normalization, largely reflects the CTC content, which is expected to correlate to prognosis. mCq, the surrogate value for CTC content in this study, was strongly associated to a short time to development of CRPC ($p = 0.0005$) (Table 3). Other genes were not associated to time to CRPC.

Table 3. Cox regression analysis for survival.

Parameter	Time to CRPC		Cancer-Specific Survival	
	p-Value	Hazard Ratio (95% Confidence Interval)	*p*-Value	Hazard Ratio (95% Confidence Interval)
Average Cq	**0.0005**	1.22 (1.09–1.38)	**0.0003**	1.24 (1.10–1.39)
EPCAM	**0.012**	1.32 (1.06–1.65)	**0.002**	1.52 (1.16–1.97)
MDK	0.061	1.11 (0.99–1.25)	**0.008**	1.25 (1.06–1.47)
TWIST	0.537	0.96 (0.85–1.09)	**0.050**	0.87 (0.75–1.00)

Bold indicates *p*-values ≤ 0.05.

Similarly, both *EPCAM* ($p = 0.002$) and mCq ($p = 0.0003$) were significantly associated to CSS with Cox regression statistics. In addition, a high *MDK* (Midkine) expression in CTCs was associated to a poor CSS ($p = 0.008$). The EMT-associated gene *TWIST1* was inversely associated to CSS ($p = 0.05$). No other genes were associated to CSS. None of the genes identified as significant in the univariable analysis remained prognostic when analyzed together in multivariable analysis.

No parameters were significantly associated with time to CRPC using Kaplan–Meier statistics. Using non-parametric Kaplan–Meier statistics to verify the associations with CSS, *EPCAM* ($p = 0.046$), mCq ($p = 0.0005$), and *MDK* ($p = 0.0003$) were confirmed to have statistically significant associations (Figure 2), while *TWIST1* did not ($p = 0.252$). In addition, the EMT-related gene *CDH2* and the androgen-regulated gene *AGR2* were found to be significantly associated with CSS ($p = 0.036$, and $p = 0.048$, respectively) (Figure 2). The expression of the highly prostate-cancer-related genes *AR*, *AR-V7* and *FOLH1* was not associated to survival.

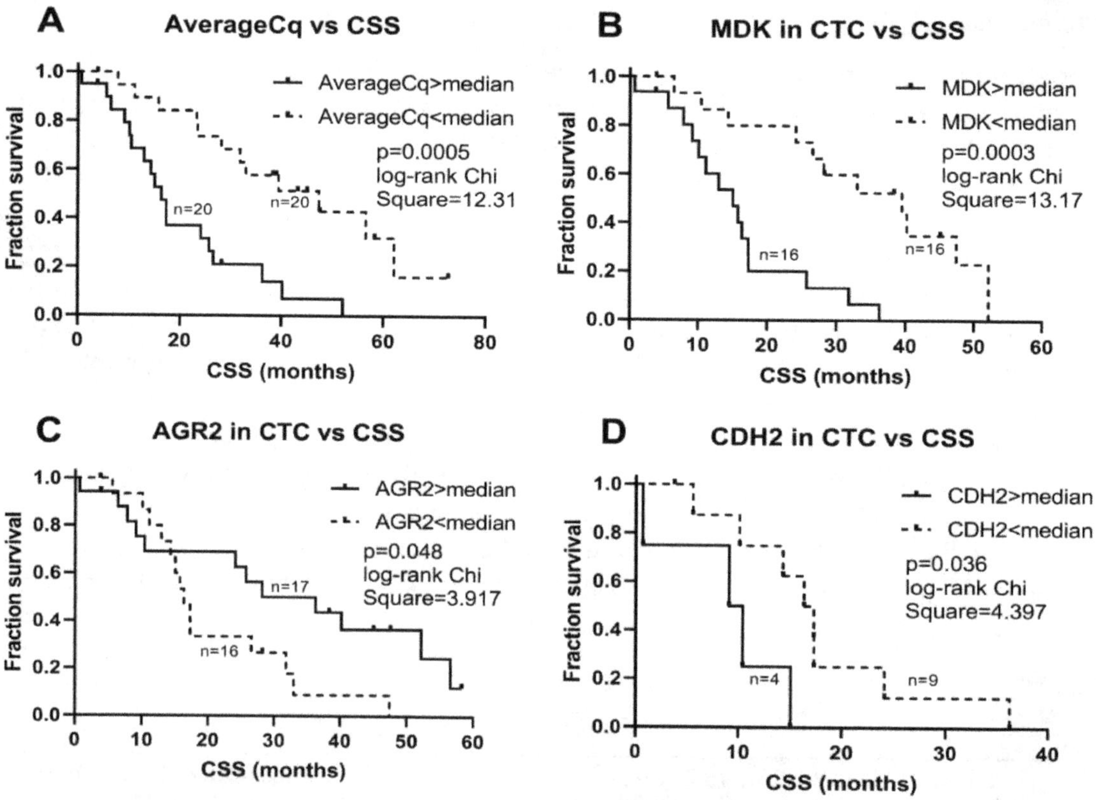

Figure 2. Survival plots of dichotomized CTC acquired parameters. Graphs illustrate Kaplan–Meier statistics of (**A**) average signal strength (mCq) as a proxy for CTC content, (**B**) midkine (MDK) gene expression in CTCs (normalized to AverageCq), (**C**) AGR2 expression in CTCs (normalized to Average Cq), and (**D**) N-cadherin (CDH2) expression in CTCs (normalized to Average Cq), in relation to cancer-specific survival (CSS). Parameters are grouped as higher or lower than their respective median values.

3. Discussion

In CRPC, CTCs have been shown to be a good biomarker for prognosis, based on their enumeration, and treatment-predictive purposes, based on their expression of AR-V7 [12,19]. Their expression pattern also reflects the phenotype of skeletal metastases [15]. However, the potential of CTCs as a source for phenotypic information also in the hormone-sensitive stage of the disease has not been extensively investigated. The present study provides data emphasizing the value of CTC characterization in hormone-sensitive PC both for phenotypic investigation of disease progression and for biomarker purposes.

This is the first study that investigates consecutive CTC samples during the progression from hormone-sensitive to castration-resistant prostate cancer. In the pairwise analysis of these samples, we identified a number of genes, the expression of which changed during the development of CRPC. In line with previous studies on tumor tissue, the CTC expression of genes in the androgen signaling pathway was increased in CRPC compared to before start of ADT. Expression of both AR and its constitutively active splice variant AR-V7 are increased in CRPC tissue [3,4,20] reflecting an increased androgen sensitivity and AR signaling in the CRPC state. Increased expression of the steroidogenic enzyme AKR1C3 has been reported in CRPC tissue [4,21], pointing out conversion of adrenal steroid precursors as a resistance mechanism to ADT and promoting intratumoral androgen synthesis enabling sustained AR signaling despite castrate levels of testosterone in the circulation. SRD5A1 converts testosterone to dihydrotestosterone (DHT) enabling more potent activation of the AR. Although previous studies have showed the increased importance of SRD5A1 compared to SRD5A2 in advanced PC [22] as well as increased expression of SRD5A1 in metastatic PC compared to primary tumors [3,21,23,24], this study

informed consent for inclusion before they participated in the study. The study was conducted in accordance with the Declaration of Helsinki, and the protocol was approved by the Ethics Committee of Gothenburg (936-12).

4.2. CTC Isolation and cDNA Synthesis

CTCs were isolated from blood and detected using the AdnaTest ProstateCancerSelect/Detect (Qiagen Hannover GmbH, Langenhagen, Germany) as previously described [41]. Briefly, patient blood samples were collected immediately prior to surgery in AdnaCollect tubes and kept refrigerated (4 °C) for no more than 24 h until CTC isolation. CTCs were captured with EPCAM and HER2 antibody-conjugated magnetic beads and lysed before mRNA was isolated with oligo-dT-conjugated magnetic beads and transcribed into cDNA.

4.3. Gene Expression Profiling

Two μL of cDNA samples were preamplified using the TATAA PreAmp Primer Mix and TATAA PreAmp GrandMaster® Mix (Cat. No. #TA05, TATAA Biocenter, Gothenburg, Sweden). Preamplification was performed in a thermocycler (T100, BioRad, Hercules, CA, USA). Preamplification with no template control and human gDNA sample (concentration 0.5 ng/μL, TATAA Biocenter, Gothenburg, Sweden) were also included. The preamplified samples were spun down (to pellet the magnetic beads), and a fraction of the supernatant was moved to a separate tube and diluted 10×. The diluted samples were analyzed with 47 assays specifically designed for this study (now available in the GrandPerformance CTC Assay Panel (TATAA Biocenter, Gothenburg, Sweden)) and ValidPrime™ assay (TATAA Biocenter, Gothenburg, Sweden). For assay details see Table S1. The qPCR was performed using TATAA Probe GrandMaster® Mix Low ROX (TATAA Biocenter, Gothenburg, Sweden) and GE 96.96 Dynamic Array™ Sample & Assay Loading Reagent Kit (P/N 85000802-R, Fluidigm, South San Francisco, CA, USA). Preamplification no template control (preAmp NTC) and no template control (NTC) for the qPCR were included. The qPCR was performed on BioMark (Fluidigm, South San Francisco, CA, USA) using the 96.96 Dynamic Array™ IFC (Integrated Fluidic Circuit). All samples (including NTCs and gDNA) were analyzed in duplicates.

4.4. Definition of CTC Load as Basis for Normalization and Handling of Missing Signals

Cycle of quantification (Cq) values for the individual genes were correlated to Cq values for CD45 as a marker for leucocyte contamination. Genes with a *p*-value less than 0.05 and a Spearman correlation coefficient above 0.5 (*AKT2, ALDH1A1, ANXA2R, CD44, GAPDH, GUSB, MYC,* and *RUNX2*) were removed from further analyses on the assumption that they largely represent gene expression in the contaminating leucocyte population.

For all other genes, missing signals (i.e., not detected with PCR) were imputed with $Cq_{max,g} + 1$, where $Cq_{max,g}$ is the highest detectable Cq value for gene g, and an average Cq (mCq_k) was calculated for each individual sample k. Using these mCq values, a ΔCq was calculated for all detected signals (ΔCq = Cq−mCq). Originally missing signals were reimputed with $\Delta Cq_{max,g} + 1$, i.e., one cycle higher than the highest identified ΔCq for that specific gene, g. A missing Cq value means that there were no detectable levels of the gene in the sample. There can of course be several different reasons for this; either there are insufficient cells in the sample, or the expression of the gene is zero or very low. To ensure that missing signals were not falsely interpreted as a low gene expression in samples where the CTC content was too low to enable signal detection, a cut-off was set for each gene, g, based on the sample, h, with the highest mCq in which the gene could be detected. For gene g, the cut off was calculated as: cut off-$mCq_g = mCq_{h,g} - ((Cq_{h,g} - Cq_{max,g})/2)$, where $mCq_{h,g} - Cq_{max,g}$ is the mCq for sample h and $Cq_{h,g}$ is the Cq for gene g in sample h. Using this cut off, a missing value was only imputed and included in the analysis if its sample mCq was lower or equal to the cut-off. The cut-off calculations were based on approximately 230 CTC samples from prostate cancer patients, most of them not included in the analysis in the present study. A step-wise example of the calculations can be found as a supplementary file (Table S3).

4.5. Statistical Analysis

Differences between gene expression in CTCs before ADT and at CRPC relapse were statistically evaluated using the Wilcoxon test, a paired rank-based non-parametric test. Survival analysis was performed using univariate Cox regression analysis and Kaplan–Meier analysis. A stricter cut-off was used for survival analysis than for the pair-wise comparisons. Samples with imputed values were only included in survival analysis if their mCq was one cycle lower (i.e., had a higher CTC content) than the mCq for the sample with the highest mCq in which the gene could be detected ($mCq_{h,g}$). Cox regression analysis was performed on continuous data, and Kaplan–Meier log-rank analysis was performed on patients categorized based on expression levels above or below the median expression level of the specific gene. Genes for which more than 50% of values were imputed were excluded from Cox regression analysis (Table S2).

5. Conclusions

We show that liquid biopsies in the form of CTCs harbor treatment prognostic information regarding CSS in mHSPC. Furthermore, CTCs can provide deeper knowledge on treatment-induced expression changes, enabling identification of novel therapeutic targets.

Author Contributions: Conceptualization: K.W., A.J., J.-E.D.; Methodology: K.W., K.L., A.J., E.F.; Validation: K.W., E.F.; Formal Analyses: K.W., E.F.; Investigation: A.J., K.W., K.L.; Resources: J.-E.D., K.W., A.J., K.L.; Writing-original draft preparation: K.W., A.J.; Writing-review editing: J.-E.D., K.L., A.J., K.W., E.F.; Visualization: K.W.; Project Administration: A.J., K.W.; Funding acquisition: K.W., J.-E.D., A.J. All authors have read and agreed to the published version of the manuscript.

Acknowledgments: The authors gratefully acknowledge the lab technicians Anita Fae at the department of Urology, Sahlgrenska Academy, and Jennifer Pettersson at the TATAA Biocenter for their excellent technical assistance, as well as Robert Sjöback and Jens Björkman (TATAA Biocenter) for setting up the CTC analyses. We also acknowledge Christina Hansson, Marianne Sanderoth and Jenny Holm Fernström at the Urology Clinic at Sahlgrenska University Hospital for their valued assistance with the blood sampling and patient logistics.

References

1. James, N.D.; Spears, M.R.; Clarke, N.W.; Dearnaley, D.P.; De Bono, J.S.; Gale, J.; Hetherington, J.; Hoskin, P.J.; Jones, R.J.; Laing, R.; et al. Survival with newly diagnosed metastatic prostate cancer in the "docetaxel era": Data from 917 patients in the control arm of the stampede trial (mrc pr08, cruk/06/019). *Eur. Urol.* **2015**, *67*, 1028–1038. [CrossRef] [PubMed]
2. Locke, J.A.; Guns, E.S.; Lubik, A.A.; Adomat, H.H.; Hendy, S.C.; Wood, C.A.; Ettinger, S.L.; Gleave, M.E.; Nelson, C.C. Androgen levels increase by intratumoral de novo steroidogenesis during progression of castration-resistant prostate cancer. *Cancer Res.* **2008**, *68*, 6407–6415. [CrossRef] [PubMed]
3. Montgomery, R.B.; Mostaghel, E.A.; Vessella, R.; Hess, D.L.; Kalhorn, T.F.; Higano, C.S.; True, L.D.; Nelson, P.S. Maintenance of intratumoral androgens in metastatic prostate cancer: A mechanism for castration-resistant tumor growth. *Cancer Res.* **2008**, *68*, 4447–4454. [CrossRef] [PubMed]
4. Stanbrough, M.; Bubley, G.J.; Ross, K.; Golub, T.R.; Rubin, M.A.; Penning, T.M.; Febbo, P.G.; Balk, S.P. Increased expression of genes converting adrenal androgens to testosterone in androgen-independent prostate cancer. *Cancer Res.* **2006**, *66*, 2815–2825. [CrossRef]
5. Titus, M.A.; Schell, M.J.; Lih, F.B.; Tomer, K.B.; Mohler, J.L. Testosterone and dihydrotestosterone tissue levels in recurrent prostate cancer. *Clin. Cancer Res.* **2005**, *11*, 4653–4657. [CrossRef] [PubMed]
6. Gregory, C.W.; Johnson, R.T., Jr.; Mohler, J.L.; French, F.S.; Wilson, E.M. Androgen receptor stabilization in recurrent prostate cancer is associated with hypersensitivity to low androgen. *Cancer Res.* **2001**, *61*, 2892–2898. [PubMed]

7. Hu, R.; Dunn, T.A.; Wei, S.; Isharwal, S.; Veltri, R.W.; Humphreys, E.; Han, M.; Partin, A.W.; Vessella, R.L.; Isaacs, W.B.; et al. Ligand-independent androgen receptor variants derived from splicing of cryptic exons signify hormone-refractory prostate cancer. *Cancer Res.* **2009**, *69*, 16–22. [CrossRef]
8. Linja, M.J.; Savinainen, K.J.; Saramaki, O.R.; Tammela, T.L.; Vessella, R.L.; Visakorpi, T. Amplification and overexpression of androgen receptor gene in hormone-refractory prostate cancer. *Cancer Res.* **2001**, *61*, 3550–3555.
9. Taplin, M.E.; Bubley, G.J.; Shuster, T.D.; Frantz, M.E.; Spooner, A.E.; Ogata, G.K.; Keer, H.N.; Balk, S.P. Mutation of the androgen-receptor gene in metastatic androgen-independent prostate cancer. *N. Engl. J. Med.* **1995**, *332*, 1393–1398. [CrossRef]
10. Barrie, S.E.; Potter, G.A.; Goddard, P.M.; Haynes, B.P.; Dowsett, M.; Jarman, M. Pharmacology of novel steroidal inhibitors of cytochrome p450(17) alpha (17 alpha-hydroxylase/c17-20 lyase). *J. Steroid Biochem. Mol. Biol.* **1994**, *50*, 267–273. [CrossRef]
11. Tran, C.; Ouk, S.; Clegg, N.J.; Chen, Y.; Watson, P.A.; Arora, V.; Wongvipat, J.; Smith-Jones, P.M.; Yoo, D.; Kwon, A.; et al. Development of a second-generation antiandrogen for treatment of advanced prostate cancer. *Science* **2009**, *324*, 787–790. [CrossRef] [PubMed]
12. Antonarakis, E.S.; Lu, C.; Wang, H.; Luber, B.; Nakazawa, M.; Roeser, J.C.; Chen, Y.; Mohammad, T.A.; Chen, Y.; Fedor, H.L.; et al. Ar-v7 and resistance to enzalutamide and abiraterone in prostate cancer. *N. Engl. J. Med.* **2014**, *371*, 1028–1038. [CrossRef] [PubMed]
13. Scher, H.I.; Lu, D.; Schreiber, N.A.; Louw, J.; Graf, R.P.; Vargas, H.A.; Johnson, A.; Jendrisak, A.; Bambury, R.; Danila, D.; et al. Association of ar-v7 on circulating tumor cells as a treatment-specific biomarker with outcomes and survival in castration-resistant prostate cancer. *JAMA Oncol.* **2016**, *2*, 1441–1449. [CrossRef] [PubMed]
14. Ritch, C.R.; Cookson, M.S. Advances in the management of castration resistant prostate cancer. *BMJ* **2016**, *355*, i4405. [CrossRef]
15. Josefsson, A.; Larsson, K.; Mansson, M.; Bjorkman, J.; Rohlova, E.; Ahs, D.; Brisby, H.; Damber, J.E.; Welen, K. Circulating tumor cells mirror bone metastatic phenotype in prostate cancer. *Oncotarget* **2018**, *9*, 29403–29413. [CrossRef]
16. Bu, H.; Bormann, S.; Schafer, G.; Horninger, W.; Massoner, P.; Neeb, A.; Lakshmanan, V.K.; Maddalo, D.; Nestl, A.; Sultmann, H.; et al. The anterior gradient 2 (agr2) gene is overexpressed in prostate cancer and may be useful as a urine sediment marker for prostate cancer detection. *Prostate* **2011**, *71*, 575–587. [CrossRef]
17. Chanda, D.; Lee, J.H.; Sawant, A.; Hensel, J.A.; Isayeva, T.; Reilly, S.D.; Siegal, G.P.; Smith, C.; Grizzle, W.; Singh, R.; et al. Anterior gradient protein-2 is a regulator of cellular adhesion in prostate cancer. *PLoS ONE* **2014**, *9*, e89940. [CrossRef]
18. Kani, K.; Malihi, P.D.; Jiang, Y.; Wang, H.; Wang, Y.; Ruderman, D.L.; Agus, D.B.; Mallick, P.; Gross, M.E. Anterior gradient 2 (agr2): Blood-based biomarker elevated in metastatic prostate cancer associated with the neuroendocrine phenotype. *Prostate* **2013**, *73*, 306–315. [CrossRef]
19. De Bono, J.S.; Scher, H.I.; Montgomery, R.B.; Parker, C.; Miller, M.C.; Tissing, H.; Doyle, G.V.; Terstappen, L.W.; Pienta, K.J.; Raghavan, D. Circulating tumor cells predict survival benefit from treatment in metastatic castration-resistant prostate cancer. *Clin. Cancer Res.* **2008**, *14*, 6302–6309. [CrossRef]
20. Hornberg, E.; Ylitalo, E.B.; Crnalic, S.; Antti, H.; Stattin, P.; Widmark, A.; Bergh, A.; Wikstrom, P. Expression of androgen receptor splice variants in prostate cancer bone metastases is associated with castration-resistance and short survival. *PLoS ONE* **2011**, *6*, e19059. [CrossRef]
21. Jernberg, E.; Thysell, E.; Bovinder Ylitalo, E.; Rudolfsson, S.; Crnalic, S.; Widmark, A.; Bergh, A.; Wikstrom, P. Characterization of prostate cancer bone metastases according to expression levels of steroidogenic enzymes and androgen receptor splice variants. *PLoS ONE* **2013**, *8*, e77407. [CrossRef] [PubMed]
22. Titus, M.A.; Gregory, C.W.; Ford, O.H., 3rd; Schell, M.J.; Maygarden, S.J.; Mohler, J.L. Steroid 5alpha-reductase isozymes i and ii in recurrent prostate cancer. *Clin. Cancer Res.* **2005**, *11*, 4365–4371. [CrossRef] [PubMed]
23. Hofland, J.; van Weerden, W.M.; Dits, N.F.; Steenbergen, J.; van Leenders, G.J.; Jenster, G.; Schroder, F.H.; de Jong, F.H. Evidence of limited contributions for intratumoral steroidogenesis in prostate cancer. *Cancer Res.* **2010** *70*, 1256–1264. [CrossRef] [PubMed]
24. Mitsiades, N.; Sung, C.C.; Schultz, N.; Danila, D.C.; He, B.; Eedunuri, V.K.; Fleisher, M.; Sander, C.; Sawyers, C.L.; Scher, H.I. Distinct patterns of dysregulated expression of enzymes involved in androgen

synthesis and metabolism in metastatic prostate cancer tumors. *Cancer Res.* **2012**, *72*, 6142–6152. [CrossRef] [PubMed]

25. Gallee, M.P.; Visser-de Jong, E.; van der Korput, J.A.; van der Kwast, T.H.; ten Kate, F.J.; Schroeder, F.H.; Trapman, J. Variation of prostate-specific antigen expression in different tumour growth patterns present in prostatectomy specimens. *Urol. Res.* **1990**, *18*, 181–187. [CrossRef] [PubMed]
26. Stege, R.H.; Tribukait, B.; Carlstrom, K.A.; Grande, M.; Pousette, A.H. Tissue psa from fine-needle biopsies of prostatic carcinoma as related to serum psa, clinical stage, cytological grade, and DNA ploidy. *Prostate* **1999**, *38*, 183–188. [CrossRef]
27. Chen, C.L.; Mahalingam, D.; Osmulski, P.; Jadhav, R.R.; Wang, C.M.; Leach, R.J.; Chang, T.C.; Weitman, S.D.; Kumar, A.P.; Sun, L.; et al. Single-cell analysis of circulating tumor cells identifies cumulative expression patterns of emt-related genes in metastatic prostate cancer. *Prostate* **2013**, *73*, 813–826. [CrossRef]
28. Markou, A.; Lazaridou, M.; Paraskevopoulos, P.; Chen, S.; Swierczewska, M.; Budna, J.; Kuske, A.; Gorges, T.M.; Joosse, S.A.; Kroneis, T.; et al. Multiplex gene expression profiling of in vivo isolated circulating tumor cells in high-risk prostate cancer patients. *Clin. Chem.* **2018**, *64*, 297–306. [CrossRef]
29. Sun, Y.; Wang, B.E.; Leong, K.G.; Yue, P.; Li, L.; Jhunjhunwala, S.; Chen, D.; Seo, K.; Modrusan, Z.; Gao, W.Q.; et al. Androgen deprivation causes epithelial-mesenchymal transition in the prostate: Implications for androgen-deprivation therapy. *Cancer Res.* **2012**, *72*, 527–536. [CrossRef]
30. Tanaka, H.; Kono, E.; Tran, C.P.; Miyazaki, H.; Yamashiro, J.; Shimomura, T.; Fazli, L.; Wada, R.; Huang, J.; Vessella, R.L.; et al. Monoclonal antibody targeting of n-cadherin inhibits prostate cancer growth, metastasis and castration resistance. *Nat. Med.* **2010**, *16*, 1414–1420. [CrossRef]
31. Ahrens-Fath, I.; Politz, O.; Geserick, C.; Haendler, B. Androgen receptor function is modulated by the tissue-specific ar45 variant. *FEBS J.* **2005**, *272*, 74–84. [CrossRef] [PubMed]
32. Goodman, O.B., Jr.; Symanowski, J.T.; Loudyi, A.; Fink, L.M.; Ward, D.C.; Vogelzang, N.J. Circulating tumor cells as a predictive biomarker in patients with hormone-sensitive prostate cancer. *Clin. Genitourin. Cancer* **2011**, *9*, 31–38. [CrossRef] [PubMed]
33. Nordin, A.; Wang, W.; Welen, K.; Damber, J.E. Midkine is associated with neuroendocrine differentiation in castration-resistant prostate cancer. *Prostate* **2013**, *73*, 657–667. [CrossRef] [PubMed]
34. Zhang, J.S.; Gong, A.; Cheville, J.C.; Smith, D.I.; Young, C.Y. Agr2, an androgen-inducible secretory protein overexpressed in prostate cancer. *Genes Chromosomes Cancer* **2005**, *43*, 249–259. [CrossRef]
35. Gravdal, K.; Halvorsen, O.J.; Haukaas, S.A.; Akslen, L.A. A switch from e-cadherin to n-cadherin expression indicates epithelial to mesenchymal transition and is of strong and independent importance for the progress of prostate cancer. *Clin. Cancer Res.* **2007**, *13*, 7003–7011. [CrossRef]
36. Jennbacken, K.; Tesan, T.; Wang, W.; Gustavsson, H.; Damber, J.E.; Welen, K. N-cadherin increases after androgen deprivation and is associated with metastasis in prostate cancer. *Endocr. Relat. Cancer* **2010**, *17*, 469–479. [CrossRef]
37. Linder, A.; Hagberg Thulin, M.; Damber, J.E.; Welen, K. Analysis of regulator of g-protein signalling 2 (rgs2) expression and function during prostate cancer progression. *Sci. Rep.* **2018**, *8*, 17259. [CrossRef]
38. Cottard, F.; Asmane, I.; Erdmann, E.; Bergerat, J.P.; Kurtz, J.E.; Ceraline, J. Constitutively active androgen receptor variants upregulate expression of mesenchymal markers in prostate cancer cells. *PLoS ONE* **2013**, *8*, e63466. [CrossRef]
39. Cottard, F.; Madi-Berthelemy, P.O.; Erdmann, E.; Schaff-Wendling, F.; Keime, C.; Ye, T.; Kurtz, J.E.; Ceraline, J. Dual effects of constitutively active androgen receptor and full-length androgen receptor for n-cadherin regulation in prostate cancer. *Oncotarget* **2017**, *8*, 72008–72020. [CrossRef]
40. Josefsson, A.; Damber, J.E.; Welen, K. Ar-v7 expression in circulating tumor cells as a potential prognostic marker in metastatic hormone-sensitive prostate cancer. *Acta Oncol.* **2019**, *58*, 1660–1664. [CrossRef]
41. Josefsson, A.; Linder, A.; Flondell-Site, D.; Canesin, G.; Stiehm, A.; Anand, A.; Bjartell, A.; Damber, J.E.; Welen, K. Circulating tumor cells as a marker for progression-free survival in metastatic castration-naïve prostate cancer. *Prostate* **2017**, *77*, 849–858. [CrossRef] [PubMed]

5

Leukocyte-Derived Extracellular Vesicles in Blood with and without EpCAM Enrichment

Afroditi Nanou *, Leonie L. Zeune and Leon W.M.M. Terstappen *

Department of Medical Cell BioPhysics, Faculty of Science and Technology, University of Twente, 7522 NH Enschede, The Netherlands

* Correspondence: a.nanou@utwente.nl (A.N.); l.w.m.m.terstappen@utwente.nl (L.W.M.M.T.).

Abstract: Large tumor-derived Extracellular Vesicles (tdEVs) detected in blood of metastatic prostate, breast, colorectal, and non-small cell lung cancer patients after enrichment for Epithelial Cell Adhesion Molecule (EpCAM) expression and labeling with 4′,6-diamidino-2-phenylindole (DAPI), phycoerythrin-conjugated antibodies against Cytokeratins (CK-PE), and allophycocyanin-conjugated antibody against the cluster of differentiation 45 (CD45-APC), are negatively associated with the overall survival of patients. Here, we investigated whether, similarly to tdEVs, leukocyte-derived EVs (ldEVs) could also be detected in EpCAM-enriched blood. Presence of ldEVs and leukocytes in image data sets of EpCAM-enriched samples of 25 healthy individuals and 75 metastatic cancer patients was evaluated using the ACCEPT software. Large ldEVs could indeed be detected, but in contrast to the 20-fold higher frequency of tdEVs as compared to Circulating Tumor Cells (CTCs), ldEVs were present in a 5-fold lower frequency as compared to leukocytes. To evaluate whether these ldEVs pre-exist in the blood or are formed during the CellSearch procedure, the blood of healthy individuals without EpCAM enrichment was labelled with the nuclear dye Hoechst and fluorescently tagged monoclonal antibodies recognizing the leukocyte-specific CD45, platelet-specific CD61, and red blood cell-specific CD235a. Fluorescence microscopy imaging using a similar setup as the CellSearch was performed and demonstrated the presence of a similar population of ldEVs present at a 3-fold lower frequency as compared to leukocytes.

Keywords: leukocyte-derived extracellular vesicles; immunofluorescence imaging; EpCAM enrichment; CellSearch; EasyCount slides; ACCEPT

---

1. Introduction

During the last decades, Extracellular Vesicles (EVs) have emerged as promising disease biomarkers bearing similar membrane and cargo composition as their originating cells [1–3]. Importantly, for nucleic acid analysis, the membrane encapsulated nucleic acid cargo is protected from enzymatic degradation, and consequently, it can circulate for a longer time compared to cell-free DNA (cfDNA) [4,5]. In the case of cancer, the presence of nucleic acids (DNA, mRNA, and miRNA) within tumor-derived EVs (tdEVs) and proteins within or on tdEV membranes could provide information of the predisposition of the tumor to metastasize in specific organs and guide treatment monitoring of patients to block metastasis and cancer progression [5–9]. It has been demonstrated that EVs in biofluids of cancer patients are significantly elevated when compared to the respective numbers of healthy individuals [10,11]. However, to our knowledge, there is no data available in regards to the composition of the redundant EVs in the blood of cancer patients. The recent *in vivo* studies of Ricklefs et al. using brain tumors expressing the green fluorescent protein (GFP) in mice showed that less than 0.5% of the total circulating EVs were GFP+ [10]. That finding implies that more cell types secrete EVs in response to the present tumor contributing to the final EV pool detected in biofluids of cancer patients. Furthermore, the pre-analytical steps of sample processing determine the EV populations to be analyzed and could lead to biased conclusions. The majority of research groups is only interested in exosomes that constitute the smallest subclass of EVs as they consider them products of active cell secretion; therefore, they are using differential centrifugation steps to get rid of other EV subclasses, collect the exosome fraction as a pellet from the final ultracentrifugation step and label them with antibodies recognizing generic exosome-enriched biomarkers, mainly tetraspanins, such as the clusters of differentiation CD81, CD9, and CD63 to identify them [10,12]. Nevertheless, EV subclasses of larger size (microvesicles, oncosomes, and apoptotic bodies) have been reported to be bioactive with a wide spectrum of functions depending on their cells of origin [13]. Importantly, Vagner et al. reported the presence of DNA in large tdEVs reflecting the genetic aberrations of the tumor; a finding that highlights their promising potential in the liquid biopsy field [9]. Padda et al. also demonstrated that the majority of prostate-specific membrane antigen (PSMA) expressing EVs in plasma of prostate cancer patients derive directly from the plasma membrane and have a larger size [14]; hence, these clinically important populations are missed by solely the exosome analysis. Very few studies have investigated the isolation and downstream characterization of specific tdEVs from patient samples using immuno-affinity techniques [15,16]. Recently, we showed that large tdEVs, immunomagnetically isolated based on their EpCAM expression together with Circulating Tumor Cells (CTCs) by the CellSearch system from the blood of metastatic prostate, breast, colorectal, and non-small cell lung cancer patients have equivalent prognostic power to CTCs [16,17]. These observations were enabled through the availability of the open-source ACCEPT image analysis program, which allows for the exploration and enumeration in a single level of all different classes of objects detected in the fluorescence images in an automated, fast and reproducible manner, free of the subjectivity and bias of different operators [18,19]. However, it is not clear whether our previously reported large tdEVs are a result of the fragmentation of CTCs during the immunomagnetic EpCAM enrichment and washing steps that the CellSearch system is using or whether they pre-exist in the blood samples of cancer patients. Their rare frequency in combination with the abundance of blood cells and EVs of different origins prevent us from addressing that question by labeling of blood samples without any pre-enrichment steps and subsequent enumeration from fluorescence images. In this study, we identified in the digitally stored CellSearch images some CD45+, DAPI-, CK- objects of similar size to tdEVs that we baptized leukocyte-derived Extracellular Vesicles (ldEVs). We addressed the question of whether large ldEVs pre-exist in the blood of individuals without EpCAM enrichment or they are by-products of cell fragmentation by the CellSearch procedure. Towards that direction, we labeled blood samples of healthy individuals with the nuclear dye Hoechst and fluorophore-conjugated antibodies against the leukocyte-specific CD45, the platelet-specific CD61, and the red blood cell-specific CD235a without any pre-enrichment or pre-analytical steps. The samples were imaged using a fluorescence microscope with a 10×/0.45 numerical aperture (NA) objective to enable fair comparison of the image datasets acquired by the CellTracks Analyzer II of the CellSearch system [20].

2. Materials and Methods

2.1. Immunofluorescence Image Data Sets of EpCAM-Enriched Cells and Extracellular Vesicles of 25 Healthy Individuals and 75 Metastatic Cancer Patients

One-hundred digitally stored CellSearch image data sets corresponding to EpCAM-enriched blood samples of 25 healthy individuals, 25 metastatic prostate (CRPC), 25 colorectal (mCRC), and 25 non-small cell lung (NSCLC) cancer patients before the initiation of a new therapy, were used for this analysis. The EpCAM-enriched leukocytes and large leukocyte-derived EVs present in these fluorescence images were enumerated. The CRPC and mCRC patients had participated in the retrospective IMMC38 (***NCT00133900***) and CAIRO 2 (***NCT00208546***) clinical studies, respectively.

Briefly, the EpCAM+ Circulating Tumor Cells (CTCs) and tdEVs were positively selected by ferrofluids conjugated to an antibody recognizing the extracellular epitope of Epithelial Cell Adhesion Molecule (EpCAM, clone VU1D9) from 7.5 mL of the blood of cancer patients using the CellSearch system (Menarini Silicon Biosystems, Huntingdon Valley PA, USA), as previously described [21]. Following EpCAM immunomagnetic enrichment, the suspension was incubated with the nuclear dye DAPI and antibodies against the epithelial-specific cytokeratins 8, 18, and 19 (clone C11) conjugated to phycoerythrin (PE) and an antibody against the leukocyte-specific cluster of differentiation CD45 conjugated to allophycocyanin (APC). The suspension was transferred to a cartridge placed within a magnest that allowed for a homogeneous distribution of the ferrofluids and the EpCAM-enriched objects on a focal plane [22]. The cartridges were imaged using a semi-automated 10×/0.45 NA objective fluorescence microscope, the CellTracks Analyzer II, as previously described [21].

2.2. *Blood Samples of 10 Healthy Individuals*

Blood samples from 10 anonymous healthy individuals were collected in ethylenediaminetetraacetic acid (EDTA) tubes after written informed consent from the Experimental Centre for Technical Medicine (ECTM) donor service (University of Twente, Enschede, The Netherlands). The frequencies of white blood cells, red blood cells, and platelets were assessed using a hematology analyzer (Beckman Coulter, Brea, CA, USA). The samples were processed on the same day of the drawing.

2.3. *Immunofluorescence Imaging of Cells and Extracellular Vesicles in Whole Blood Samples*

10–20 μL of EDTA blood samples of 10 different healthy individuals were 10× diluted in 0.2 μm filtered 1% *w/v* bovine serum albumin (BSA) in phosphate- buffered saline (PBS) solution. Blood was incubated with the nuclear dye Hoechst 33342 (Invitrogen, cat. # H3570), the fluorescently tagged monoclonal antibodies CD45-PerCP (clone HI30 Invitrogen, cat. # MHCD4531) recognizing leukocytes, CD235a-Alexa Fluor® 647 (clone YTH89.1, bio-rad, MCA 506A647) antibodies recognizing erythrocytes and CD61-Alexa Fluor® 488 (clone Y2.51, bio-rad, cat. # MCA 2588A488) antibodies recognizing platelets. The final concentrations used were 4.0 μg/mL Hoechst, 0.5 μg/mL CD45-PerCP, 2.5 μg/mL CD235a-Alexa 647, and 0.6 μg/mL CD61-Alexa 488. The samples were incubated with the antibodies at 37 °C for 1–2 h and stored at 4 °C until further processing. Subsequently, the samples were further diluted to a final dilution of 500×. 10 μL of the diluted sample (corresponding to 0.02 μL of undiluted blood) were loaded in a well of EasyCount™ Slide-6™ (Immunicon Corp., Huntingdon Valley, PA, USA). Four–six technical replicates of samples were used to assess the reproducibility of the measurements. Image data sets of 55–65 frames/channel were acquired to cover the whole surface of each well using a semi-automated inverted fluorescence microscope (Eclipse Ti-E, Nikon Instruments, Amsterdam, The Netherlands) equipped with a 10×/0.45 numerical aperture (NA) objective, a camera (Orca flash 4.0 LT, C11440, Hamamatsu, Almere, The Netherlands) and fluorescence filter cubes (DAPI, FITC, PerCP, APC filter sets for the detection of Hoechst, CD61-Alexa 488, CD45-PerCP and CD235a-Alexa 647, respectively). The operator determined three corners of the surface to be scanned and adjusted the focus at four points distributed throughout each well. The exposure times used for the imaging were 20 ms for DAPI, 400 ms for PerCP, 500 ms for FITC, 1000 ms for APC, and 500 ms for brightfield. Few images were obtained using a 60×/0.70 NA objective and the same exposure times for comparison. However, only the images obtained with the 10× objective were used as an input for the enumeration of the different populations to allow a fair comparison with the images of the CellTracks Analyzer II. An example of a frame acquired with the 10× and 60× objectives is shown in Supplementary Figure S1.

2.4. *Automated Enumeration of Objects in Immunofluorescence Images Using the Open-Source ACCEPT Software*

All immunofluorescence image data sets, obtained with the 10×/0.45 NA objective microscopes, were processed with the open-source software for Automated CTC Classification, Enumeration and Phenotyping (ACCEPT) (http://github.com/LeonieZ/ACCEPT). The software detects all present objects, larger than four pixels, and extracts for each of them 10 morphological and fluorescence signal intensity measurements per fluorescence channel [18]. The user can design linear gates based on these features to define the classes of their interest and enumerate the objects falling within them [16,23]. The application of the same gates for all different samples allows the elimination of inter- and intra- operator variations leading subsequently, to a more objective consensus [24].

For the CellSearch generated images, gates for the enumeration of leukocyte-derived Extracellular Vesicles (ldEVs) and leukocytes were applied. The gates are summarized in Table 1. For the image data sets corresponding to the EasyCount Slides-6, gates for the enumeration of red blood cells, leukocytes, platelets and ldEVs were used and are summarized in Table 1.

Table 1. ACCEPT gates used for the automated enumeration of leukocytes, leukocyte-derived Extracellular Vesicles (ldEVs), platelets, and red blood cells in blood A with EpCAM enrichment and B without EpCAM enrichment.

		A. EpCAM Enrichment		**B. No Enrichment**	
Leukocytes	DAPI/Hoechst [a]	Mean Intensity	>30	Mean Intensity	>30
		Max Intensity	>50	Max Intensity	>50
		Size	>16 μm^2	Size	>16 μm^2
	CD45	Mean Intensity	>30	Mean Intensity	>30
		Max Intensity	>50	Max Intensity	>50
				Size	≤400 μm^2
	CK	Standard Deviation	≤5	n/a [b]	
	CD61	n/a [b]		Standard Deviation	≤5
	CD235a	n/a [b]		Standard Deviation	≤5
	Extra channel	Standard Deviation	≤5	Standard Deviation	≤5
ldEVs	DAPI/Hoechst [a]	Standard Deviation	≤5	Standard Deviation	≤5
	CD45	Mean Intensity	>30	Mean Intensity	>30
		Max Intensity	>50	Max Intensity	>50
		Perimeter	>5 pixels	Perimeter	>5 pixels
		Size	≤150	Size	≤150 μm^2
		Eccentricity	≤0.85	Eccentricity	≤0.85
	CK	Standard Deviation	≤5	n/a [b]	
	CD61	n/a [b]		Standard Deviation	≤5
	CD235a	n/a [b]		Standard Deviation	≤5
	Extra channel	Standard Deviation	≤5	Standard Deviation	≤5
Platelets	CD45	n/a [b]		Standard Deviation	≤5
	CD61			Mean Intensity	>30
				Max Intensity	>50
				Perimeter	>5 pixels
				Size	≤150 μm^2
				Eccentricity	≤0.85
	CD235a			Standard Deviation	≤5
				Standard Deviation	≤5
	Extra Channel			Standard Deviation	≤5
Red blood cells	Hoechst	n/a [b]		Standard Deviation	≤5
	CD45			Standard Deviation	≤5
	CD61			Standard Deviation	≤5
	CD235a			Mean Intensity	>30
				Max Intensity	>50
				Perimeter	>5 pixels
	Extra Channel			Standard Deviation	≤5

[a] DAPI was used in EpCAM-enriched blood and Hoechst in the blood without pre-enrichment, [b] n/a: not applicable.

3. Results

3.1. Detection of ldEVs in EpCAM-Enriched Blood Samples of Healthy Individuals and Metastatic Cancer Patients

After careful examination of the immunofluorescence images of the CellSearch cartridges, CD45+, DAPI-, CK- objects, that resemble in size our previously reported CD45-, DAPI-, CK+ tdEVs [16,25], could be observed. We baptized these objects leukocyte-derived Extracellular Vesicles (ldEVs). Examples of single ldEV events in EpCAM-enriched samples by the CellSearch system are shown in Figure 1 next to some examples of leukocytes as a reference to their size and CD45 phenotype. The observation of the presence of these ldEVs in the CellSearch cartridges raised questions about their formation: are they fragments of leukocytes formed during the CellSearch procedure or do they pre-exist in the blood circulation?

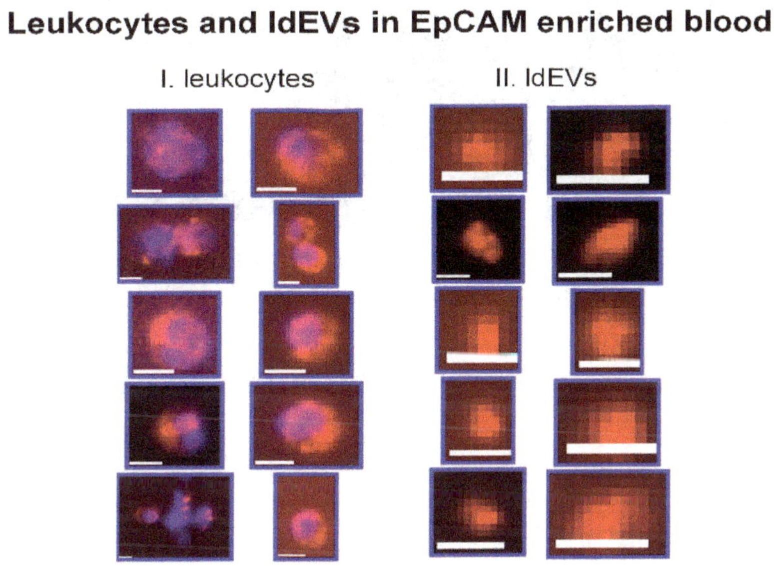

Figure 1. Thumbnails of I. leukocytes and II. Leukocyte-derived Extracellular Vesicles (ldEVs) detected in EpCAM-enriched blood samples. The red color represents CD45 and blue represents DAPI. Scale bars indicate 6.4 µm.

3.2. Detection of Cell and Extracellular Vesicle Classes in the Blood Of Healthy Individuals without EpCAM Enrichment

In order to address the aforementioned question, blood samples of healthy individuals were labeled with Hoechst, CD45-PerCP, CD61-Alexa 488, and CD235a-Alexa 647 and were imaged with a similar fluorescence microscope as the CellTracks Analyzer II. No pre-enrichment or washing steps were used in order to minimize the cell fragmentation or activation. The inclusion of the aforementioned antibodies allowed the detection of four different classes of objects in the whole blood of healthy individuals, namely leukocytes, platelets, red blood cells, and ldEVs, as shown in Figure 2. Leukocytes are defined as nucleated CD45+, CD61-, CD235a- cells of a size between 7 and 20 µm (Panel A); leukocyte-derived Extracellular Vesicles (ldEVs) as CD45+, CD61-, CD235- objects without a nucleus and of undefined size as shown in the respective brightfield image (Panel B); platelets as CD45-, CD61+, CD235a- objects without a nucleus of size between 2 and 5 µm (Panel C) and red blood cells as CD45-, CD61-, CD235a+ cells without a nucleus and a size range between 6 and 10 µm (Panel D). In Panel C, three platelets are shown, of which the middle one is clearly smaller and with a lower expression of CD61; examination at higher magnification would allow for better identification of the smaller size platelets, but no discrimination could be made between small platelets and larger platelet-derived EVs. The presence of ldEVs in blood samples without pre-enrichment (Panel B) confirmed their pre-existence in whole blood.

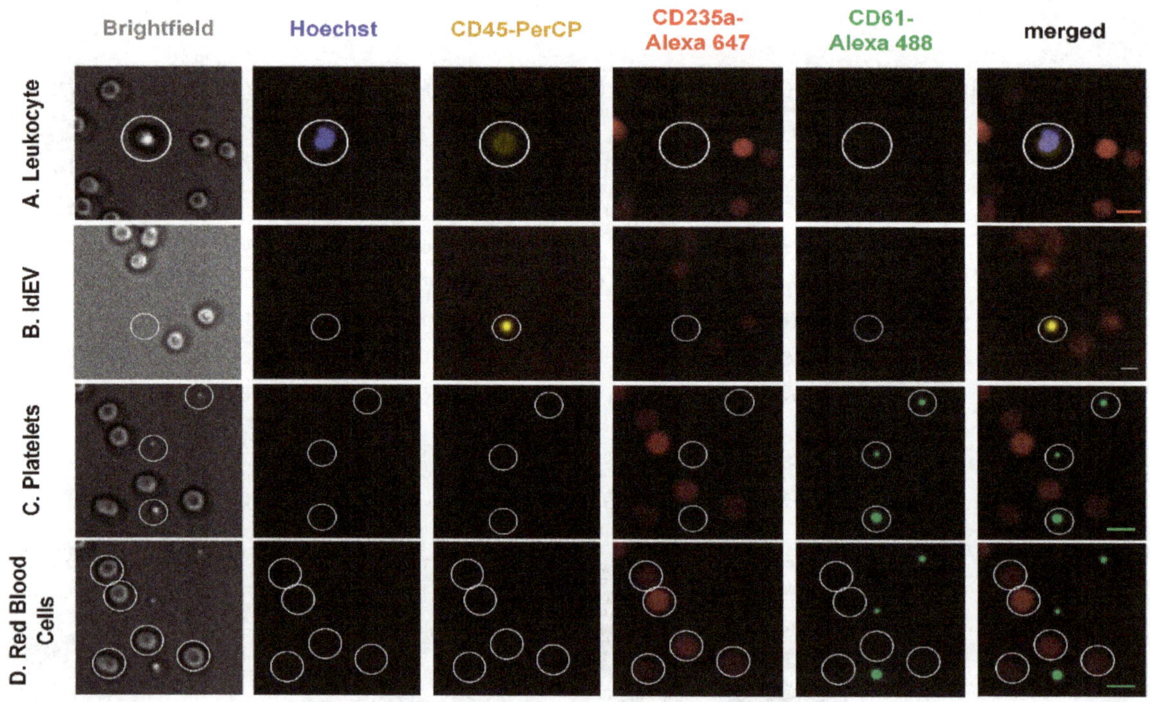

Figure 2. Bright field and immunofluorescence (IF) images of leukocytes, ldEVs, platelets and red blood cells in blood samples without EpCAM enrichment. Scale bars in the merged IF images indicate 10 μm.

3.3. ACCEPT Gates for the Automated Enumeration of Different Classes in the Blood with and without EpCAM Enrichment.

In order to acquire the absolute counts of the different classes from each data set of the healthy individuals in a fast and unbiased manner, we processed all data sets with the open-source ACCEPT software. Based on the aforementioned characteristics of the different classes, we developed linear gates to enumerate the objects falling within each class automatically. The gates are summarized in Table 1. Three examples of objects per class are shown in Figure 3 (Panel A). The objects that fall into each class are depicted in blue dots in the scatter plots (Panel B) showing the mean Hoechst intensity versus the mean CD45-PerCP intensity and the mean CD61-Alexa 488 intensity versus the mean CD235a-Alexa 647 intensity. Objects falling in the "leukocyte" gate are double-positive for CD45 and Hoechst and negative for CD61 and CD235a, (Panel B1); ldEVs are positive only for CD45 (Panel B2); platelets are only positive for CD61 (Panel B3) and red blood cells are only positive for CD235a (Panel B4).

In order to achieve a fair comparison between the leukocyte and ldEV counts detected in the blood with and without EpCAM enrichment, very similar ACCEPT gates were developed for the automated enumeration of the two classes applied in the different image data sets. The gates can be found in Table 1.

The size threshold of $20 \times 20\ \mu m^2$ in the case of the "leukocyte" gate that was applied in blood samples with no enrichment was removed in the respective gate of EpCAM-enriched samples because within the CellSearch cartridges, there are many leukocytes present in close proximity to each other, as shown in Figure 1, that are segmented as one object by the ACCEPT software. Therefore, the inclusion of such a parameter in the EpCAM-enriched samples would lead to an even higher underestimation of this population compared to the underestimation already introduced by cell clusters counted as one object. On the other hand, the removal of that parameter in case of blood samples with no enrichment (Figure 3), where it is very rare to find two or more leukocytes in close proximity, leads to the inclusion of artefacts and an overestimation of leukocytes.

These results do not deviate a lot from the previously reported ldEV frequencies (median value: 356, interquartile range: 268–529) of Simak et al. in the plasma of healthy donors; ldEVs were larger than 200 nm and were defined as CD45+, CD105-, CD235a- by flow cytometry [36]. The use of solely one specific but weakly expressed inclusion marker, namely CD45 to define them using either approaches results in the underestimation of the whole circulating ldEV population; a point also stressed out by Lacroix et al. [37]. Further investigation of ldEVs in terms of their size distribution and surface marker expression using established techniques in the EV field, such as nanoparticle tracking analysis, electron microscopy, flow cytometry, and surface plasmon resonance imaging, would lead to their better profiling [10,38,39]. Eventually, a similar test to the hematology analyzer having as an output the EV populations (of platelet, erythrocyte, leukocyte, endothelial, and epithelial origin) in the biofluids of individuals could serve as an important diagnostic tool in clinical practice, since EVs have been associated with numerous pathophysiological conditions, such as thrombogenicity, inflammation, angiogenesis, and cancer [26,40–44].

Our study has several limitations. Although the extrapolated counts of leukocytes, platelets, and red blood cells per μL of blood correlated to the respective values of the hematology analyzer (Figure 5), the detected frequencies of all the cell populations were consistently lower as compared to the respective ones measured by the hematology analyzer. The particularly lower detection of red blood cells could be explained by the overlap and aggregation of more than one red blood cell segmented as one object by the open-source ACCEPT software and the large range in the distribution of the fluorescence intensity of CD235-APC in the fluorescence images (Supplementary Figure S1, Panel A) through which part of the red blood cells fall outside the applied gate. The underestimation of the leukocytes could be explained by the lower expression of CD45 by the granulocytes that consist to 60–70% of white blood cells. The detection of the smaller platelet population and their secreted EVs is limited by our approach, because of the use of a 0.45 NA objective that results in a resolution of 0.64 μm/pixel. The abundance and high signal of the red blood cells prevented us from the detection and enumeration of low signal-to-background ratio red blood cell-derived EVs. Importantly, the Hoechst 33342 labeled nucleic acids in platelets could be detected by flow cytometry but not with our microscopy set-up. This observation implies that the zero DAPI signal detected with ACCEPT inside our previously reported tdEVs does not rule out the presence of nucleic acids within them. This is an important finding encouraging the further characterization of tdEVs that are immunomagnetically isolated based on their EpCAM expression [16,17,25]. That would come into agreement with the findings of Vagner et al. that the DNA cargo of large EVs reflect the genotype of prostate cancer patients [9]. The use of a membrane permeable dye, binding to both DNA and RNA, with a higher sensitivity, like SYTO13 [45] could also facilitate the detection of nucleic acids within the isolated EVs.

Interestingly, platelets have a similar size to ldEVs based on the immunofluorescence images of CD61 and CD45, respectively, with a minimum area of the detected objects being 9 μm^2 based on our observations from the ACCEPT scatter plots. That area corresponds to a circular object of an approximate radius of 1.7 μm. However, it was not possible to confirm the size of ldEVs from the respective bright field images, because opposite to platelets, the contrast between the background and ldEV intensity was inadequate to detect them (Figure 2 and Supplementary Figure S2) suggesting that their physical properties (absorption coefficient, scattering coefficient, scattering anisotropy, refractive index) differ from the ones of platelets.

It is worth mentioning that ldEVs were found in 5–6 lower frequencies compared to leukocytes in EpCAM enriched samples, whereas tdEVs in our previous studies were detected in 10–20 times higher frequencies compared to CTCs [16]. That observation could be explained after taking into consideration some technical and biological facts. From a technical perspective, our study was limited by the resolution of a 10×/0.45 NA objective fluorescence microscope, implying that only the larger EVs with a larger than 1 μm diameter or the ones with a high expression of inclusion markers could be detected. CK is the inclusion marker used for the detection of tdEVs, whereas the detection of ldEVs is accomplished by the inclusion of CD45. Since the CK expression is intracellular and proportional to the volume instead of the surface, as in the case of the CD45 expression of ldEVs, CK is easier to detect in smaller tdEVs than CD45 in similar size ldEVs. Consequently, the CD45 expression may be present in more particles in blood samples but not exceeding the detection limit to be considered positive. Further characterization of the size distribution and the surface marker expression profile could elaborate on the detection limits of our technique. From a biological perspective, tdEVs could be found in higher frequencies either because of the increased apoptosis and fragmentation of CTCs in the blood circulation or because of different EV secretion pattern between normal and cancerous cells. Regarding the first hypothesis, the lifespan of neutrophils is around 24 h [46], whereas the circulation lifetime of CTCs has been estimated to be 1 to 2.4 h [47]; that; however, does not imply that CTCs are fragmented and cleared by the blood. On the contrary, *in vivo* animal studies showed the trap of more than 80% of viable CTCs by the liver and lung, that serve as "filter" organs, and the survival of CTCs for the prolonged time in a dormant state [48,49]. The survival of CTCs in the bloodstream is further supported by studies on their mechanical phenotype. Atomic force microscopy studies on cell lines suggest that cells with increased metastatic potential are more deformable (as expressed by Young's modulus), compared to less metastatic or non-malignant cells [50,51]. These results were further confirmed in clinical samples from pleural effusions, where metastatic cells had lower stiffness compared to benign cells from the same effusions and leukocytes [52,53]. Interestingly, Sun et al. demonstrated that deformable cancer cells engulf neighboring ones with higher stiffness via entosis further encouraging the increased survivorship of CTCs [54]. Regarding the second hypothesis of different EV secretion pattern of CTCs and leukocytes, it is well known that cancer cells have reprogrammed metabolism and acquired traits that promote their survival and growth [55,56]. Recent findings of independent research groups converge into the survival of tumor cells regardless of their phenotypic characteristics of possible apoptosis, such as caspases activation, amoeboid phenotype and membrane blebbing [57,58]. Instead of undergoing apoptosis, cells with such traits have a more tumorigenic and invasive phenotype [59]. Hence, tumor cells may actively secrete EVs similar in size to apoptotic bodies, but without special receptors on their surface to be recognized and ingested by macrophages for their clearance as in case of healthy cells (e.g., white blood cells).

5. Conclusions

In conclusion, the relative frequencies of large (above 1 μm) leukocyte-derived Extracellular Vesicles (ldEVs) to leukocytes are similar in EpCAM-enriched blood samples of healthy individuals and cancer patients (1:6 to 1:5) as in the blood of healthy individuals without EpCAM enrichment (1:3), implying their pre-existence in the blood circulation rather than their formation from activated or apoptotic leukocytes using the CellSearch system. Furthermore, the Hoechst signal of platelets could not be detected using a similar fluorescence microscope as the CellTracks Analyzer II. These two findings have important implications for our previously reported tumor-derived Extracellular Vesicles (tdEVs), that were immunomagnetically co-isolated with CTCs based on their EpCAM expression from metastatic cancer patients [17]. Firstly, tdEVs are most likely not a result of CTC fragmentation during the CellSearch procedure and secondly, the presence of undetectable nucleic acids within ldEVs and tdEVs should not be excluded but instead further investigated. No conclusions could be drawn in regards to the smaller ldEV population, namely exosomes, since they are not expected to be detected with our imaging setup. Last but not least, our results do not allow for comparison of ldEVs between healthy individuals and cancer patients since the available image data sets of patients corresponded to only EpCAM enriched samples.

Author Contributions: Conceptualization—L.W.M.M.T. and A.N.; methodology—A.N., L.L.Z., and L.W.M.M.T.; Validation—A.N.; formal analysis, A.N.; Investigation—A.N.; Resources—A.N., L.L.Z., L.W.M.M.T.; Data curation—A.N.; Writing—original draft preparation—A.N.; Writing—review and editing—A.N., L.L.Z., L.W.M.M.T.; Visualization—A.N.; Supervision—L.W.M.M.T.; Project administration—L.W.M.M.T.; funding acquisition—L.W.M.M.T.

Acknowledgments: The authors acknowledge the Experimental Centre for Technical Medicine (ECTM) of the University of Twente for providing us blood samples of healthy individuals. We would also like to thank all the patients and healthy individuals, the blood samples of whom were used for the accomplishment of the present study. Special acknowledgments to C. Breukers and J. Weersink for their technical support with the inverted fluorescence microscope whenever needed.

References

1. Van Niel, G.D.; Angelo, G.; Raposo, G. Shedding light on the cell biology of extracellular vesicles. *Nat. Rev. Mol. Cell. Biol.* **2018**, *19*, 213–228. [CrossRef]
2. Fais, S.O.; Driscoll, L.; Borras, F.E.; Buzas, E.; Camussi, G.; Cappello, F.; Carvalho, J.; Cordeiro da Silva, A.; Del Portillo, H.; El Andaloussi, S. Evidence-based clinical use of nanoscale extracellular vesicles in nanomedicine. *ACS Nano* **2016**, *10*, 3886–3899. [CrossRef]
3. Yanez-Mo, M.; Siljander, P.R.; Andreu, Z.; Zavec, A.B.; Borras, F.E.; Buzas, E.I.; Buzas, K.; Casal, E.; Cappello, F.; Carvalho, J.; et al. Biological properties of extracellular vesicles and their physiological functions. *J. Extracell. Vesicles* **2015**, *4*, 27066. [CrossRef]
4. Jia, S.; Zhang, R.; Li, Z.; Li, J. Clinical and biological significance of circulating tumor cells, circulating tumor DNA, and exosomes as biomarkers in colorectal cancer. *Oncotarget* **2017**, *8*, 55632–55645. [CrossRef]
5. Krug, A.K.; Enderle, D.; Karlovich, C.; Priewasser, T.; Bentink, S.; Spiel, A.; Brinkmann, K.; Emenegger, J.; Grimm, D.G.; Castellanos-Rizaldos, E.; et al. Improved EGFR mutation detection using combined exosomal RNA and circulating tumor DNA in NSCLC patient plasma. *Ann. Oncol. Off. J. Eur. Soc. Med. Oncol.* **2018**, *29*, 2143. [CrossRef]
6. Costa-Silva, B.; Aiello, N.M.; Ocean, A.J.; Singh, S.; Zhang, H.; Thakur, B.K.; Becker, A.; Hoshino, A.; Mark, M.T.; Molina, H.; et al. Pancreatic cancer exosomes initiate pre-metastatic niche formation in the liver. *Nat. Cell Biol.* **2015**, *17*, 816–826. [CrossRef]
7. Hoshino, A.; Costa-Silva, B.; Shen, T.L.; Rodrigues, G.; Hashimoto, A.; Tesic Mark, M.; Molina, H.; Kohsaka, S.; Di Giannatale, A.; Ceder, S.; et al. Tumour exosome integrins determine organotropic metastasis. *Nature* **2015**, *527*, 329–335. [CrossRef]
8. Kosaka, N.; Yoshioka, Y.; Fujita, Y.; Ochiya, T. Versatile roles of extracellular vesicles in cancer. *J. Clin. Investig.* **2016**, *126*, 1163–1172. [CrossRef]
9. Vagner, T.; Spinelli, C.; Minciacchi, V.R.; Balaj, L.; Zandian, M.; Conley, A.; Zijlstra, A.; Freeman, M.R.; Demichelis, F.; De S Posadas, E.M. Large extracellular vesicles carry most of the tumour DNA circulating in prostate cancer patient plasma. *J. Extracell. Vesicles* **2018**, *7*, 1505403. [CrossRef]
10. Ricklefs, F.L.; Maire, C.L.; Reimer, R.; Duhrsen, L.; Kolbe, K.; Holz, M.; Schneider, E.; Rissiek, A.; Babayan, A.; Hille, C.; et al. Imaging flow cytometry facilitates multiparametric characterization of extracellular vesicles in malignant brain tumours. *J. Extracell. Vesicles.* **2019**, *8*, 1588555. [CrossRef]
11. Konig, L.; Kasimir-Bauer, S.; Bittner, A.K.; Hoffmann, O.; Wagner, B.; Santos Manvailer, L.F.; Kimmig, R.; Horn, P.A.; Rebmann, V. Elevated levels of extracellular vesicles are associated with therapy failure and disease progression in breast cancer patients undergoing neoadjuvant chemotherapy. *Oncoimmunology* **2017**, *7*, e1376153. [CrossRef]
12. Kanwar, S.S.; Dunlay, C.J.; Simeone, D.M.; Nagrath, S. Microfluidic device (ExoChip) for on-chip isolation, quantification and characterization of circulating exosomes. *Lab Chip* **2014**, *14*, 1891–1900. [CrossRef]
13. Slomka, A.; Urban, S.K.; Lukacs-Kornek, V.; Zekanowska, E.; Kornek, M. Large Extracellular Vesicles: Have We Found the Holy Grail of Inflammation? *Front. Immunol.* **2018**, *9*, 2723. [CrossRef]
14. Padda, R.S.; Deng, F.K.; Brett, S.I.; Biggs, C.N.; Durfee, P.N.; Brinker, C.J.; Williams, K.C.; Leong, H.S. Nanoscale flow cytometry to distinguish subpopulations of prostate extracellular vesicles in patient plasma. *Prostate* **2019**, *79*, 592–603. [CrossRef]
15. Reategui, E.; van der Vos, K.E.; Lai, C.P.; Zeinali, M.; Atai, N.A.; Aldikacti, B.; Floyd, F.P.A.H.K., Jr.; Thapar, V.;

Hochberg, F.H.; Sequist, L.V.; et al. Engineered nanointerfaces for microfluidic isolation and molecular profiling of tumor-specific extracellular vesicles. *Nat. Commun.* **2018**, *9*, 175. [CrossRef]

16. Nanou, A.; Coumans, F.A.W.; van Dalum, G.; Zeune, L.L.; Dolling, D.; Onstenk, W.; Crespo, M.; Fontes, M.S.; Rescigno, P.; Fowler, G.; et al. Circulating tumor cells, tumor-derived extracellular vesicles and plasma cytokeratins in castration-resistant prostate cancer patients. *Oncotarget* **2018**, *9*, 19283–19293. [CrossRef]
17. Nanou, A.; Zeune, L.L.; de Wit, S.; Miller, C.M.; Punt, C.J.A.; Groen, H.J.M.; Hayes, D.F.; de Bono, J.S.L.W.M.M.T. *Tumor-Derived Extracellular Vesicles in Blood of Metastatic Breast, Colorectal, Prostate and Non-Small Cell Lung Cancer Patients Associate with Worse Survival*; American Association for Cancer Research: Philadelphia, PA, USA, 2019.
18. Zeune, L.; van Dalum, G.; Terstappen, L.W.M.M.; van Gils, S.A.; Brune, C. Multiscale Segmentation via Bregman Distances and Nonlinear Spectral Analysis. *SIAM J. Imaging Sci.* **2017**, *10*, 111–146. [CrossRef]
19. Zeune, L. *Automated CTC Classification, Enumeration and Pheno Typing: Where Marh Meets Biology. Medical Cell BioPhysics*; University of Twente: Overijssel, The Netherland, 2019.
20. Coumans, F.; Terstappen, L. Detection and Characterizati of Circulating Tumor Cells by the CellSearch Approach. *Methods Mol. Biol.* **2015**, *1347*, 263–278.
21. Allard, W.J.; Matera, J.; Miller, M.C.; Repollet, M.; Connelly, M.C.; Rao, C.; Tibbe, A.G.; Uhr, J.W.; Terstappen, L.W. Tumor cells circulate in the peripheral blood of all major carcinomas but not in healthy subjects or patients with nonmalignant diseases. *Clin. Cancer Res.* **2004**, *10*, 6897–6904. [CrossRef]
22. Tibbe, A.G.; de Grooth, B.G.; Greve, J.; Dolan, G.J.; Rao, C.; Terstappen, L.W. Magnetic field design for selecting and aligning immunomagnetic labeled cells. *Cytometry* **2002**, *47*, 163–172. [CrossRef]
23. De Wit, S.; Zeune, L.L.; Hiltermann, T.J.N.; Groen, H.J.M.; Dalum, G.V.; Terstappen, L.W.M.M. Classification of Cells in CTC-Enriched Samples by Advanced Image Analysis. *Cancers* **2018**, *10*, 377. [CrossRef]
24. Zeune, L.L.; de Wit, S.; Berghuis, A.M.S.M.J.I.J.; Terstappen, L.; Brune, C. How to Agree on a CTC: Evaluating the Consensus in Circulating Tumor Cell Scoring. *Cytom. A* **2018**, *93*, 1202–1206. [CrossRef]
25. De Wit, S.; Rossi, E.; Weber, S.; Tamminga, M.; Manicone, M.; Swennenhuis, J.F. Groothuis-Oudshoorn, C.G.M.; Vidotto, R.; Facchinetti, A.; Zeune, L.L.; et al. Single tube liquid biopsy for advanced non-small cell lung cancer. *Int. J. Cancer* **2019**, *144*, 3127–3137. [CrossRef]
26. Van der Pol, E.; Boing, A.N.; Harrison, P.; Sturk, A.; Nieuwland, R. Classification, functions, and clinical relevance of extracellular vesicles. *Pharmacol. Rev.* **2012**, *64*, 676–705. [CrossRef]
27. Heijnen, H.F.; Schiel, A.E.; Fijnheer, R.; Geuze, H.J.; Sixma, J.J. Activated platelets release two types of membrane vesicles: Microvesicles by surface shedding and exosomes derived from exocytosis of multivesicular bodies and alpha-granules. *Blood* **1999**, *94*, 3791–3799.
28. Pitanga, T.N.; de Aragao Franca, L.; Rocha, V.C.; Meirelles, T.; Borges, V.M.; Goncalves, M.S.; Pontes-de-Carvalho, L.C.; Noronha-Dutra, A.A.; dos-Santos, W.L. Neutrophil-derived microparticles induce myeloperoxidase-mediated damage of vascular endothelial cells. *BMC Cell Biol.* **2014**, *15*, 21. [CrossRef]
29. Baka, Z.; Senolt, L.; Vencovsky, J.; Mann, H.; Simon, P.S.; Kittel, A.; Buzas, E.; Nagy, G. Increased serum concentration of immune cell derived microparticles in polymyositis/dermatomyositis. *Immunol. Lett.* **2010**, *128*, 124–130. [CrossRef]
30. Canellini, G.; Rubin, O.; Delobel, J.; Crettaz, D.; Lion, N.; Tissot, J.D. Red blood cell microparticles and blood group antigens: An analysis by flow cytometry. *Blood Transfus.* **2012**, *10*, s39–s45.
31. Wheway, J.; Latham, S.L.; Combes, V.; Grau, G.E.R. Endothelial microparticles interact with and support the proliferation of T cells. *J. Immunol.* **2014**, *193*, 3378–3387. [CrossRef]
32. Deregibus, M.C.; Cantaluppi, V.; Calogero, R.; Lo Iacono, M.; Tetta, C.; Biancone, L.; Bruno, S.; Bussolati, B.; Camussi, G. Endothelial progenitor cell derived microvesicles activate an angiogenic program in endothelial cells by a horizontal transfer of mRNA. *Blood* **2007**, *110*, 2440–2448. [CrossRef]
33. Ruhen, O.; Meehan, K. Tumor-Derived Extracellular Vesicles as a Novel Source of Protein Biomarkers for Cancer Diagnosis and Monitoring. *Proteomics* **2019**, *19*, e1800155. [CrossRef]
34. Nanou, A.; Crespo, M.; Flohr, P.; De Bono, J.S.; Terstappen, L. Scanning Electron Microscopy of Circulating Tumor Cells and Tumor-Derived Extracellular Vesicles. *Cancers* **2018**, *10*, 416. [CrossRef]
35. Rikkert, L.G.; van der Pol, E.; van Leeuwen, T.G.; Nieuwland, R.; Coumans, F.A.W. Centrifugation affects the purity of liquid biopsy-based tumor biomarkers. *Cytometry A* **2018**, *93*, 1207–1212. [CrossRef]
36. Simak, J.; Holada, K.; Risitano, A.M.; Zivny, J.H.; Young, N.S.; Vostal, J.G. Elevated circulating endothelial

membrane microparticles in paroxysmal nocturnal haemoglobinuria. *Br. J. Haematol.* **2004**, *125*, 804–813. [CrossRef]

37. Lacroix, R.; Robert, S.; Poncelet, P.; Dignat-George, F. Overcoming limitations of microparticle measurement by flow cytometry. *Semin. Thromb. Hemost.* **2010**, *36*, 807–818. [CrossRef]
38. Van der Pol, E.; Coumans, F.A.; Grootemaat, A.E.; Gardiner, C.; Sargent, I.L.; Harrison, P.; Sturk, A.; van Leeuwen, T.G.; Nieuwland, R. Particle size distribution of exosomes and microvesicles determined by transmission electron microscopy, flow cytometry, nanoparticle tracking analysis and resistive pulse sensing. *J. Thromb. Haemost.* **2014**, *12*, 1182–1192. [CrossRef]
39. Gool, E.L.; Stojanovic, I.; Schasfoort, R.B.M.; Sturk, A.; van Leeuwen, T.G.; Nieuwland, R.; Terstappen, L.; Coumans, F.A.W. Surface Plasmon Resonance is an Analytically Sensitive Method for Antigen Profiling of Extracellular Vesicles. *Clin. Chem.* **2017**, *63*, 1633–1641. [CrossRef]
40. Halim, A.T.A.; Ariffin, N.A.F.M.; Azlan, M. Review: The Multiple Roles of Monocytic Microparticles. *Inflammation* **2016**, *39*, 1277–1284. [CrossRef]
41. Tissot, J.D.; Canellini, G.; Rubin, O.; Angelillo-Scherrer, A.; Delobel, J.; Prudent, M.; Lion, N. Blood microvesicles: From proteomics to physiology. *Transl. Proteom.* **2013**, *1*, 38–52. [CrossRef]
42. Burnier, L.; Fontana, P.; Kwak, B.R.; Angelillo-Scherrer, A. Cell-derived microparticles in haemostasis and vascular medicine. *Thromb. Haemost.* **2009**, *101*, 439–451. [CrossRef]
43. Julich-Haertel, H.; Urban, S.K.; Krawczyk, M.; Willms, A.; Jankowski, K.; Patkowski, W.; Kruk, B.; Krasnodebski, M.; Ligocka, J.; Schwab, R.; et al. Cancer-associated circulating large extracellular vesicles in cholangiocarcinoma and hepatocellular carcinoma. *J. Hepatol.* **2017**, *67*, 282–292. [CrossRef] [PubMed]
44. Ullal, A.J.; Pisetsky, D.S.; Reich, C.F., III. Use of SYTO 13, a fluorescent dye binding nucleic acids, for the detection of microparticles in in vitro systems. *Cytom. Part A J. Int. Soc. Anal. Cytol.* **2010**, *7*, 294–301. [CrossRef] [PubMed]
45. McCracken, J.M.; Allen, L.A. Regulation of human neutrophil apoptosis and lifespan in health and disease. *J. Cell Death* **2014**, *7*, 15–23. [CrossRef] [PubMed]
46. Meng, S.; Tripathy, D.; Frenkel, E.P.; Shete, S.; Naftalis, E.Z.; Huth, J.F.; Beitsch, P.D.; Leitch, M.; Hoover, S.; Euhus, D.; et al. Circulating tumor cells in patients with breast cancer dormancy. *Clin. Cancer Res.* **2004**, *10*, 8152–8162. [CrossRef] [PubMed]
47. Luzzi, K.J.; MacDonald, I.C.; Schmidt, E.E.; Kerkvliet, N.; Morris, V.L.; Chambers, A.F.; Groom, A.C. Multistep nature of metastatic inefficiency: Dormancy of solitary cells after successful extravasation and limited survival of early micrometastases. *Am. J. Pathol.* **1998**, *153*, 865–873. [CrossRef]
48. Cameron, M.D.; Schmidt, E.E.; Kerkvliet, N.; Nadkarni, K.V.; Morris, V.L.; Groom, A.C.; Chambers, A.F.; MacDonald, I.C. Temporal progression of metastasis in lung: Cell survival, dormancy, and location dependence of metastatic inefficiency. *Cancer Res.* **2000**, *60*, 2541–2546. [PubMed]
49. Li, Q.S.; Lee, G.Y.H.; Ong, C.N.; Lim, C.T. AFM indentation study of breast cancer cells. *Biochem. Biophys. Res. Commun.* **2008**, *374*, 609–613. [CrossRef]
50. Zhang, W.; Kai, K.; Choi, D.S.; Iwamoto, T.; Nguyen, Y.H.; Wong, H.; Landis, M.D.; Ueno, N.T.; Chang, J.; Qin, L. Microfluidics separation reveals the stem-cell-like deformability of tumor-initiating cells. *Proc. Natl. Acad. Sci. USA* **2012**, *109*, 18707–18712. [CrossRef]
51. Cross, S.E.; Jin, Y.S.; Rao, J.; Gimzewski, J.K. Nanomechanical analysis of cells from cancer patients. *Nat. Nanotechnol.* **2007**, *2*, 780–783. [CrossRef]
52. Gossett, D.R.; Tse, H.T.K.; Lee, S.A.; Ying, Y.; Lindgren, A.G.; Yang, O.O.; Rao, J.; Clark, A.T.; Di Carlo, D. Hydrodynamic stretching of single cells for large population mechanical phenotyping. *Proc. Natl. Acad. Sci. USA* **2012**, *109*, 7630–7635. [CrossRef]
53. Sun, Q.; Luo, T.; Ren, Y.; Florey, O.; Shirasawa, S.; Sasazuki, T.; Robinson, D.N.; Overholtzer, M. Competition between human cells by entosis. *Cell Res.* **2014**, *24*, 1299–1310. [CrossRef] [PubMed]
54. Hanahan, D.; Weinberg, R.A. Hallmarks of cancer: The next generation. *Cell* **2011**, *144*, 646–674. [CrossRef] [PubMed]
55. Hanahan, D.; Weinberg, R.A. The hallmarks of cancer. *Cell* **2000**, *100*, 57–70. [CrossRef]
56. Jinesh, G.G.; Choi, W.; Shah, J.B.; Lee, E.K.; Willis, D.L.; Kamat, A.M. Blebbishields, the emergency program for cancer stem cells: Sphere formation and tumorigenesis after apoptosis. *Cell Death Differ.* **2013**, *20*, 382–395. [CrossRef] [PubMed]

57. Di Vizio, D.; Morello, M.; Dudley, A.C.; Schow, P.W.; Adam, R.M.; Morley, S.; Mulholland, D.; Rotinen, M.; Hager, M.H.; Insabato, L.; et al. Large oncosomes in human prostate cancer tissues and in the circulation of mice with metastatic disease. *Am. J. Pathol.* **2012**, *181*, 1573–1584. [CrossRef] [PubMed]
58. Di Vizio, D.; Kim, J.; Hager, M.H.; Morello, M.; Yang, W.; Lafargue, C.J.; True, L.D.; Rubin, M.A.; Adam, R.M.; Beroukhim, R.; et al. Oncosome formation in prostate cancer: Association with a region of frequent chromosomal deletion in metastatic disease. *Cancer Res.* **2009**, *69*, 5601–5609. [CrossRef] [PubMed]
59. Reis-Sobreiro, M.; Chen, J.F.; Novitskaya, T.; You, S.; Morley, S.; Steadman, K.; Gill, N.K.; Eskaros, A.; Rotinen, M.; Chu, C.Y.; et al. Emerin Deregulation Links Nuclear Shape Instability to Metastatic Potential. *Cancer Res.* **2018**, *78*, 6086–6097. [CrossRef] [PubMed]

6

Molecular and Kinetic Analyses of Circulating Tumor Cells as Predictive Markers of Treatment Response in Locally Advanced Rectal Cancer Patients

Bianca C. Troncarelli Flores [1,†], Virgilio Souza e Silva [2,†], Emne Ali Abdallah [1], Celso A.L. Mello [2], Maria Letícia Gobo Silva [3], Gustavo Gomes Mendes [4], Alexcia Camila Braun [1], Samuel Aguiar Junior [5,6] and Ludmilla Thomé Domingos Chinen [1,6,*]

1 International Research Center, A.C. Camargo Cancer Center, São Paulo 01508-010, Brazil
2 Department of Medical Oncology, A.C. Camargo Cancer Center, São Paulo 01509-900, Brazil
3 Department of Radiotherapy, A.C. Camargo Cancer Center, São Paulo 01509-900, Brazil
4 Department of Radiology, A.C. Camargo Cancer Center, São Paulo 01509-900, Brazil
5 Department of Pelvic Surgery, A.C. Camargo Cancer Center, São Paulo 01509-900, Brazil
6 National Institute for Science and Technology in Oncogenomics and Therapeutic Innovation, São Paulo 01509-900, Brazil
* Correspondence: ludmilla.chinen@accamargo.org.br
† These authors have contributed equally to the work.

Abstract: Neoadjuvant chemoradiation (NCRT) followed by total mesorectal excision is the standard treatment for locally advanced rectal cancer (LARC). To justify a non-surgical approach, identification of pathologic complete response (pCR) is required. Analysis of circulating tumor cells (CTCs) can be used to evaluate pCR. We hypothesize that monitoring of thymidylate synthase (TYMS) and excision repair protein, RAD23 homolog B (RAD23B), can be used to predict resistance to chemotherapy/radiotherapy. Therefore, the aims of this study were to analyze CTCs from patients with LARC who underwent NCRT plus surgery for expression of TYMS/RAD23B and to evaluate their predictive value. Blood samples from 30 patients were collected prior to NCRT (S1) and prior to surgery (S2). CTCs were isolated and quantified by ISET®, proteins were analyzed by immunocytochemistry, and *TYMS* mRNA was detected by chromogenic in situ hybridization. CTC counts decreased between S1 and S2 in patients exhibiting pCR ($p = 0.02$) or partial response ($p = 0.01$). Regarding protein expression, TYMS was absent in 100% of CTCs from patients with pCR ($p = 0.001$) yet was expressed in 83% of non-responders at S2 ($p < 0.001$). Meanwhile, RAD23B was expressed in CTCs from 75% of non-responders at S1 ($p = 0.01$) and in 100% of non-responders at S2 ($p = 0.001$). Surprisingly, 100% of non-responders expressed *TYMS* mRNA at both timepoints ($p = 0.001$). In addition, TYMS/RAD23B was not detected in the CTCs of patients exhibiting pCR ($p = 0.001$). We found 83.3% of sensitivity for TYMS mRNA at S1 ($p = 0.001$) and 100% for TYMS ($p = 0.064$) and RAD23B ($p = 0.01$) protein expression at S2. Thus, *TYMS* mRNA and/or TYMS/RAD23B expression in CTCs, as well as CTC kinetics, have the potential to predict non-response to NCRT and avoid unnecessary radical surgery for LARC patients with pCR.

Keywords: locally advanced rectal cancer; circulating tumor cells; RAD23B; thymidylate synthase; chemoradioresistance

---

1. Introduction

Colorectal carcinoma is one of the most commonly occurring neoplasms in the Western world. Moreover, despite improvements in treatment, colorectal carcinoma remains an important cause of cancer-related deaths worldwide [1]. Rectal carcinoma, which accounts for approximately 30% of all primary colorectal cancers, is characterized by an anatomy and natural history that are distinct from other colonic tumors [2,3].

Neoadjuvant chemoradiation (NCRT) followed by total mesorectal excision is the standard treatment for locally advanced rectal carcinoma (LARC). NCRT is generally recommended for patients with cT3/T4 disease or lymph node involvement [4].

The absence of viable tumor cells in surgical specimens after NCRT is defined as pathological complete response (pCR). It occurs in 10–30% of patients [5] and it is associated with a better prognosis. Capecitabin and 5-fluorouracil (5-FU) are currently the most widely used radiosensitizers for NCRT. In some patients, chemoradiation induces pCR, although radical surgery is recommended for most patients [6,7].

Accurate identification of patients with pCR is necessary before pursuing a non-surgical approach. Many studies have evaluated the accuracy of the digital rectal exam and images in identifying patients with pCR [8,9]. However, the sensitivity and specificity of these methods are too low to accurately predict pCR. Molecular analyses of gene expression have also been performed and have been shown to be unsuitable for identifying patients with pCR who can be followed without radical surgery [10]. Recently, promising data have supported the use of a "watch and wait" strategy for patients who have no signs of a viable tumor in a digital rectal exam, rectoscopy (with or without biopsy), or imaging. The likelihood of tumor regrowth is minimal for these patients, and most could be cured by salvage surgery [11].

In addition, considering that the majority of patients will not respond to NCRT, it is extremely important to identify these patients, in an attempt to optimize the response to these patients. As there are no randomized phase-III studies to date, comparing this strategy with isolated induction chemoradiotherapy, it becomes fundamental to identify a biomarker capable of selecting patients for these approach [12–14].

For organ-preserving treatment strategies, the ability to identify tissues or blood biomarkers that predict NCRT response/non-response is very important. Circulating tumor cells (CTCs) are a real-time source of biomarkers, which have shown promise in facilitating the detection and monitoring of pCR and non-responder patients. It has been proposed that the identification and analysis of CTCs would facilitate investigations to understand intrinsic tumor features and characteristics of patients with pCR or non-responders so that individualized treatment strategies can be applied [15]. To date, use of CTC counts has been approved by the US Food and Drug Administration as a prognostic tool for metastatic prostate, colon, and breast cancers [16–18]. Additionally, CTC kinetics can be used to monitor tumor response to systemic treatment [19,20]. Meanwhile, molecular characterization of CTCs has facilitated studies of biomarkers, including proteins, gene expression, and chromosomal translocations [21,22].

Since only a small proportion of patients with LARC experience complete response following chemoradiation, an improved understanding of the molecular mechanisms underlying resistance to chemotherapy and radiation therapy resistance is essential. For chemotherapy, the main target of 5-FU is the enzyme thymidylate synthase (TYMS). TYMS expression analysis has been used to predict individual response to NCRT and has exhibited good prognostic value for rectal cancer recurrence [23,24]. Radiation therapy is an effective component of neoadjuvant treatment and it induces genetic damage. Consequently, the ultraviolet excision repair protein, RAD23 homolog B (RAD23B), which is part of the nucleotide excision repair process [25,26], would potentially be induced by the genetic damage introduced by radiation therapy. Recently, RAD23B was found to be associated with breast cancer relapse risk [27].

The aim of the present study was to explore the role of CTCs in patients undergoing NCRT followed by surgery for treatment of LARC. In addition, the predictive values of TYMS and RAD23B before and after NCRT were evaluated.

2. Methods

2.1. Patients and Treatment

This prospective study was conducted at the A. C. Camargo Cancer Center (São Paulo, Brazil) and was approved by the local ethics committee (2141/15C). Written informed consent was obtained from all patients prior to enrolment. Patients met the inclusion criteria for this study if they had a diagnosis of rectal cancer, as confirmed by biopsy pathology; had locally advanced disease staged as cT3–cT4 or N0–N+; and were candidates for chemoradiation therapy per institutional protocol. Patients were excluded if they had evidence of distant metastasis; a history of any surgery (e.g., colostomy) within two weeks prior to the initiation of treatment; or were taking anticoagulants at the time of the study. Cancer stage was determined with pelvic magnetic resonance imaging and chest and abdominal computed tomography.

Blood samples were collected at baseline or prior to NCRT (S1), and then prior to surgery (S2). Venous blood was collected from the antecubital vein and these samples were stored at room temperature for a maximum of 4 h prior to analysis.

Radiation therapy was applied with a three-dimensional (3D) conformal technique. A 45-Gy dose was applied in 25 fractions over the entire pelvis. In addition, a 5.4-Gy radiation boost was administered to the primary tumor and involved lymph nodes in three fractions, for a total of 50.4 Gy applied in 28 fractions. Chemotherapy regimens consisted of either intravenous 5-FU administered at a dose of 1000 mg/m^2 on days 1–5 during weeks 1 and 5 of radiation therapy; or oral capecitabin administered at a dose of 1650 mg/m^2/d during the entire radiation treatment period. Each patient's physician determined which regimen was appropriate.

The evaluation of the response was determined by the comparative analysis of the baseline images in S1 with the images before surgery (S2). In addition, we evaluated the pathologic response in comparison to the clinical staging established by baseline images.

2.2. Isolation and Quantification of CTCs and Protein Analysis of TYMS and RAD

ISET® (Isolation by size of epithelial tumors) was used to quantify and analyze CTCs. Briefly, peripheral blood samples were collected from patients into EDTA tubes (8.0 mL BD Vacutainer, Franklin Lakes, NJ, USA) and then were homogenized at room temperature for up to 4 h. Samples were then prepared as described previously [28]. Briefly, ISET membrane spots were cut out and subjected to immunocytochemistry (ICC) with an anti-RAD23B antibody (1:100 CSB-PA019260LA01HU; CusaBio, Wuhan, People's Republic of China), an anti-TYMS antibody (1:230 WH0007298M1; Sigma-Aldrich, St. Louis, MO, USA), and an anti-CD45 antibody (1:200 HPA000440; Sigma-Aldrich).

Selected spots from the ISET membranes were additionally subjected to a 24-well dual color ICC assay (Polink DS-RR-Hu/Ms A Kit; GBI Labs, Bothell, WA, USA). Briefly, antigen retrieval was performed by using a DakoTM Antigen Retrieval Solution (Dako, Santa Clara, CA, USA). Cells were then hydrated with 1X Tris-buffered saline (TBS) for 10 min and then permeabilized with Triton X-100 for 5 min. Next, cells were rinsed with 1X TBS and incubated with 3% hydrogen peroxide in the dark for 15 min to block endogenous peroxides. Immobilized cells on the membrane spots were incubated overnight with primary antibodies diluted in 10% fetal calf serum in TBS. To amplify primary antibody signals, the spots were incubated for 30 min with rabbit horseradish peroxidase (HRP) polymer (GBI Labs), then with 3,3-diaminobenzidine (DAB) for 10 min. After amplification, the spots were incubated with a second primary antibody for 2 h, a rabbit AP polymer (GBI Labs) for 30 min, and then GBI-Permanent Red (GBI Labs) for 10 min. The latter reagent was freshly prepared according to the manufacturer's instructions. After staining the cells with haematoxylin, specimens were examined with light microscopy (Research System Microscope BX61; Olympus, Waltham, MA, USA). CTCs were counted per 1 mL blood, as previously described by Krebs et al. [29]. CTCs were characterized according to five criteria: Negativity for CD45 staining; nucleus size >12 μm, hyperchromatic and irregular nuclei; visible cytoplasm; and a nuclear to cytoplasm ratio >80% [30]. CTCs were considered positive for TYMS or RAD23B expression if at least one cell in a specimen stained for these markers on ICC analysis.

2.3. CTC Isolation and Immunostaining Control

Negative controls were healthy donor blood filtered by ISET®. Positive controls included healthy donor blood spiked with HCT-8 colorectal carcinoma cells. For the ICC reaction and TYMS antibody control, leukocytes from healthy filtered blood were used. According to the Human Protein Atlas (http://www.proteinatlas.org/), the latter express TYMS. As a positive control for the RAD23B antibody, HCT-8 cells were spiked in healthy donor blood and filtered on ISET®. To create a negative control for ICC, the same cell line was used without primary antibodies in order to ensure exclusion of cross-reactivity. To confirm that analyzed CTCs were not leukocytes, staining with an anti-CD45 antibody was performed.

2.4. Chromogenic In situ Hybridization Assay for TYMS

TYMS mRNA was detected in intact cells by using a chromogenic in situ hybridization (CISH) assay employing RNAscope Technology (ACDbio, Newark, CA, USA), according to the manufacturer's protocol. Cytology methods standardized by our group were also used. Briefly, one spot was cut out for each patient sample and these were placed in individual wells of a 24-well plate. The membrane spots were hydrated with 1X TBS for 5 min before being incubated in 1% formaldehyde solution for 5 min at room temperature. The spots were then rinsed 2× with distilled water before applying 5–8 drops of RNAscope Hydrogen Peroxide (ACDbio) to each spot. After incubating the samples in a humid chamber for 10 min at room temperature, the spots were washed 2× with distilled water and mounted on slides. To each slide, 3 drops of cytology pepsin were applied. After 10 min at room temperature the spots were returned to the 24-well plate, washed 2× with distilled water, then rapidly dehydrated in successive 1-min incubations with 70%, 85%, and 100% ethanol solutions. After the spots were left to air dry on the slides, 3 drops of a TYMS-specific probe were added to each spot. After 2 h at 40 °C in a HybEZ oven hybridizer (ACDbio), drops of Amp1-6 solutions were added, according to the manufacturer's protocol, with washes performed with 1X TBS. Finally, the cells were incubated with chromogen and 50% haematoxylin then placed on slides with aqueous mounting media and coverslips. The samples were inspected with brightfield microscopy (BX61-Olympus; Olympus). To classify *TYMS* mRNA expression, an absence or presence of staining was classified as negative or positive, respectively.

2.5. Statistical Analysis

Initially, a descriptive analysis was performed to obtain absolute (n) and relative (%) frequency distributions. To evaluate possible associations between the variables of interest, a contingency table was constructed from sample data. Chi-square tests of independence or Fisher's exact test were used, as appropriate. The level of significance was set at 5% and all statistical analyses were performed by using the SPSS program for Windows, version 25.

3. Results

3.1. Patients

Thirty patients with rectal cancer were enrolled in this study between April 2016 and January 2018. Clinical characteristics of these patients are listed in Table 1. The mean age of this cohort was 56 years (range, 34–72) and 60% of the patients were male. For 67% of the patients, their rectal tumors were located 7 cm or less from the anal verge.

The baseline (prior to NCRT) T stage was cT2 for 4 patients (13%), cT3 for 21 patients (70%), and cT4 for 5 patients (17%). Baseline N stage was cN0 for 22 patients (73%) and cN+ for 8 patients (27%). The mean time to surgery completion following NCRT was 77 days. Pathologic T stage was pT0 for 6 patients (20%), pT1 for 5 patients (17%), pT2 for 7 patients (23%), pT3 for 9 patients (30%), and pT4 for 2 patients (7%). Pathologic N stage was pN0 for 21 patients (70%), pN1 for 6 patients (20%), and pN2 for 2 patients (7%). One patient (3%) exhibited disease progression during NCRT. Complete pathologic response after preoperative therapy was detected in 6 patients (20%), while 7 patients (23%) had their tumors down-staged (ypT1-2N0).

Table 1. Patient characteristics.

Characteristics	N (%)
Average age (min–max), years	56 (34–72)
Gender	
Male	18 (60)
Female	12 (40)
Tumor distance from the anal verge	
≤7 cm	20 (67)
>7 cm	10 (33)
Clinical T baseline stage	
T2	04 (13)
T3	21 (70)
T4	05 (17)
Clinical N baseline stage	
N0	22 (73)
N+	08 (27)
Pathological T stage	
T0	06 (20)
T1–T2	12 (40)
T3–T4	11 (37)
DP	1 (3)
Pathological N stage	
N0	21 (70)
N1–N2	08 (27)
DP	01 (3)
Average time (min–max) of completion of RDT for surgery (days)	77 (50–143)

Abbreviations: DP: Disease Progression; RDT: radiotherapy.

All patients were treated with 25 fractions of 45 Gy to the pelvis with a 3D conformal technique. In addition, a 5.4-Gy radiation boost was applied to the primary tumor. There were no treatment delays or interruptions that lasted more than two days. Twenty-six patients (86.7%) received intravenous 5-FU and four patients (13.3%) received capecitabine. The main treatment toxicities were grade 1 or 2 diarrhea in 10 patients (33%) and grade 1 or 2 oral mucositis in 8 patients (27%). No grade 3 or 4 toxicities were reported. There were also no differences in adverse events between the two chemotherapy groups.

3.2. CTCs

CTCs were detected at S1 in all 30 patients of our cohort, and at S2 in 24 patients (Figure 1A,D). The mean CTC concentrations were 6 cells/mL at S1 and 3.5 cells/mL at S2. Kinetic analyses showed that CTC levels were increased in 3 patients, decreased in 22 patients, and remained unchanged in 5 patients. The mean CTC count per mL was 3.1 for those with pCR, 2.5 for those exhibiting a partial response (PR), and 2.9 for those with non-responsive tumors. Patients exhibiting pCR and PR showed a decrease in CTC kinetics (calculated as: CTC baseline [CTC1] × CTC post-CRT [CTC2]) during treatment ($P = 0.02$ and $P = 0.01$, respectively; Table 2).

Table 2. Kinetic counts of circulating tumor cells (CTCs) between baseline (CTC1) and post-neoadjuvant chemoradiation (NCRT) (CTC2) time points.

9	Patient ID	CTCs/mL before NCRT	CTCs/mL after NCRT	Kinetics of CTC1 vs. CTC2	
CR	8	3	1	>	$P = 0.02$
	11	4	1	>	
	18	1	0	>	
	21	4	0	>	
	25	4	2	>	
	27	3	2	>	
PR	3	5	5	=	$P = 0.01$
	4	3	2	>	
	7	3	2	>	
	9	1	0	>	
	10	0	1	<	
	13	1	1	=	
	15	2	0	>	
	16	2	0	>	
	23	2	1	>	
	24	6	3	>	
	29	3	0	>	
	30	2	2	=	
NR	1	3	4	<	$P = 0.07$
	2	3	2	>	
	5	1	1	=	
	6	1	1	=	
	12	1	4	<	
	14	2	1	>	
	17	3	1	>	
	19	4	2	>	
	20	5	1	>	
	22	8	4	>	
	26	2	1	>	
	28	2	1	>	

Abbreviations: NCRT: Neoadjuvant chemoradiotherapy; CR: Complete response; PR: Partial response; NR: No response.

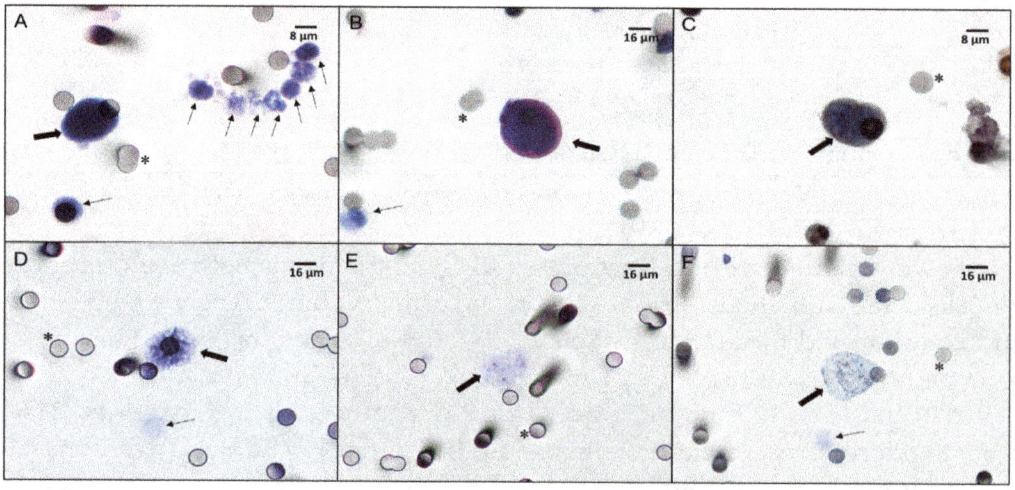

Figure 1. Immunostaining (**A–C**) and chromogenic in situ hybridization (CISH) (**D–F**) of CTCs from locally advanced rectal cancer (LARC) patients. (**A**) CTCs and leukocytes visualized with haematoxylin-eosin staining (×40 magnification). (**B**) CTCs stained with an anti-thymidylate synthase (TYMS) antibody, visualized with Permanent Red, and counterstained with haematoxylin (×40 magnification). (**C**) CTCs stained with an anti-RAD23B antibody, visualized with 3,3-diaminobenzidine (DAB), and counterstained with haematoxylin (×60 magnification). (**D**) CTCs negative for *TYMS* mRNA and counterstained with haematoxylin (×40 magnification). (**E**) CTCs with a low *TYMS* mRNA signal (normal expression) counterstained with haematoxylin (×40 magnification). (**F**) CTCs with a high *TYMS* mRNA signal (overexpression) counterstained with haematoxylin (×40 magnification). All images were analyzed on a Research System Microscope BX61 (Olympus, Tokyo, Japan) coupled to a digital camera (SC100–Olympus). Thick arrows indicate CTCs, thin arrows indicate leukocytes, and asterisks indicate 8 µm pores of the ISET® membranes.

3.3. *TYMS and RAD23B*

At baseline, TYMS-positive CTCs were detected in 7 patients (23.5%) by ICC (Figure 1B) and in 21 patients (70%) by CISH (Figure 1E,F). RAD23B-positive CTCs were detected in 13 patients (43.3%) by ICC (Figure 1C).

After chemoradiation, TYMS-positive CTCs were detected in 10 patients (41.6%) by ICC, while *TYMS* mRNA was detected in 16 patients (61.5%) by CISH. RAD23B-positive CTCs were detected in 14 patients (58.3%).

Baseline *TYMS* mRNA and RAD23B-positive CTCs were associated with poor clinical response. For example, all 12 patients with non-responsive tumors had *TYMS* mRNA detected in their CTCs by CISH. In contrast, 83.5% of patients with pCR did not have CTCs expressing *TYMS* mRNA ($P = 0.001$; Table 3). At S1, RAD23B-positive CTCs were detected in 33% of patients with pCR, in 16% of patients with PR, and in 75% of patients exhibiting no response ($P = 0.01$). Thus, there was no association between detection of TYMS expression in CTCs by ICC and response type in S1 (Table 3). To confirm the value of these proteins as biomarkers, we compared these found with the most used clinical parameter, which is pathological response of the primary tumor. We found, for TYMS mRNA, 83.3% sensitivity, 83.3% specificity, and 95.2% positive predictive value (PPV) ($p = 0.001$) (supplementary table).

Table 3. Expression profiles of RAD23B and TYMS proteins and *TYMS* mRNA.

			CR	*PR*	*NR*	
Before NCRT	CISH (TYMS)	+	16.5	66.5	100	$P = 0.001$
		–	83.5	33.5	0	
	TYMS (protein)	+	16.5	25	25	$P = 1.0$
		–	83.5	75	75	
	RAD (protein)	+	33.5	16.5	75	$P = 0.01$
		–	66.5	83.5	25	
After NCRT	CISH (TYMS)	+	25	30	100	$P = 0.001$
		–	75	70	0	
	TYMS (protein)	+	0	0	83.5	$P = 0.0001$
		–	100	100	16.5	
	RAD (protein)	+	0	25	100	$P = 0.0001$
		–	100	75	0	

Abbreviations: NCRT: Neoadjuvant chemoradiotherapy; CR: Complete response; PR: Partial response; NR: No response.

Post-chemoradiation analyses (S2) showed that expression of TYMS and RAD23B in CTCs strongly correlates with poor response. For example, TYMS-positive CTCs were not detected in the patients with pCR or pPR, yet they were detected in 83% of non-responsive patients ($P < 0.001$; Table 3). Furthermore, *TYMS* mRNA expression at S2 correlated with response, with *TYMS* mRNA detected in all of the patients exhibiting no response ($P = 0.001$). We found, for TYMS protein, 100% sensitivity, 50% specificity, and 100% positive predictive value (PPV) ($p = 0.064$). For RAD23B, the values were: 100% sensitivity, 70% specificity, and 100% positive predictive value (PPV) ($p = 0.01$) (supplementary table).

Lastly, we found that expression of TYMS and RAD23B in CTCs was strongly predictive of response type. For example, after NCRT, $TYMS^-/RAD23B^-$ CTCs were detected in 100% of the pCR patients, in 83.5% of the PR patients, and in none of the NR patients. However, $TYMS^+/RAD23B^+$ CTCs were not detected in patients with pCR or PR, and yet were detected in 83.5% of patients with no response ($P = 0.001$). At S1, CTCs expressing TYMS and RAD23B did not correlate with response type ($P = 0.1$; Table 4). Meanwhile, expression of *TYMS* mRNA determined which patients did not respond to chemoradiotherapy at S1 and S2 (Figure 2).

Table 4. Correlation between RAD and TYMS protein expression profiles.

Profile	*CR %*	*PR %*	*NR %*	
TYMS–/RAD–	50	66.5	16.5	Before NCRT $P = 0.1$
TYMS+/RAD+	0	8.5	16.5	
TYMS+/RAD– TYMS–/RAD+	50	25	67	
TYMS–/RAD–	100	83.5	0	After NCRT $P = 0.001$
TYMS+/RAD+	0	0	83.5	
TYMS+/RAD– TYMS–/RAD+	0	16.5	16.5	

Abbreviations: NCRT: Neoadjuvant chemoradiotherapy; CR: Complete response; PR: Partial response; NR: No response.

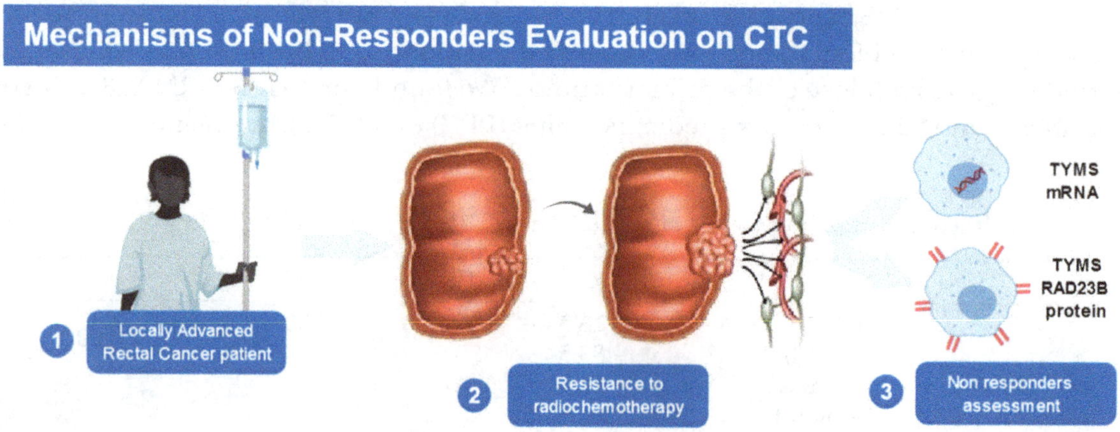

Figure 2. Scheme demonstrating both methodologies used to select responders and non-responders to neoadjuvant chemoradiotherapy: mRNA detected by CISH and protein expression detected by immunocytochemistry (ICC).

4. Discussion

LARC treatment generally consists of NCRT followed by total mesorectal excision. For high-risk patients, postoperative adjuvant chemotherapy may additionally be considered [31]. The ability to identify patients who have undergone complete eradication of a tumor following NCRT is crucial. In the current study, we prospectively demonstrated a strong correlation between expression of TYMS and RAD23B by CTCs in patients with LARC and pCR following NCRT.

Among the 50–60% of patients with LARC who respond to NCRT (i.e., their tumors are down-staged following treatment), many exhibit improved survival. It has been reported that pCR after NCRT is associated with improved cancer outcome and significantly decreased rates of local recurrence. Conversely, the 40% non-response rate after NCRT [31] represents heterogeneity in response to standard treatment [32,33]. For the latter, new therapeutic strategies, avoidance of toxicity associated with ineffective treatment, and novel neoadjuvant chemotherapeutic options are needed [34,35]. In addition, the identification of biomarkers would facilitate the creation of individualized treatment plans.

The ability to identify tumors that do not respond to radiotherapy is useful for helping patients avoid radiation side effects such as fibrosis, fecal and bladder incontinence, diarrhea, dysuria, and myelosuppression [36,37]. This knowledge would also facilitate discussions regarding alternate approaches, such as more potent neoadjuvant multi-agent chemotherapy strategies or a rationale for foregoing radiation therapy. There are several studies that have discussed these considerations, although robust biomarkers that will predict non-responders of NCRT with high accuracy still need to be identified [37–39]. In the present study, we were able to identify NCRT non-responders at S1 by detecting *TYMS* mRNA. Thus, it is possible that detection of *TYMS* mRNA in CTCs could represent a valuable tool in identifying non-responder patients prior to the start of NCRT. Furthermore, detection of

RAD23B expression could make this patient selection process more accurate. It is well-established that RAD23B is involved in DNA repair following radiation damage, and its identification in CTCs can guide treatment plans. In the present study, 75% of non-responding tumors expressed this protein on CTCs before NCRT. Furthermore, when both RAD23B and TYMS protein expression on CTCs were detected after NCRT, 100% of patients with pCR did not express either protein. Meanwhile, 83.5% of patients with non-responsive tumors expressed both proteins and the remaining 16.5% expressed at least one of these proteins. In addition, *TYMS* mRNA expression after NCRT showed high positivity for non-responders (100%) and was not related to protein expression (Figure 3). For the seven patients who presented discordant mRNA/protein positivity at S1 (Supplementary Table S1), they exhibited a correspondence between *TYMS* gene expression and TYMS protein expression at S2. Chemotherapeutic agents have previously been associated with changes in gene expression. In addition, post-transcriptional mechanisms for blocking protein synthesis have been characterized [40–42]. For 5-FU, there are several papers that describe this correlation [43,44]. The present results support the hypothesis that CTC analysis can be a useful tool for identifying patients who will respond to chemoradiotherapy. As a result, a "watch and wait" strategy becomes an option to be considered in addition to radical surgery.

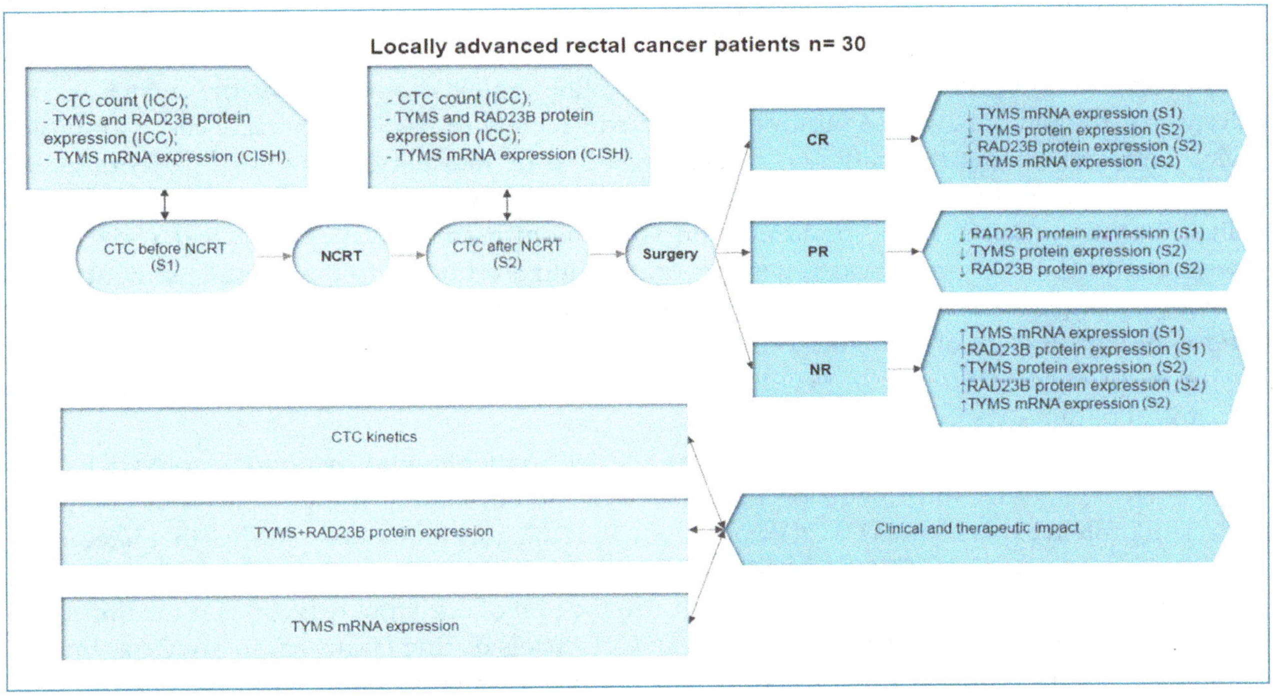

Figure 3. Summary of methodologies, analyses, and results in the present study. Patients were enrolled prior to the start of NCRT. Blood was collected to perform CTC counts and molecular analyses. At baseline, complete response (CR) correlated with low levels of *TYMS* mRNA. In contrast, NR correlated with high levels of *TYMS* mRNA and RAD23B protein expression. Blood samples were collected again during follow-up after NCRT. CTC analyses showed that CR correlated to low levels of *TYMS* mRNA and RAD23B protein expression, while NR correlated to high levels of *TYMS* mRNA and RAD23B/TYMS protein expression. CTC kinetics also correlated to pathological response. Based on these results, *TYMS* mRNA and RAD23B and TYMS protein expression appear to have a clinical and therapeutic impact in LARC patients. Abbreviations: CR: Complete response; LARC: Locally advanced rectal cancer; NCRT: Neoadjuvant chemoradiotherapy; NR: No response; PR: Partial response.

Our analysis of RAD23B and TYMS expression showed similar profiles for these two proteins. Based on the different protein patterns for TYMS in relation to pCR and PR at S1 and S2, we decided to further examine the mRNA expression of *TYMS* with RNA hybridization assays performed in situ with CTCs. All of the CTCs from non-responding patients expressed *TYMS* mRNA at both S1 and

S2. Thus, this assay exhibited high sensitivity and specificity for identifying LARC patients who are predicted to be non-responders to radiochemotherapy.

After NCRT, TYMS expression exhibited a strong correlation with *TYMS* mRNA (kappa value 0.6; $P = 0.003$) in absence of response to treatment. This second assay at S2 cost less to perform besides the possibility of performing more than one protein per time on the CTCs isolated by ISET® methodology. Furthermore, we previously showed that elevated TYMS expression in CTCs was associated with poor prognosis among patients with metastatic colorectal cancer [26].

TYMS, a downstream target molecule of 5-FU, plays a key role in DNA synthesis. The enzyme catalyzes deoxyuridine monophosphate methylation to produce deoxythymidine triphosphate, and subsequently, thymidylate. Increased TYMS expression is thought to be responsible for 5-FU resistance in patients with colorectal cancer [42,43]. Meanwhile, RAD23B is a member of the nucleotide excision repair system, which stabilizes the xeroderma pigmentosum complementation group C (XPC) protein and potentiates its interaction with damaged DNA [45]. XPC subsequently initiates nucleotide excision repair. It is not clear why increased production of RAD23B protein was observed in CTCs from non-responders to NCRT in the present study. A possible hypothesis is that increased levels of this repair protein during treatment make it difficult to eliminate cancer cells by NCRT, a strategy that is based on inducing apoptosis as a result of DNA damage.

A treatment approach that deserves discussion in the context of our study is the "watch and wait" strategy. For the patients whose tumors respond completely to NCRT (as evidenced by radiologic, clinical, and endoscopic evaluations), a "watch and wait" strategy may allow patients to avoid the morbidity, mortality, and functional consequences of radical surgical treatment. A recently published meta-analysis of 880 patients who were monitored with a "watch and wait" strategy showed that the strategy can be safely incorporated into a multidisciplinary management plan for the treatment of patients with rectal cancer who achieve pCR after neoadjuvant treatment [12,46–48]. Therefore, identification of a biomarker that can predict pCR to NCRT treatment prior to treatment, or in the early stages of treatment, would be of great clinical utility in combination with commonly used clinical parameters.

In the present study, it was observed that 100% of patients who responded to NCRT did not express TYMS or RAD23B proteins on their CTCs at the beginning of their follow-up monitoring. This lack of protein expression in CTCs is consistent with imaging and kinetic studies that have showed a reduction or elimination of CTCs post NCRT. Furthermore, we observed that CTC kinetics correlated with disease outcome for our patients with LARC. To the best of our knowledge, this is the first study to demonstrate this result. The observed decrease in CTC levels during treatment in our cases involving complete or partial responses further support the use of CTC analyses to predict response to NCRT.

In summary, our results provide valuable data regarding two potential biomarkers of chemoradiation resistance in patients undergoing neoadjuvant treatment for LARC. Despite the small sample size of our study, CTC kinetics, as well as *TYMS* mRNA and/or RAD23B/TYMS protein expression in CTCs, were found to strongly correlate with pCR. Further studies are needed to validate these findings with a larger patient cohort. If CTC analysis proves useful in predicting pCR with high accuracy, many patients may be spared radical surgery for rectal cancer treatment. In addition, biomarker and kinetic analyses of CTCs may identify potential non-responders to treatment with NCRT, thereby identifying a need to evaluate other forms of therapy for these patients.

Author Contributions: B.C.T.F.: conception/design, data analysis and interpretation, collection and/or assembly of data, manuscript writing; V.S.e.S.: conception/design, provision of study material or patients, collection and/or assembly of data, manuscript writing; E.A.A.: data analysis and interpretation; C.A.L.M.: provision of study material or patients; M.L.G.S.: data analysis and interpretation; G.G.M.: data analysis and interpretation; A.C.B.: data analysis and interpretation; S.A.J.: conception/design; L.T.D.C.: conception/design, data analysis and interpretation, manuscript writing, final approval of manuscript.

Acknowledgments: We thank the Public Ministry (TAC-MP- PAJ no. 000968.2012.10.000/0), the National Institute for Science and Technology in Oncogenomics and Therapeutic Innovation (INCT FAPESP/CNPq 2014/50943-1), and the São Paulo Research Foundation (2016/18786-9) for financial support for this study.

References

1. Siegel, R.L.; Miller, K.D.; Jemal, A. Cancer statistics, 2018. *CA Cancer J. Clin.* **2018**, *68*, 7–30. [CrossRef] [PubMed]
2. Cravo, M.; Rodrigues, T.; Ouro, S.; Ferreira, A.; Féria, L.; Maio, R. Management of rectal cancer: Times they are changing. *GE Port. J. Gastroenterol.* **2014**, *21*, 192–200. [CrossRef]
3. Gaertner, W.B.; Kwaan, M.R.; Madoff, R.D.; Melton, G.B. Rectal cancer: An evidence-based update for primary care providers. *World J. Gastroenterol.* **2015**, *21*, 7659–7671. [CrossRef] [PubMed]
4. Li, Y.; Wang, J.; Ma, X.; Tan, L.; Yan, Y.; Xue, C.; Hui, B.; Liu, R.; Ma, H.; Ren, J. A Review of Neoadjuvant Chemoradiotherapy for Locally Advanced Rectal Cancer. *Int. J. Biol. Sci.* **2016**, *12*, 1022–1031. [CrossRef] [PubMed]
5. Ryan, J.E.; Warrier, S.K.; Lynch, A.C.; Ramsay, R.G.; Phillips, W.A.; Heriot, A.G. Predicting pathological complete response to neoadjuvant chemoradiotherapy in locally advanced rectal cancer: a systematic review. *Colorectal Dis.* **2016**, *18*, 234–246. [CrossRef] [PubMed]
6. Zou, X.-C.; Wang, Q.-W.; Zhang, J.-M. Comparison of 5-FU-based and Capecitabine-based Neoadjuvant Chemoradiotherapy in Patients With Rectal Cancer: A Meta-analysis. *Clin. Colorectal Cancer* **2017**, *16*, e123–e139. [CrossRef] [PubMed]
7. Schmoll, H.-J.; Haustermans, K.; Price, T.J.; Nordlinger, B.; Hofheinz, R.; Daisne, J.-F.; Janssens, J.; Brenner, B.; Schmidt, P.; Reinel, H.; et al. Preoperative chemoradiotherapy and postoperative chemotherapy with capecitabine +/- oxaliplatin in locally advanced rectal cancer: Final results of PETACC-6. *JCO* **2018**, *36*, 3500. [CrossRef]
8. Pozo, M.E.; Fang, S.H. Watch and wait approach to rectal cancer: A review. *World J. Gastrointest. Surg.* **2015**, *7*, 306–312. [CrossRef]
9. Vojtíšek, R.; Korčáková, E.; Mařan, J.; Šorejs, O.; Fínek, J. Neoadjuvant chemoradiotherapy of the rectal carcinoma - The correlation between the findings on the restaging multiparametric 3T MRI scanning and the surgical findings. *Rep. Pract. Oncol. Radiother.* **2017**, *22*, 265–276. [CrossRef]
10. Molinari, C.; Matteucci, F.; Caroli, P.; Passardi, A. Biomarkers and Molecular Imaging as Predictors of Response to Neoadjuvant Chemoradiotherapy in Patients With Locally Advanced Rectal Cancer. *Clin. Colorectal Cancer* **2015**, *14*, 227–238. [CrossRef]
11. van der Valk, M.J.M.; Hilling, D.E.; Bastiaannet, E.; Meershoek-Klein Kranenbarg, E.; Beets, G.L.; Figueiredo, N.L.; Habr-Gama, A.; Perez, R.O.; Renehan, A.G.; van de Velde, C.J.H.; et al. Long-term outcomes of clinical complete responders after neoadjuvant treatment for rectal cancer in the International Watch & Wait Database (IWWD): an international multicentre registry study. *Lancet* **2018**, *391*, 2537–2545. [PubMed]
12. Fernandez-Martos, C.; Garcia-Albeniz, X.; Pericay, C.; Maurel, J.; Aparicio, J.; Montagut, C.; Safont, M.J.; Salud, A.; Vera, R.; Massuti, B.; et al. Chemoradiation, surgery and adjuvant chemotherapy versus induction chemotherapy followed by chemoradiation and surgery: long-term results of the Spanish GCR-3 phase II randomized trial†. *Ann. Oncol.* **2015**, *26*, 1722–1728. [CrossRef] [PubMed]
13. Cercek, A.; Roxburgh, C.S.D.; Strombom, P.; Smith, J.J.; Temple, L.K.F.; Nash, G.M.; Guillem, J.G.; Paty, P.B.; Yaeger, R.; Stadler, Z.K.; et al. Adoption of Total Neoadjuvant Therapy for Locally Advanced Rectal Cancer. *JAMA Oncol.* **2018**, *4*, e180071. [CrossRef] [PubMed]
14. Yamashita, K.; Matsuda, T.; Hasegawa, H.; Mukohyama, J.; Arimoto, A.; Tanaka, T.; Yamamoto, M.; Matsuda, Y.; Kanaji, S.; Nakamura, T.; et al. Recent advances of neoadjuvant chemoradiotherapy in rectal cancer: Future treatment perspectives. *Ann. Gastroenterol. Surg.* **2019**, *3*, 24–33. [CrossRef] [PubMed]
15. Sun, W.; Li, G.; Wan, J.; Zhu, J.; Shen, W.; Zhang, Z. Circulating tumor cells: A promising marker of predicting tumor response in rectal cancer patients receiving neoadjuvant chemo-radiation therapy. *Oncotarget* **2016**, *7*, 69507–69517. [CrossRef] [PubMed]
16. Cristofanilli, M.; Hayes, D.F.; Budd, G.T.; Ellis, M.J.; Stopeck, A.; Reuben, J.M.; Doyle, G.V.; Matera, J.; Allard, W.J.; Miller, M.C.; et al. Circulating tumor cells: a novel prognostic factor for newly diagnosed metastatic breast cancer. *J. Clin. Oncol.* **2005**, *23*, 1420–1430. [CrossRef]

17. Cohen, S.J.; Punt, C.J.A.; Iannotti, N.; Saidman, B.H.; Sabbath, K.D.; Gabrail, N.Y.; Picus, J.; Morse, M.; Mitchell, E.; Miller, M.C.; et al. Relationship of circulating tumor cells to tumor response, progression-free survival, and overall survival in patients with metastatic colorectal cancer. *J. Clin. Oncol.* **2008**, *26*, 3213–3221. [CrossRef]
18. León-Mateos, L.; Vieito, M.; Anido, U.; López López, R.; Muinelo Romay, L. Clinical Application of Circulating Tumour Cells in Prostate Cancer: From Bench to Bedside and Back. *Int. J. Mol. Sci.* **2016**, *17*, 1580. [CrossRef]
19. Souza E Silva, V.; Chinen, L.T.D.; Abdallah, E.A.; Damascena, A.; Paludo, J.; Chojniak, R.; Dettino, A.L.A.; de Mello, C.A.L.; Alves, V.S.; Fanelli, M.F. Early detection of poor outcome in patients with metastatic colorectal cancer: tumor kinetics evaluated by circulating tumor cells. *Onco. Targets Ther.* **2016**, *9*, 7503–7513. [CrossRef]
20. Yan, W.-T.; Cui, X.; Chen, Q.; Li, Y.-F.; Cui, Y.-H.; Wang, Y.; Jiang, J. Circulating tumor cell status monitors the treatment responses in breast cancer patients: a meta-analysis. *Sci Rep* **2017**, *7*, 43464. [CrossRef]
21. Benini, S.; Gamberi, G.; Cocchi, S.; Garbetta, J.; Alberti, L.; Righi, A.; Gambarotti, M.; Picci, P.; Ferrari, S. Detection of circulating tumor cells in liquid biopsy from Ewing sarcoma patients. *Cancer Manag. Res.* **2018**, *10*, 49–60. [CrossRef] [PubMed]
22. Pantel, K.; Alix-Panabières, C. Liquid biopsy and minimal residual disease–latest advances and implications for cure. *Nat. Rev. Clin. Oncol.* **2019**. [CrossRef]
23. Kuremsky, J.G.; Tepper, J.E.; McLeod, H.L. Biomarkers for response to neoadjuvant chemoradiation for rectal cancer. *Int. J. Radiat. Oncol. Biol. Phys.* **2009**, *74*, 673–688. [CrossRef] [PubMed]
24. Conradi, L.-C.; Bleckmann, A.; Schirmer, M.; Sprenger, T.; Jo, P.; Homayounfar, K.; Wolff, H.A.; Rothe, H.; Middel, P.; Becker, H.; et al. Thymidylate synthase as a prognostic biomarker for locally advanced rectal cancer after multimodal treatment. *Ann. Surg. Oncol.* **2011**, *18*, 2442–2452. [CrossRef] [PubMed]
25. Watkins, J.F.; Sung, P.; Prakash, L.; Prakash, S. The Saccharomyces cerevisiae DNA repair gene RAD23 encodes a nuclear protein containing a ubiquitin-like domain required for biological function. *Mol. Cell. Biol.* **1993**, *13*, 7757–7765. [CrossRef] [PubMed]
26. Schauber, C.; Chen, L.; Tongaonkar, P.; Vega, I.; Lambertson, D.; Potts, W.; Madura, K. Rad23 links DNA repair to the ubiquitin/proteasome pathway. *Nature* **1998**, *391*, 715–718. [CrossRef] [PubMed]
27. Pérez-Mayoral, J.; Pacheco-Torres, A.L.; Morales, L.; Acosta-Rodríguez, H.; Matta, J.L.; Dutil, J. Genetic polymorphisms in RAD23B and XPC modulate DNA repair capacity and breast cancer risk in Puerto Rican women. *Mol. Carcinog.* **2013**, *52*, E127–E138. [CrossRef]
28. Abdallah, E.A.; Fanelli, M.F.; Buim, M.E.C.; Machado Netto, M.C.; Gasparini Junior, J.L.; Souza E Silva, V.; Dettino, A.L.A.; Mingues, N.B.; Romero, J.V.; Ocea, L.M.M.; et al. Thymidylate synthase expression in circulating tumor cells: a new tool to predict 5-fluorouracil resistance in metastatic colorectal cancer patients. *Int. J. Cancer* **2015**, *137*, 1397–1405. [CrossRef]
29. Krebs, M.G.; Hou, J.-M.; Sloane, R.; Lancashire, L.; Priest, L.; Nonaka, D.; Ward, T.H.; Backen, A.; Clack, G.; Hughes, A.; et al. Analysis of circulating tumor cells in patients with non-small cell lung cancer using epithelial marker-dependent and -independent approaches. *J. Thorac. Oncol.* **2012**, *7*, 306–315. [CrossRef]
30. Khoja, L.; Backen, A.; Sloane, R.; Menasce, L.; Ryder, D.; Krebs, M.; Board, R.; Clack, G.; Hughes, A.; Blackhall, F.; et al. A pilot study to explore circulating tumour cells in pancreatic cancer as a novel biomarker. *Br. J. Cancer* **2012**, *106*, 508–516. [CrossRef]
31. Sung, S.; Son, S.H.; Park, E.Y.; Kay, C.S. Prognosis of locally advanced rectal cancer can be predicted more accurately using pre- and post-chemoradiotherapy neutrophil-lymphocyte ratios in patients who received preoperative chemoradiotherapy. *PLoS ONE* **2017**, *12*, e0173955. [CrossRef] [PubMed]
32. Powell, A.A.; Talasaz, A.H.; Zhang, H.; Coram, M.A.; Reddy, A.; Deng, G.; Telli, M.L.; Advani, R.H.; Carlson, R.W.; Mollick, J.A.; et al. Single Cell Profiling of Circulating Tumor Cells: Transcriptional Heterogeneity and Diversity from Breast Cancer Cell Lines. *PLoS ONE* **2012**, *7*, e33788. [CrossRef] [PubMed]
33. Greenbaum, A.; Martin, D.R.; Bocklage, T.; Lee, J.-H.; Ness, S.A.; Rajput, A. Tumor Heterogeneity as a Predictor of Response to Neoadjuvant Chemotherapy in Locally Advanced Rectal Cancer. *Clin. Colorectal Cancer* **2019**. [CrossRef] [PubMed]
34. Franke, A.J.; Parekh, H.; Starr, J.S.; Tan, S.A.; Iqbal, A.; George, T.J. Total Neoadjuvant Therapy: A Shifting Paradigm in Locally Advanced Rectal Cancer Management. *Clin. Colorectal Cancer* **2018**, *17*, 1–12. [CrossRef] [PubMed]
35. Garcia-Aguilar, J.; Chen, Z.; Smith, D.D.; Li, W.; Madoff, R.D.; Cataldo, P.; Marcet, J.; Pastor, C. Identification

of a biomarker profile associated with resistance to neoadjuvant chemoradiation therapy in rectal cancer. *Ann. Surg.* **2011**, *254*, 486–492, discussion 492-493. [CrossRef]

36. Santos, M.D.; Silva, C.; Rocha, A.; Nogueira, C.; Castro-Poças, F.; Araujo, A.; Matos, E.; Pereira, C.; Medeiros, R.; Lopes, C. Predictive clinical model of tumor response after chemoradiation in rectal cancer. *Oncotarget* **2017**, *8*, 58133–58151. [CrossRef]
37. Sun, Y.; Wu, X.; Zhang, Y.; Lin, H.; Lu, X.; Huang, Y.; Chi, P. Pathological complete response may underestimate distant metastasis in locally advanced rectal cancer following neoadjuvant chemoradiotherapy and radical surgery: Incidence, metastatic pattern, and risk factors. *Eur. J. Surg. Oncol.* **2019**. [CrossRef]
38. Gotanda, K.; Hirota, T.; Matsumoto, N.; Ieiri, I. MicroRNA-433 negatively regulates the expression of thymidylate synthase (TYMS) responsible for 5-fluorouracil sensitivity in HeLa cells. *BMC Cancer* **2013**, *13*, 369. [CrossRef]
39. Kwon, M.J.; Soh, J.S.; Lim, S.-W.; Kang, H.S.; Lim, H. HER2 as a limited predictor of the therapeutic response to neoadjuvant therapy in locally advanced rectal cancer. *Pathol.–Res. Pract.* **2019**, *215*, 910–917. [CrossRef]
40. Kaneno, R.; Shurin, G.V.; Kaneno, F.M.; Naiditch, H.; Luo, J.; Shurin, M.R. Chemotherapeutic agents in low noncytotoxic concentrations increase immunogenicity of human colon cancer cells. *Cell Oncol. (Dordr)* **2011**, *34*, 97–106. [CrossRef]
41. Lee, S.-C.; Xu, X.; Lim, Y.-W.; Iau, P.; Sukri, N.; Lim, S.-E.; Yap, H.L.; Yeo, W.-L.; Tan, P.; Tan, S.-H.; et al. Chemotherapy-induced tumor gene expression changes in human breast cancers. *Pharmacogenet. Genomics* **2009**, *19*, 181–192. [CrossRef] [PubMed]
42. Negrei, C.; Hudita, A.; Ginghina, O.; Galateanu, B.; Voicu, S.N.; Stan, M.; Costache, M.; Fenga, C.; Drakoulis, N.; Tsatsakis, A.M. Colon Cancer Cells Gene Expression Signature As Response to 5- Fluorouracil, Oxaliplatin, and Folinic Acid Treatment. *Front. Pharmacol.* **2016**, *7*, 172. [CrossRef] [PubMed]
43. Tang, M.; Lu, X.; Zhang, C.; Du, C.; Cao, L.; Hou, T.; Li, Z.; Tu, B.; Cao, Z.; Li, Y.; et al. Downregulation of SIRT7 by 5-fluorouracil induces radiosensitivity in human colorectal cancer. *Theranostics* **2017**, *7*, 1346–1359. [CrossRef]
44. Liang, X.; Shi, H.; Yang, L.; Qiu, C.; Lin, S.; Qi, Y.; Li, J.; Zhao, A.; Liu, J. Inhibition of polypyrimidine tract-binding protein 3 induces apoptosis and cell cycle arrest, and enhances the cytotoxicity of 5- fluorouracil in gastric cancer cells. *Br. J. Cancer* **2017**, *116*, 903–911. [CrossRef] [PubMed]
45. Ng, J.M.Y.; Vermeulen, W.; van der Horst, G.T.J.; Bergink, S.; Sugasawa, K.; Vrieling, H.; Hoeijmakers, J.H.J. A novel regulation mechanism of DNA repair by damage-induced and RAD23-dependent stabilization of xeroderma pigmentosum group C protein. *Genes Dev.* **2003**, *17*, 1630–1645. [CrossRef] [PubMed]
46. Xu, W.; Jiang, H.; Zhang, F.; Gao, J.; Hou, J. MicroRNA-330 inhibited cell proliferation and enhanced chemosensitivity to 5-fluorouracil in colorectal cancer by directly targeting thymidylate synthase. *Oncol. Lett.* **2017**, *13*, 3387–3394. [CrossRef]
47. Bunick, C.G.; Miller, M.R.; Fuller, B.E.; Fanning, E.; Chazin, W.J. Biochemical and structural domain analysis of xeroderma pigmentosum complementation group C protein. *Biochemistry* **2006**, *45*, 14965–14979. [CrossRef] [PubMed]
48. Rushworth, D.; Mathews, A.; Alpert, A.; Cooper, L.J.N. Dihydrofolate Reductase and Thymidylate Synthase Transgenes Resistant to Methotrexate Interact to Permit Novel Transgene Regulation. *J. Biol. Chem.* **2015**, *290*, 22970–22976. [CrossRef]

7

CTCs 2020: Great Expectations or Unreasonable Dreams

Elisabetta Rossi [1,2] and Francesco Fabbri [3,*]

1 Department of Surgery, Oncology and Gastroenterology, Oncology Section, University of Padova, 35124 Padova, Italy
2 Veneto Institute of Oncology IOV-IRCCS, 35128 Padua, Italy
3 Istituto Scientifico Romagnolo per lo Studio e la Cura dei Tumori (IRST) IRCCS, 47014 Meldola, Italy
* Correspondence: francesco.fabbri@irst.emr.it

Abstract: Circulating tumor cells (CTCs) are cellular elements that can be scattered into the bloodstream from primary cancer, metastasis, and even from a disseminated tumor cell (DTC) reservoir. CTCs are "seeds", able to give rise to new metastatic lesions. Since metastases are the cause of about 90% of cancer-related deaths, the significance of CTCs is unquestionable. However, two major issues have stalled their full clinical exploitation: rarity and heterogeneity. Therefore, their full clinical potential has only been predicted. Finding new ways of studying and using such tremendously rare and important events can open new areas of research in the field of cancer research, and could drastically improve tumor companion diagnostics, personalized treatment strategies, overall patients management, and reduce healthcare costs.

Keywords: CTC; heterogeneity; liquid biopsy; liquid surgery; clinical utility

---

1. Introduction

In 2004, Cristofanilli and colleagues reported, for the first time, a trial regarding the level of circulating tumor cells (CTCs) in metastatic breast cancer. The trial results indicated this marker was a useful predictor of progression-free survival (PFS) and overall survival (OS) [1]. In 2007, the American Society of Clinical Oncology (ASCO) cited CTCs and disseminated tumor cells (DTCs) in recommendations on tumor markers. Ten years later, the American Joint Committee of Cancer (AJCC) proposed a new category for TNM staging in breast cancer M0(i+). The M0(i+) is defined as the presence of CTC or DTC, respectively, in the blood or in the bone marrow, in case of absence of clinical or radiological evidence of distant metastases. Fifteen years after the first Cristofanilli study, this marker was validated as a prognostic marker in a number of clinical settings, and few oncology topics have been so intensely investigated. [2–6]. However, due to their biological features, the technical hurdles regarding their investigation, and the lack of methodological standardization that make it difficult to compare different CTC studies [7–11], CTCs have still not reached the Olympus of the "clinical utility value". Their predictive value, which we have been researching for decades, is still far from being fully demonstrated. Therefore, old promises should be carefully reconsidered and new directions should be undertaken. A first step in this direction was reported at the beginning of this year. Bidard and colleagues showed, for the first time, results on actual CTC clinical utility. They observed that CTC count, in ER+HER- metastatic breast cancer (MBC) patients, may be used to objectively choose the best therapy [12]. In addition, as other authors stated, another important challenge is how to trigger a paradigm shift in oncology research: anti-cancer personalized treatments should also strive to hit CTCs, and not only target the solid tumor compartment [13]. Recently, Gkountela and colleagues investigated CTC clusters. They observed that CTC clustering could direct the DNA methylation pattern. Specific FDA-approved drugs may take apart CTC clusters, triggering methylation remodeling and metastasis suppression [14]. This is a clear example of how it may be possible to bridge the gap between a potentially important marker, although still limited (i.e., the current CTC status), and a very useful theranostic approach, i.e., the new potential clinical utility of CTCs. Hence, although a lot of work has to be done, the aim is clear: Demonstrating and validating real CTC clinical utility, and looking for new CTC-targeting theranostic approaches.

2. Being Different, to Be Stronger

CTCs are a major cause of tumor relapse [15]. Disseminated into the bloodstream, CTCs are the "seeds", able to give rise to new metastatic lesions. Hence, their significance is unquestionable. Despite the highly inefficient mechanism of spreading [16], some of them are able to survive into the bloodstream, resist biophysical and cell-mediated insults, and reach their final destination. There, they take root in the pre-metastatic niche, most probably already prepared and fueled by tumor cell-derived messengers. In the niche, CTCs lay dormant for an indefinite period, until some still little known signals trigger their lethal awakening [17–19]. CTCs are extremely rare and heterogeneous elements [20–22]. Although their quantification in the peripheral blood of cancer patients is quite unpredictable, even in patients with advanced disease, they have been found down to 1–10 cells/per mL of peripheral blood of metastatic breast, prostate, and colon cancer patients. Their number may be even lower in other cancers and/or in nonmetastatic setting [23–26]. Secondly, and equally important, CTCs are dynamically heterogeneous. Dynamic heterogeneity (DH) can be described as the characteristic of the tumor to evolve through space and time [20,27–29]. This event can produce solid masses composed of many clones, which may be different in transcriptomic, proteomic, and functional makeup. DH, and in particular that of the most aggressive tumors such as triple negative breast cancers (TNBCs), can be quite striking. It has been shown that no two single cells from TNBCs have an identical genomic profile [28]. CTC subpopulation onset and spreading may follow the solid tumor clonality and adaptability. Indeed, CTCs may acquire DH traits with time, from their onset, starting from a more epithelial-like phenotype and transitioning to a more mesenchymal-like state and vice versa, through processes known as epithelial-to-mesenchymal and mesenchymal-to-epithelial transitions (EMT and MET) [30–35]. In addition, EMT can be observed in distinct transitional states. These EMT states are fairly different and can be detected using cell surface markers and single-cell RNA sequencing. Indeed, EMT-hybrid CTC could have the highest metastatic potential with different degrees of aggressiveness due to their capabilities and proteomic and transcriptomic features [36,37]. These elements suggest that studying only a few biomarkers at a few time points, e.g., at first diagnosis and/or relapse, could only offer a very limited actionable vision of the disease. Monitoring the tumor evolution and progression through a timely and accurate multi-marker detection is a crucial investigation opportunity. CTCs lend themselves to this possibility, being a potential mirror/reflection of the evolving and progressing tumor. CTCs, indeed, are often genotypically, genomically, phenotypically, and functionally different. Some have stem cell features (e.g., CD44- or ABC-G2-positivity), others mesenchymal-like characteristics (e.g., N-Cadherin), others still, hybrids phenotypes [29,38–40]. With an appropriate technical set-up (Gallerani et al., unpublished), we could detect CTCs of all three types, at different sampling times, in the blood of esophageal cancer patients. Despite our current inability to see if other populations were present in the blood, our data agree with previously published reports and confirms that some CTC subgroups may be more dangerous than others may. Furthermore, some CTCs express endothelial markers and they reproduce vascular mimicry (VM), a phenomenon present in several human cancers associated with aggressive diseases. Williamson et al. demonstrated, in small cell lung cancer (SCLC), that these rare cells (VE-cadherin CTC) exhibit essentially the same copy number of gains and losses present in other CTCs and in ctDNA from the same patient, a highly related CNA profile and typical of SCLC [41]. We should also note that several CTC populations that coexist in the bloodstream could establish a relation with other normal blood cells. Recently, several authors demonstrated a crosstalk between different CTCs subpopulations and the immune system [42,43]. It is well-known that CTCs may express the receptor programmed death-ligand 1 (PD-L1) on their membrane [44].

PD-L1 is probably able to suppress the immune response against CTCs, helping their survival in the bloodstream [45,46]. Aceto and colleagues proposed a model where the association between CTCs and neutrophils supports the cell cycle progression during the blood trip [47]. The CTCs with the ability to survive in the bloodstream and to interact with leukocytes and platelets increase the possibility of forming metastasis. To this end, CTCs use several strategies to crosstalk with leukocytes and platelet:

(a) High expression of the immunosuppressive molecule PD-L1, which prevents T cell-mediated destruction [44,48–50];

(b) The expression of CD47, which provides a 'don't eat me' signal [51–53];

(c) An altered expression of the apoptotic FAS and/or FASL proteins that may induce the apoptosis of T cells [54] or protect tumor cells from FAS-mediated apoptosis [55];

(d) The interaction with platelets, which induces EMT-like features in CTCs [56], and promotes tumor cell arrest and extravasation [57];

(e) Platelets promote the survival of CTCs during metastasis by conferring resistance to shear stress and attacks from NK cells [58,59];

(f) Platelet–CTC interaction can lead to the transfer of platelet MHC-I to tumor cells, thereby preventing the identification of NK cells and aiding the CTCs to escape from the cytolytic activity mediated by NK cells [60].

Furthermore, a $CD14^{+}CD11c^{+}CD45^{+}$ myeloid subpopulation has been observed inside circulating tumor microemboli (CTM) or free in the bloodstream. Adams and colleagues named these cells 'cancer-associated macrophage-like cells' (CAMLs). Typically, these cells are giant (30–300 μm in length) with big multiple or polylobulated nuclei (14–64 μm in diameter) [61]. Adams et al. showed that CAMLs could be detected in the peripheral blood of patients with breast, prostate, pancreas, and lung cancer in percentages ranging from 81% to 97% of the total number of patients, contrary to healthy subjects, in which they are totally absent [61,62]. CAMLs also express epithelial markers, such as EpCAM and/or CK8/18/19. We still need to clarify if they directly express these epithelial markers (expression levels depend on the differentiation stage), or if they phagocytize material of epithelial origin. CAMLs seem to originate in the primary tumor and increase in blood samples from patients responding to radiotherapy, chemotherapy, or other treatments [63]. However, CAMLs have also been shown to actively interact with CTCs, or have a proangiogenic activity, as CD146 and TIE2 markers suggest [61]. In support of a pro-tumor role of these cells, in metastatic breast cancer patients, the presence of EpCAM+ CAMLs correlates with a shorter OS and PFS [64]. We could suppose that CAMLs might have an active role in helping CTC intravasation, extravasation, or survival in the bloodstream, thus participating in the metastatic process. The high degree of heterogeneity and complex relationships between CTC subpopulations and other blood "resident" cells, makes studying this biomarker more complex. It offers the opportunity to act both with drugs that affect CTCs, but also with drugs that limit the formation of clusters, as already mentioned [14], or with anticoagulant agents or those acting by stimulating the immune system to recognize CTCs [65,66].

Another pivotal aspect is the contribution of circulating cancer stem cells (C-CSCs), or stem-like CTCs. Cancer stem cells are cellular elements thought to propel cancer progression and to be responsible for the rooting of the disease in primary and metastatic sites, and hence, for local and distant recurrence. Although not fully determined, C-CSCs may derive from CSCs and have been identified in numerous types of cancer [40,67,68]. If CTCs can be seen as the overall population of tumor "seeds", almost an epiphenomenon of cancer spreading, C-CSCs can be described as the actual first cause of metastasis, the proverbial needle in a haystack, since such events are even rarer than general CTCs. Furthermore, a close relation between EMT and C-CSCs has been shown, and an enhanced metastatic competence and a high drug resistance capability have also been described for these cells [69,70]. Despite being quite controversial elements, difficult to characterize specifically and unequivocally, C-CSCs may be considered key players in the genesis of cancer metastasis, and may clearly fulfill the model by Paget [16,71–75]. Specific C-CSC markers are difficult to find, but in general, some of them are fairly well established, e.g., CD133, CD44 (including different isoforms),

ALDH1, and ABC transporters. A number of these markers have also been investigated in CTCs, also aimed at finding diagnostic, prognostic, and predictive indicators. None have been found, validated, and progressively translated into the clinical setting [25]. In 2017, Whang and colleagues showed that CTCs could act as potential precursor cells of metastasis and not only act in the intermediate step of metastatic cascade. Probably due to a stress response, CTCs could evolve in cells that were more aggressive and became tumor-initiating cells, probably different from the CSC-like cells present in the primary tumor [76]. The metastatic potential of EpCAM-positive CTCs has been confirmed using xenograft models in several papers [52,77,78]. A subset of CTCs showed a stem cell-like phenotype, these C-CSC have been detected in different kinds of cancer such as lung cancer with expression of BMI1, hepatocarcinoma where C-CSCs are CD90+CXCR4+, colon rectal cancer with CD44v9+ cells. In breast cancer one or more EMT markers, such as *TWIST*, *AKT2*, *PI3K-alpha*, *ALDH1*, were detectable on CTCs. This CTC subset highlights therapy resistance and a poor prognosis [68,79].

As already mentioned earlier, important papers have recently been published. Mirroring C-CSC potentialities, CTC-clusters have been demonstrated to be 23 to 50 times more capable of giving rise to actual metastases than single CTCs [80]. This capability has been explained by uncovering the link between the CTC cluster state and an increased accessibility of stemness-related transcription factors. CTC clustering is related to a DNA methylation pattern that promotes stemness and metastasis [14]. Again, the same group published a study on the striking importance of the cooperation between "normal" and cancer cells. The association of CTCs with neutrophils induces a proliferative boost, making them more competent in metastasis formation [47,81,82].

In conclusion, a number of elements may induce a CTC to gain a CSC-like cluster-state with specific cellular, molecular, and functional phenotypes and capabilities. More than ten years ago, the exposome was described as the overall environmental complement to the genome in determining the risk of a disease. It can be seen as the totality of exposures throughout one's lifespan [83]. Hence, it can be inferred that in addition to genetic predisposition, every molecule or element that can prompt tumor onset and growth can be seen as exposome, or, more specifically, cancer exposome [Fabbri et al., under review]. Consequently, cancer exposome could include inflammation-related molecules, "normal" cellular elements associated to cancer progression (e.g., CAMs, CAFs, and platelets), environmental factors, and other molecules, i.e., elements known to be connected to tumor development, such as induction of EMT, C-CSCs, and progression. It becomes clear that the cancer exposome could contributes to the onset of CSC-like CTCs. Only a focused and multidisciplinary approach may truly help in finding innovative paths to unravel the enigmatic Gordian knot that may well represent cancer metastasis. In the light of recent results and these exposome-related concepts, we have to strive to find innovative treatments to suppress the spread of cancer.

3. Clinical Data: Hopes of Utility

In clinical practice, after pre- and analytical standardizations and diagnostic accuracy evaluations, a test has to demonstrate that it leads to health benefits [84] before being introduced in the decisional path.

Regarding CTCs, the multivariate analysis demonstrated that the CTCs count was the strongest prognostic biomarkers for patients survival [3,85]. For this reason, CTCs could be used to stratify the disease, but this biomarker is rarely used in clinical practice. Budd and colleagues studied 138 MBC patients, who underwent CT scans before and after start treatment. They compared the results with the CTC count obtained at baseline and after four weeks from the start of therapy, and demonstrated that CTC count is more reproducible than radiological response. This suggests that CTCs are a superior surrogate endpoint [86]. In metastatic breast cancer, the first trial designed to study the clinical utility of CTC count, SWOG S0500, was not able to find a survival difference for patients stratified using CTC count. This negative result was because the study suffered from a design fault and not a failure of CTC test [87,88]. In fact, cancers that show early resistance to first line chemotherapy are likely to be resistant to second line chemotherapy, whenever the second line is started. Moreover, it had already been reported that early changes of first line chemotherapy never demonstrated a gain in OS, whatever the technique used to guide the early change (functional imaging, etc.) [89]. Therefore, SWOG S0500 should not be regarded as the final proof that CTCs are not useful in patients' management. However, because of these negative results, the 2015 American Society of Clinical Oncology, in its clinical practice guidelines, did not recommend the use of CTC count in women with metastatic BC [90]. In Europe, two other clinical utility trials based on CTC count are currently ongoing:

- The "CirCe01" trial (NCT01349842), similarly to SWOG0500, assesses early changes of CTC count during treatment in metastatic patients; patients were enrolled before the start of third line of chemotherapy (CT) and followed with the CTC test throughout the successive lines of CT.
- The "STIC CTC" trial (NCT01710605) investigated the clinical utility of the prognostic value of baseline CTC count. In this trial, patients were randomized in two arms: In the first arm, clinically driven patient's treatment choose between CT and hormone therapy (HT) (CTC count not disclosed); in the second arm, CTC count driven treatment chooses patients. In fact, such as first-line treatment, patients with CTC ≥5/7.5 mL received CT, whereas patients with CTC <5/7.5 mL received HT.

As already discussed, this last trial demonstrated the clinical utility of CTC count in first line therapy in ER+HER2-MBC. Although results are awaiting a definitive publication, Bidard and colleagues have shown that in the majority of patients, the treatment chosen based on the CTC count was the same as the treatment chosen based on the clinic. On the contrary, in discrepant cases, CTC count may be more reliable for either escalating (i.e., considering CT in patients with high CTC count) or de-escalating (i.e., considering HT in patients with low CTC count) first line therapies [12]. SWOG0500 and CirCe01 have similar designs. Both studies evaluated CTCs at baseline and after treatment. In particular, CirCe01 evaluated CTCs number after every new line of CT in patients randomized in the CTC arm. The number of therapy lines represent the main difference between these two studies. In fact, CirCe01-enrolled patients about to start third line therapy (a population with a high chemo-resistance prevalence). On the contrary, SWOG500 enrolled patients just before first line CT. Moreover, CirCe01 trials, in nonrandomized run-in phase, also wanted to establish a CTC threshold to be used in the randomized part of the study. The STIC CTC, a randomized trial, focused on evaluating the health economic interest of taking into account CTCs to determine the kind of first line treatment for HR–positive MBC. Major limitations of this trial were a) the lack of standardized clinical criteria for recommending CT in the clinically driven arm, and b) that it was conducted prior to the widespread use of CDK4/6 inhibitors, now regarded as the standard of care for HR–positive MBC. The SWOG0500 was the first trial with a goal regarding the clinical utility of CTCs. However, the heterogeneity of first line treatments allowed in this trial might have variably influenced CTC behavior. Changing treatment to a different CT agent after the rise of early chemo-resistance may be not significantly effective, and this most probably influenced the results of the SWOG0500 [91]. A similar problem, given the design of the study, could have occurred in CirCE01. The STIC CTC trial had a different study design and, as preliminary results presented at the San Antonio Breast Cancer Symposium showed, it suggested that the CTC count could allow for identifying patients in whom the dose of therapy can be scaled. If the preliminary results of this study are confirmed, this could lead to a benefit for patients who will be able to see reduced side effects of the treatment and, in the long term, an economic saving for the national health systems.

Although still not incorporating recently proposed drugs, e.g., cyclin-dependent kinase 4/6 inhibitors, and indicating to escalate therapy, while it is usually suggested to try to de-escalate regimens in this clinical setting, the STIC CTC trial is the first investigation that showed an actual CTC-based clinical benefit. Hopefully, this huge step should be the first of a long series, prompting further investigations and highlighting the pivotal importance of clearly and specifically designed future clinical studies not to fall into previous misleading results.

Several studies evaluated the impact of CTC count and CTC characterization in monitoring treatment. Indeed, in metastatic castration-resistant prostate cancer (mCRPC), CTC elimination (CTC0) after short-term treatment proved to be an early response endpoint [92]. In this study, 3000 men were enrolled in five phase III trials. The response measures were evaluated regardless of the specific intervention and the changes in CTC status from CTC-positive (T baseline) to CTC-negative (after 13 weeks). This change was shown to be strongly associated with longer survival. Interestingly, the percentage change from baseline of prostate specific antigen (PSA), which is more widely used as end-point, did not discriminate survival at the same level of CTC0. In four out of five trials, the treatments included HT; HT can induce a PSA-level modulation, independent of an effect on cell killing. This HT-based effect limits the role of post-therapy PSA change measures as a reliable indicator of efficacy. In this study, CTCs allowed researchers to determine treatment response seven to eight weeks earlier than determined by standard RECIST criteria and PSA. For these reasons, CTC number, a biomarker that is not affected by modulations in androgen receptor signaling, should be included in clinical management of patients. It is important to note that this result could allow for the prevention of drug toxicity and cause a significant health savings [12,93–95].

4. Challenging the Current Paradigm: Liquid Surgery Premises and Hopes

So far, systemic therapies have aimed at preventing relapse after the removal of the primary tumor and delaying the already present metastatic tumor, confidently eliminating therapy-resistant clones and minimal residual disease. To contrast cancer dissemination, preventing metastasis in the first place, we think it is necessary to expand the conventional line of thinking that aims to hit an already widespread disease, and start new paths that anticipate its diffusion, e.g., directly and therapeutically targeting CTCs.

The investigation of the CTCs, and in general everything that can be included in the term liquid biopsy (LB), has aimed at simple, fast, and cost-effective monitoring of disease status or response to treatment. LB is less troublesome than tissue biopsy, thanks to the ease of access to body fluids, in particular when taking a tissue biopsy is often clinically impossible or not recommended [96]. Moreover, as already suggested, the tumor may change and tissue biopsies may not appropriately reflect the complex profile of a tumor disease, because of its dynamic heterogeneity (DH), which can only be addressed by taking biopsies from different tumor areas at different time points. Hence, LB may offer a more comprehensive cross-section of the disease.

LB can be a crucial alternative and/or addition to conventional and tissue-based diagnostic procedures. Targeting CTCs in the blood of patients could control or avoid metastasis spreading and the risk of relapse, just like the surgical removal of solid tumor can limit the disease or even cure a cancer patient. Therefore, two major lines of research could be followed to deplete CTCs as much as possible, striving not to leave any cells with actual metastasis-generating properties. One possible approach is to examine, in depth, CTC biology in order to identify actionable targets that could lead to their depletion and/or to make them harmless, as already cited [14,97]. A second possibility is to find new paths that can exploit already well-known aspects of CTCs: Their biological features, i.e., surface marker expression, dimension, clusterization potential, and their ecosystem, the blood flow.

Both ways are necessary to identify better-tailored CTC-based anticancer treatments for individual patients. The small number of CTCs per mL of peripheral blood is the major challenging physical limit. Although it may appear to be a simple solution, it is increasingly clear that it is necessary to extend the volumes of blood that should be analyzed and/or treated. Then, new approaches to overcome this obstacle, i.e., in vivo CTC collection and/or count devices, and apheresis-based procedures, are under investigation. The GILUPI CellCollector (CC) allows collecting CTC directly from the blood, in vivo. It is a functionalized medical wire covered with anti-EpCAM antibodies that, when inserted in the cubital vein of a patient for 30 min, could trap CTCs from up to one liter and a half of blood. Studies reported an increased number of CTCs detected compared to standard methods [98–100]. Recently, Didzdar and colleagues investigated the prognostic impact of CTCs rescued with CC in colorectal cancer (CRC) [101], but in contrast to other studies, they did not report prognostic significance of CC-CTC. The authors estimated that recovering CTCs with the CC is similar to extracting tumor cells from a blood volume range of 0.33–18 mL, a different volume compared to what was suggested in previous studies (volume range 1–3 L). Also, in our experience, the median number of CTCs retrieved from the blood of metastatic non-small lung cancer (NSCLC) patients was quite low (around three cells per CC device/patient (Fabbri et al., unpublished data). So, despite being a feasible [102] and promising idea, it seems that the CC has not maintained what it had predicted. In June 2019, Galanzha et al. exploited a photacoustic (PA) flow cytometry (PAFC) platform (cytophone platform) to count circulating melanoma cells in vivo. This method allows a noninvasive assessment of a large volume of blood using the PA effect as a transformation of absorbent laser pulse energy into sound through thermoelastic expansion of light objects. The probe is placed on the patients' skin above the selected vein. The authors utilized 18 melanoma patients as the training set and another 10 new melanoma patients as the validation set. This system was able to detect circulating single melanoma cells, circulating clusters of melanoma cells, and it could discriminate between CMC cluster and circulating blood clots (CBCs). The PAFC detection limit achieved in vivo is about 1 CTC/1 L. The PACF-based CTC count needs further studies to confirm the great potential of this noninvasive method [103].

To avoid the bottleneck due to the low number of CTCs, the CTCTrap consortium and the Nick Stoecklein group used and validated the diagnostic leukapheresis (DLA) in order to enrich a higher number of CTCs. The CTCTrap efforts demonstrated that the procedure used allowed researchers to improve the number of CTCs isolated in MBC and metastatic prostate cancer (MPC), and that the procedure could be safely performed in different clinical sites [104]. The DLA product analyzed with Cell Search demonstrated a 30-fold increase in median CTC numbers detected [105] and an increase in the number of CTC-positive patients as well as M0 breast cancer patients [106]. This high number of enriched CTCs was exploited for the single cell molecular characterizations of CTCs [107] and also to culture CTCs [106]. Despite DLA enrichment, the CTC-cultured success is limited to few patients. Only in two out of eight patients were CTCs cultured from fresh and cryopreserved DLA [106]. The culture success rate was reached in samples with more than 300 CTCs, in agreement with other group results; this is probably because DLA samples with high CTC numbers and lower ratios of apoptotic CTCs were more likely to grow in culture. [106,108].

Usually, therapeutic apheresis is used for the rapid elimination of harmful or excessive blood components such as plasma proteins (plasma exchange) or for the harvest or elimination of cells (leukapheresis and platelet apheresis) and has found broad applications in a vast array of hematologic and onco-hematologic disorders. Keeping these reasoning in mind, removing CTC by filtration and adsorption, and reintroducing the blood without tumor cells back into the body would maximize CTC recovery from individual patients.

Statistically, it has been estimated that to find at least one CTC, the total blood volume (TBV) has to be analyzed, and that no less than 7.5 L of blood are necessary to detect at least 10 CTCs in almost every patient [109]. It is generally accepted that the TBV, i.e., the overall amount of fluid circulating within the arteries, capillaries, veins, and chambers of the heart, at any time in a typical adult human, is around 5 liters [110]. Interestingly, before metastasis onset, the presence of about 9 ± 6 CTC/L of blood has been mathematically predicted even in M0 breast cancer patients [111]. Scaling up this estimation, in low CTC number patients also, analyzing up to 72 L of blood would result in finding at least 200 to almost 1000 CTCs or more. Therefore, in order to treat a patient by eliminating up to all of the CTCs, it would be mandatory to perform an approach involving: a) A huge increase in the volume of blood to process, i.e., more than one TBV; b) a device that can catch all of the CTC subpopulations and that will not clog; and c) redirecting the blood into the patient after CTC removal. Although highly pioneering, we can predict that the removal of CTCs from the blood of a patient could control or even avoid metastasis spreading and the risk of relapse, just as surgical removal of a solid tumor can limit the disease or even cure a cancer patient. Such a kind of approach could be called liquid surgery (LS). A medical device able to eliminate CTCs from the blood circulation could substantially improve cancer management decreasing the overall disease burden, increasing OS and PFS, and could be used to support clinical decisions, as well as a higher DLA-based CTC retrieval [112,113]. In a long-term vision, it could even lower the need for high-cost therapeutic approaches and decrease therapy related toxicities. Preclinical tests and experimental set ups are currently running in our laboratories in order to improve and optimize such an approach. In our long-term vision, LB approaches will guide the decision of executing LS based on CTC presence in peripheral blood, exactly as a solid tumor is surgically removed after a standard tissue biopsy has revealed its presence and nature. LB could also monitor CTC levels after a LS procedure to evaluate depletion efficiency. Although reasonably invasive, the LS will be clinically feasible, not harmful, and manageable, comparable to a standard hemodialytic procedure used for kidney failure. The benefits of limiting, or even avoiding, metastasis onset have high preclinical and clinical potential [114,115] and will most probably exceed the possible drawbacks of extracorporeal circulation. Just like hemodialysis, a medical specialist will decide when the LS is needed, the therapeutic regimen, and the various parameters for the treatment. These would include frequency (e.g., treatments per week), duration of each treatment (e.g., up to 4 h or up to six months), and blood flow rates. As a maximum, up to three treatments per week, for up to 1–4 h per treatment, repeated during a period of up to six months, similar and in addition to an adjuvant therapeutic regimen, is predictable. Considering clinical parameters and patient conditions, the procedure could be repeated one or more times per patient per week, similarly to a standard chemotherapy infusion regimen. Of course, this pioneering and challenging idea has to be investigated in depth, tested, and validated, in vitro and in vivo, before becoming a reasonable treatment option that can be examined in a clinical setting. However, it remains a thrilling breakthrough hypothesis that should be put to the test.

5. Conclusions

In these last fifteen years, CTCs have been intensively studied in the field of cancer biology. CTC evaluation may provide clinically relevant and valuable information regarding cancer. The biological information gained from CTCs will be extremely important for opening new research fields, accurately targeting CTCs, and discovering new CTC-based treatment strategies. Multidisciplinary research approaches aimed at the overall "blood ecosystem", an ecosystem for the CTCs and the metastatic cascade, are increasingly mandatory to fulfill this challenge. Monitoring CTC counts during therapy is a tool that may allow the assessment of disease development in real time, even prior to overt clinical signs of relapse. Targeting CTCs should become a translational objective that could pave the way towards increasing survival outcomes and reducing distal recurrence, preventing metastasis before it occurs. In the era of tailored therapy, precision oncology increases the range of treatment options, but to date, only a relatively small number of people could benefit. Moreover, to apply precision oncology is very expensive. The high cost is due to screening patients for tailored treatments and designing drugs that would be utilized in single patients or in limited groups of patients. We should not underestimate that CTCs could improve the health system costs and savings if the CTC-based diagnostic and therapeutic application could guide (a) better patient stratification, (b) first appropriate selection of personalized regimen, and (c) early treatment discontinuation and/or switch. The theranostic use of CTCs, which may be called LS, could establish a completely new area of research in the field of cancer research and management, and trigger the further improvement of CTC-based companion diagnostics.

Author Contributions: E.R. and F.F. equally contributed to manuscript conceptualization, writing, original draft preparation, and review and editing.

References

1. Cristofanilli, M.; Budd, G.T.; Ellis, M.J.; Stopeck, A.; Matera, J.; Miller, M.C.; Reuben, J.M.; Doyle, G.V.; Allard, W.J.; Terstappen, L.W.M.M.; et al. Circulating tumor cells, disease progression, and survival in metastatic breast cancer. *N. Engl. J. Med.* **2004**, *351*, 781–791. [CrossRef]

2. Cabel, L.; Proudhon, C.; Gortais, H.; Loirat, D.; Coussy, F.; Pierga, J.Y.; Bidard, F.C. Circulating tumor cells: Clinical validity and utility. *Int. J. Clin. Oncol.* **2017**, *22*, 421–430. [CrossRef]
3. Bidard, F.-C.; Peeters, D.J.; Fehm, T.; Nolé, F.; Gisbert-Criado, R.; Mavroudis, D.; Grisanti, S.; Generali, D.; Garcia-Saenz, J.A.; Stebbing, J.; et al. Clinical validity of circulating tumour cells in patients with metastatic breast cancer: A pooled analysis of individual patient data. *Lancet Oncol.* **2014**, *15*, 406–414. [CrossRef]
4. Huang, X.; Gao, P.; Song, Y.; Sun, J.; Chen, X.; Zhao, J.; Xu, H.; Wang, Z. Meta-analysis of the prognostic value of circulating tumor cells detected with the CellSearch System in colorectal cancer. *BMC Cancer* **2015**, *15*, 202. [CrossRef]
5. Scher, H.I.; Heller, G.; Molina, A.; Attard, G.; Danila, D.C.; Jia, X.; Peng, W.; Sandhu, S.K.; Olmos, D.; Riisnaes, R.; et al. Circulating tumor cell biomarker panel as an individual-level surrogate for survival in metastatic castration-resistant prostate cancer. *J. Clin. Oncol.* **2015**, *33*, 1348–1355. [CrossRef]
6. Punnoose, E.A.; Atwal, S.; Liu, W.; Raja, R.; Fine, B.M.; Hughes, B.G.M.; Hicks, R.J.; Hampton, G.M.; Amler, L.C.; Pirzkall, A.; et al. Evaluation of circulating tumor cells and circulating tumor DNA in non-small cell lung cancer: Association with clinical endpoints in a phase II clinical trial of pertuzumab and erlotinib. *Clin. Cancer Res.* **2012**, *18*, 2391–2401. [CrossRef]
7. Rossi, G.; Ignatiadis, M. Promises and Pitfalls of Using Liquid Biopsy for Precision Medicine. *Cancer Res.* **2019**, *79*, 2798–2804. [CrossRef]
8. Bünger, S.; Zimmermann, M.; Habermann, J.K. Diversity of assessing circulating tumor cells (CTCs) emphasizes need for standardization: A CTC Guide to design and report trials. *Cancer Metastasis Rev.* **2015**, *34*, 527–545. [CrossRef]
9. Gallerani, G.; Fabbri, F. Circulating tumor cells in the adenocarcinoma of the esophagus. *Int. J. Mol. Sci.* **2016**, *17*, 1266. [CrossRef]
10. Lianidou, E.S. Circulating tumor cell isolation: A marathon race worth running. *Clin. Chem.* **2014**, *60*, 287–289. [CrossRef]
11. Parkinson, D.R.; Dracopoli, N.; Gumbs Petty, B.; Compton, C.; Cristofanilli, M.; Deisseroth, A.; Hayes, D.F.; Kapke, G.; Kumar, P.; Lee, J.S.; et al. Considerations in the development of circulating tumor cell technology for clinical use. *J. Transl. Med.* **2012**, *10*, 138. [CrossRef]
12. Bidard, F.-C.; Jacot, W.; Dureau, S.; Brain, E.; Bachelot, T.; Bourgeois, H.; Goncalves, A.; Ladoire, S.; Naman, H.; Dalenc, F.; et al. Abstract GS3-07: Clinical utility of circulating tumor cell count as a tool to chose between first line hormone therapy and chemotherapy for ER+ HER2- metastatic breast cancer: Results of the phase III STIC CTC trial. *Cancer Res.* **2019**, *79*. [CrossRef]
13. Dasgupta, A.; Lim, A.R.; Ghajar, C.M. Circulating and disseminated tumor cells: Harbingers or initiators of metastasis? *Mol. Oncol.* **2017**, *11*, 40–61. [CrossRef]
14. Gkountela, S.; Castro-Giner, F.; Szczerba, B.M.; Vetter, M.; Landin, J.; Scherrer, R.; Krol, I.; Scheidmann, M.C.; Beisel, C.; Stirnimann, C.U.; et al. Circulating Tumor Cell Clustering Shapes DNA Methylation to Enable Metastasis Seeding. *Cell* **2019**, *176*, 98–112. [CrossRef]
15. Mitra, A.; Mishra, L.; Li, S. EMT, CTCs and CSCs in tumor relapse and drug-resistance. *Oncotarget* **2015**, *10*, 10697–10711. [CrossRef]
16. Massagué, J.; Obenauf, A.C. Metastatic colonization by circulating tumour cells. *Nature* **2016**, *529*, 298–306. [CrossRef]
17. Joosse, S.A.; Gorges, T.M.; Pantel, K. Biology, detection, and clinical implications of circulating tumor cells. *EMBO Mol. Med.* **2014**, 1–12. [CrossRef]
18. Kang, Y.; Pantel, K. Tumor Cell Dissemination: Emerging Biological Insights from Animal Models and Cancer Patients. *Cancer Cell* **2013**, *23*, 573–581. [CrossRef]
19. Vishnoi, M.; Peddibhotla, S.; Yin, W.; Scamardo, A.T.; George, G.C.; Hong, D.S.; Marchetti, D. The isolation and characterization of CTC subsets related to breast cancer dormancy. *Sci. Rep.* **2015**, *5*, 17533. [CrossRef]
20. Swanton, C. Intratumor heterogeneity: Evolution through space and time. *Cancer Res.* **2012**, *72*, 4875–4882. [CrossRef]
21. Tellez-Gabriel, M.; Heymann, M.-F.; Heymann, D. Circulating Tumor Cells as a Tool for Assessing Tumor Heterogeneity. *Theranostics* **2019**, *9*, 4580–4594. [CrossRef]
22. Reinhardt, F.; Franken, A.; Meier-Stiegen, F.; Driemel, C.; Stoecklein, N.H.; Fischer, J.C.; Niederacher, D.; Ruckhaeberle, E.; Fehm, T.; Neubauer, H. Diagnostic Leukapheresis Enables Reliable Transcriptomic Profiling

of Single Circulating Tumor Cells to Characterize Inter-Cellular Heterogeneity in Terms of Endocrine Resistance. *Cancers* **2019**, *11*, 903. [CrossRef]

23. Maltoni, R.; Fici, P.; Amadori, D.; Gallerani, G.; Cocchi, C.; Zoli, M.; Rocca, A.; Cecconetto, L.; Folli, S.; Scarpi, E.; et al. Circulating tumor cells in early breast cancer: A connection with vascular invasion. *Cancer Lett.* **2015**, *10*, 43–48. [CrossRef]
24. Broncy, L.; Paterlini-br, P. Clinical Impact of Circulating Tumor Cells in Patients with Localized Prostate Cancer. *Cells* **2019**, *8*, 676. [CrossRef]
25. Strati, A.; Nikolaou, M.; Georgoulias, V.; Lianidou, E.S. Prognostic Significance of TWIST1, CD24, CD44, and ALDH1 Transcript Quantification in EpCAM-Positive Circulating Tumor Cells from Early Stage Breast Cancer Patients. *Cells* **2019**, *8*, 652. [CrossRef]
26. Magbanua, M.J.M.; Yau, C.; Wolf, D.M.; Lee, J.S.; Chattopadhyay, A.; Scott, J.H.; Bowlby-Yoder, E.; Hwang, E.S.; Alvarado, M.; Ewing, C.A.; et al. Synchronous detection of circulating tumor cells in blood and disseminated tumor cells in bone marrow predict adverse outcome in early breast cancer. *Clin. Cancer Res.* **2019**. [CrossRef]
27. Ellsworth, R.E.; Blackburn, H.L.; Shriver, C.D.; Soon-Shiong, P.; Ellsworth, D.L. Molecular heterogeneity in breast cancer: State of the science and implications for patient care. *Semin. Cell Dev. Biol.* **2017**, *64*, 65–72. [CrossRef]
28. Wang, Y.; Waters, J.; Leung, M.L.; Unruh, A.; Roh, W.; Shi, X.; Chen, K.; Scheet, P.; Vattathil, S.; Liang, H.; et al. Clonal evolution in breast cancer revealed by single nucleus genome sequencing. *Nature* **2014**, *512*, 155–160. [CrossRef]
29. Yu, M.; Bardia, A.; Wittner, B.S.; Stott, S.L.; Smas, M.E.; Ting, D.T.; Isakoff, S.J.; Ciciliano, J.C.; Wells, M.N.; Shah, A.M.; et al. Circulating breast tumor cells exhibit dynamic changes in epithelial and mesenchymal composition. *Science* **2013**, *339*, 580–584. [CrossRef]
30. Gradilone, A.; Raimondi, C.; Nicolazzo, C.; Petracca, A.; Gandini, O.; Vincenzi, B.; Naso, G.; Aglianò, A.M.; Cortesi, E.; Gazzaniga, P. Circulating tumour cells lacking cytokeratin in breast cancer: The importance of being mesenchymal. *J. Cell. Mol. Med.* **2011**, *15*, 1066–1070. [CrossRef]
31. Gorges, T.M.; Tinhofer, I.; Drosch, M.; Roese, L.; Zollner, T.M.; Krahn, T.; von Ahsen, O. Circulating tumour cells escape from EpCAM-based detection due to epithelial-to-mesenchymal transition. *BMC Cancer* **2012**, *12*, 178. [CrossRef] [PubMed]
32. De Craene, B.; Berx, G. Regulatory networks defining EMT during cancer initiation and progression. *Nat. Rev. Cancer* **2013**, *13*, 97–110. [CrossRef] [PubMed]
33. Barriere, G.; Fici, P.; Gallerani, G.; Fabbri, F.; Rigaud, M. Epithelial Mesenchymal Transition: A double-edged sword. *Clin. Transl. Med.* **2015**, *4*, 4–9. [CrossRef] [PubMed]
34. Lowes, L.E.; Allan, A.L. Circulating tumor cells and implications of the epithelial-to-mesenchymal transition. In *Advances in Clinical Chemistry*; Elsevier: Amsterdam, the Nederlands, 2018; Volume 83, pp. 121–181.
35. Francart, M.E.; Lambert, J.; Vanwynsberghe, A.M.; Thompson, E.W.; Bourcy, M.; Polette, M.; Gilles, C. Epithelial–mesenchymal plasticity and circulating tumor cells: Travel companions to metastases. *Dev. Dyn.* **2018**, *247*, 432–450. [CrossRef] [PubMed]
36. Jolly, M.K.; Mani, S.A.; Levine, H. Hybrid epithelial/mesenchymal phenotype(s): The 'fittest' for metastasis? *Biochim. Biophys. Acta Rev. Cancer* **2018**, *1870*, 151–157. [CrossRef] [PubMed]
37. Pastushenko, I.; Blanpain, C. EMT Transition States during Tumor Progression and Metastasis. *Trends Cell Biol.* **2019**, *29*, 212–226. [CrossRef] [PubMed]
38. Mocellin, S.; Keilholz, U.; Rossi, C.R.; Nitti, D. Circulating tumor cells: The "leukemic phase" of solid cancers. *Trends Mol. Med.* **2006**, *12*, 130–139. [CrossRef]
39. Krawczyk, N.; Meier-Stiegen, F.; Banys, M.; Neubauer, H.; Ruckhaeberle, E.; Fehm, T. Expression of Stem Cell and Epithelial-Mesenchymal Transition Markers in Circulating Tumor Cells of Breast Cancer Patients. *Biomed. Res. Int.* **2014**, *2014*, 1–11. [CrossRef]
40. Werner, S.; Stenzl, A.; Pantel, K.; Todenhöfer, T. Expression of epithelial mesenchymal transition and cancer stem cell markers in circulating tumor cells. In *Advances in Experimental Medicine and Biology*; Springer Nature International Publishing: Berlin, Germany, 2017; Volume 994, pp. 205–228.
41. Williamson, S.C.; Metcalf, R.L.; Trapani, F.; Mohan, S.; Antonello, J.; Abbott, B.; Leong, H.S.; Chester, C.P.E.; Simms, N.; Polanski, R.; et al. Vasculogenic mimicry in small cell lung cancer. *Nat. Commun.* **2016**, *7*, 13322. [CrossRef]

42. Buchbinder, E.I.; Desai, A. CTLA-4 and PD-1 Pathways: Similarities, Differences, and Implications of Their Inhibition. *Am. J. Clin. Oncol.* **2016**, *39*, 98–106. [CrossRef]
43. Leone, K.; Poggiana, C.; Zamarchi, R. The Interplay between Circulating Tumor Cells and the Immune System: From Immune Escape to Cancer Immunotherapy. *Diagnostics* **2018**, *8*, 59. [CrossRef]
44. Mazel, M.; Jacot, W.; Pantel, K.; Bartkowiak, K.; Topart, D.; Cayrefourcq, L.; Rossille, D.; Maudelonde, T.; Fest, T.; Alix-Panabières, C. Frequent expression of PD-L1 on circulating breast cancer cells. *Mol. Oncol.* **2015**, 1–10. [CrossRef]
45. Wang, X.; Sun, Q.; Liu, Q.; Wang, C.; Yao, R.; Wang, Y. CTC immune escape mediated by PD-L1. *Med. Hypotheses* **2016**, *93*, 138–139. [CrossRef]
46. Guibert, N.; Delaunay, M.; Lusque, A.; Boubekeur, N.; Rouquette, I.; Clermont, E.; Mourlanette, J.; Gouin, S.; Dormoy, I.; Favre, G.; et al. PD-L1 expression in circulating tumor cells of advanced non-small cell lung cancer patients treated with nivolumab. *Lung Cancer* **2018**, *120*, 108–112. [CrossRef]
47. Szczerba, B.M.; Castro-Giner, F.; Vetter, M.; Krol, I.; Gkountela, S.; Landin, J.; Scheidmann, M.C.; Donato, C.; Scherrer, R.; Singer, J.; et al. Neutrophils escort circulating tumour cells to enable cell cycle progression. *Nature* **2019**, *566*, 553–557. [CrossRef]
48. Nicolazzo, C.; Raimondi, C.; Mancini, M.; Caponnetto, S.; Gradilone, A.; Gandini, O.; Mastromartino, M.; del Bene, G.; Prete, A.; Longo, F.; et al. Monitoring PD-L1 positive circulating tumor cells in non-small cell lung cancer patients treated with the PD-1 inhibitor Nivolumab. *Sci. Rep.* **2016**, *6*, 31726. [CrossRef]
49. Yue, C.; Jiang, Y.; Li, P.; Wang, Y.; Xue, J.; Li, N.; Li, D.; Wang, R.; Dang, Y.; Hu, Z.; et al. Dynamic change of PD-L1 expression on circulating tumor cells in advanced solid tumor patients undergoing PD-1 blockade therapy. *Oncoimmunology* **2018**, *7*, e1438111. [CrossRef]
50. Kallergi, G.; Vetsika, E.-K.; Aggouraki, D.; Lagoudaki, E.; Koutsopoulos, A.; Koinis, F.; Katsarlinos, P.; Trypaki, M.; Messaritakis, I.; Stournaras, C.; et al. Evaluation of PD-L1/PD-1 on circulating tumor cells in patients with advanced non-small cell lung cancer. *Ther. Adv. Med. Oncol.* **2018**, *10*, 1758834017750121. [CrossRef]
51. Chao, M.P.; Tang, C.; Pachynski, R.K.; Chin, R.; Majeti, R.; Weissman, I.L. Extranodal dissemination of non-Hodgkin lymphoma requires CD47 and is inhibited by anti-CD47 antibody therapy. *Blood* **2011**, *118*, 4890–4901. [CrossRef]
52. Baccelli, I.; Schneeweiss, A.; Riethdorf, S.; Stenzinger, A.; Schillert, A.; Vogel, V.; Klein, C.; Saini, M.; Bäurele, T.; Wallwiener, M.; et al. Identification of a population of blood circulating tumor cells from breast cancer patients that initiates metastasis in a xenograft assay. *Nat. Biotechnol.* **2013**, *31*, 539–544. [CrossRef]
53. Steinert, G.; Schölch, S.; Niemietz, T.; Iwata, N.; García, S.A.; Behrens, B.; Voigt, A.; Kloor, M.; Benner, A.; Bork, U.; et al. Immune Escape and Survival Mechanisms in Circulating Tumor Cells of Colorectal Cancer. *Cancer Res.* **2014**, *74*, 1694–1704. [CrossRef]
54. Mego, M.; Gao, H.; Cohen, E.; Anfossi, S.; Giordano, A.; Sanda, T.; Fouad, T.; De Giorgi, U.; Giuliano, M.; Woodward, W.; et al. Circulating Tumor Cells (CTC) Are Associated with Defects in Adaptive Immunity in Patients with Inflammatory Breast Cancer. *J. Cancer* **2016**, *7*, 1095–1104. [CrossRef]
55. Hallermalm, K.; De Geer, A.; Kiessling, R.; Levitsky, V.; Levitskaya, J. Autocrine secretion of Fas ligand shields tumor cells from Fas-mediated killing by cytotoxic lymphocytes. *Cancer Res.* **2004**, *64*, 6775–6782. [CrossRef]
56. Labelle, M.; Begum, S.; Hynes, R.O. Direct signaling between platelets and cancer cells induces an epithelial-mesenchymal-like transition and promotes metastasis. *Cancer Cell* **2011**, *20*, 576–590. [CrossRef]
57. Weber, M.R.; Zuka, M.; Lorger, M.; Tschan, M.; Torbett, B.E.; Zijlstra, A.; Quigley, J.P.; Staflin, K.; Eliceiri, B.P.; Krueger, J.S.; et al. Activated tumor cell integrin $\alpha v\beta 3$ cooperates with platelets to promote extravasation and metastasis from the blood stream. *Thromb. Res.* **2016**, *140*, 27–36. [CrossRef]
58. Cravioto-Villanueva, A.; Luna-Perez, P.; Gutierrez-de la Barrera, M.; Martinez-Gómez, H.; Maffuz, A.; Rojas-Garcia, P.; Perez-Alvarez, C.; Rodriguez-Ramirez, S.; Rodriguez-Antezana, E.; Ramirez-Ramirez, L. Thrombocytosis as a Predictor of Distant Recurrence in Patients with Rectal Cancer. *Arch. Med. Res.* **2012**, *43*, 305–311. [CrossRef]
59. Leblanc, R.; Peyruchaud, O. Metastasis: New functional implications of platelets and megakaryocytes. *Blood* **2016**, *128*, 24–31. [CrossRef]
60. Placke, T.; Orgel, M.; Schaller, M.; Jung, G.; Rammensee, H.-G.; Kopp, H.-G.; Salih, H.R. Platelet-Derived MHC Class I Confers a Pseudonormal Phenotype to Cancer Cells That Subverts the Antitumor Reactivity of Natural Killer Immune Cells. *Cancer Res.* **2012**, *72*, 440–448. [CrossRef]
61. Adams, D.L.; Martin, S.S.; Alpaugh, R.K.; Charpentier, M.; Tsai, S.; Bergan, R.C.; Ogden, I.M.; Catalona, W.;

Chumsri, S.; Tang, C.-M.; et al. Circulating giant macrophages as a potential biomarker of solid tumors. *Proc. Natl. Acad. Sci. USA* **2014**, *111*, 3514–3519. [CrossRef]

62. Adams, D.L.; Adams, D.K.; Alpaugh, R.K.; Cristofanilli, M.; Martin, S.S.; Chumsri, S.; Tang, C.-M.; Marks, J.R. Circulating Cancer-Associated Macrophage-Like Cells Differentiate Malignant Breast Cancer and Benign Breast Conditions. *Cancer Epidemiol. Biomark. Prev.* **2016**, *25*, 1037–1042. [CrossRef]
63. Adams, D.L.; Adams, D.K.; He, J.; Kalhor, N.; Zhang, M.; Xu, T.; Gao, H.; Reuben, J.M.; Qiao, Y.; Komaki, R.; et al. Sequential Tracking of PD-L1 Expression and RAD50 Induction in Circulating Tumor and Stromal Cells of Lung Cancer Patients Undergoing Radiotherapy. *Clin. Cancer Res.* **2017**, *23*, 5948–5958. [CrossRef]
64. Mu, Z.; Wang, C.; Ye, Z.; Rossi, G.; Sun, C.; Li, L.; Zhu, Z.; Yang, H.; Cristofanilli, M. Prognostic values of cancer associated macrophage-like cells (CAML) enumeration in metastatic breast cancer. *Breast Cancer Res. Treat.* **2017**, *165*, 733–741. [CrossRef]
65. Lian, S.; Xie, R.; Ye, Y.; Lu, Y.; Cheng, Y.; Xie, X.; Li, S.; Jia, L. Dual blockage of both PD-L1 and CD47 enhances immunotherapy against circulating tumor cells. *Sci. Rep.* **2019**, *9*, 4532. [CrossRef]
66. Lou, X.-L.; Deng, J.; Deng, H.; Ting, Y.; Zhou, L.; Liu, Y.-H.; Hu, J.-P.; Huang, X.-F.; Qi, X.-Q. Aspirin inhibit platelet-induced epithelial-to-mesenchymal transition of circulating tumor cells (Review). *Biomed. Rep.* **2014**, *2*, 331–334. [CrossRef]
67. Van Schaijik, B.; Wickremesekera, A.C.; Mantamadiotis, T.; Kaye, A.H.; Tan, S.T.; Stylli, S.S.; Itinteang, T. Circulating tumor stem cells and glioblastoma: A review. *J. Clin. Neurosci.* **2019**, *61*, 5–9. [CrossRef]
68. Agnoletto, C.; Corrà, F.; Minotti, L.; Baldassari, F.; Crudele, F.; Cook, W.J.J.; Di Leva, G.; D'Adamo, A.P.; Gasparini, P.; Volinia, S. Heterogeneity in circulating tumor cells: The relevance of the stem-cell subset. *Cancers* **2019**, *11*, 483. [CrossRef]
69. Steinbichler, T.B.; Dudás, J.; Skvortsov, S.; Ganswindt, U.; Riechelmann, H.; Skvortsova, I.-I. Therapy resistance mediated by cancer stem cells. *Semin. Cancer Biol.* **2018**, *53*, 156–167. [CrossRef]
70. Agliano, A.; Calvo, A.; Box, C. The challenge of targeting cancer stem cells to halt metastasis. *Semin. Cancer Biol.* **2017**, *44*, 25–42. [CrossRef]
71. Paget, S. The distribution of secondary growths in cancer of the breast 1889. *Cancer Metastasis Rev.* **1989**, *8*, 98–101.
72. Fidler, I.J. The pathogenesis of cancer metastasis: The "seed and soil" hypothesis revisited. *Nat. Rev. Cancer* **2003**, *3*, 453–458. [CrossRef]
73. Koren, E.; Fuchs, Y. The bad seed: Cancer stem cells in tumor development and resistance. *Drug Resist. Updat.* **2016**, *28*, 1–12. [CrossRef]
74. Hu, Y.; Yu, X.; Xu, G.; Liu, S. Metastasis: An early event in cancer progression. *J. Cancer Res. Clin. Oncol.* **2017**, *143*, 745–757. [CrossRef]
75. Kuşoğlu, A.; Biray Avcı, Ç. Cancer stem cells: A brief review of the current status. *Gene* **2019**, *681*, 80–85. [CrossRef]
76. Luo, Y.T.; Cheng, J.; Feng, X.; He, S.J.; Wang, Y.W.; Huang, Q. The viable circulating tumor cells with cancer stem cells feature, where is the way out? *J. Exp. Clin. Cancer Res.* **2018**, *37*, 38. [CrossRef]
77. Rossi, E.; Rugge, M.; Facchinetti, A.; Pizzi, M.; Nardo, G.; Barbieri, V.; Manicone, M.; De Faveri, S.; Chiara Scaini, M.; Basso, U.; et al. Retaining the long-survive capacity of Circulating Tumor Cells (CTCs) followed by xeno-transplantation: Not only from metastatic cancer of the breast but also of prostate cancer patients. *Oncoscience* **2014**, *1*, 49–56. [CrossRef]
78. Carvalho, F.L.F.; Simons, B.W.; Antonarakis, E.S.; Rasheed, Z.; Douglas, N.; Villegas, D.; Matsui, W.; Berman, D.M. Tumorigenic potential of circulating prostate tumor cells. *Oncotarget* **2013**, *4*, 413–421. [CrossRef]
79. Aktas, B.; Tewes, M.; Fehm, T.; Hauch, S.; Kimmig, R.; Kasimir-Bauer, S. Stem cell and epithelial-mesenchymal transition markers are frequently overexpressed in circulating tumor cells of metastatic breast cancer patients. *Breast Cancer Res.* **2009**, *11*, R46. [CrossRef]
80. Aceto, N.; Bardia, A.; Miyamoto, D.T.; Donaldson, M.C.; Wittner, B.S.; Spencer, J.A.; Yu, M.; Pely, A.; Engstrom, A.; Zhu, H.; et al. Circulating tumor cell clusters are oligoclonal precursors of breast cancer metastasis. *Cell* **2014**, *158*, 1110–1122. [CrossRef]
81. Iriondo, O.; Yu, M. Unexpected Friendship: Neutrophils Help Tumor Cells En Route to Metastasis. *Dev. Cell* **2019**, *49*, 308–310. [CrossRef]
82. Yu, M. Metastasis Stemming from Circulating Tumor Cell Clusters. *Trends Cell Biol.* **2019**, *29*, 275–276. [CrossRef]

83. Wild, C.P. Complementing the genome with an exposome: The outstanding challenge of environmental exposure measurement in molecular epidemiology. *Cancer Epidemiol. Biomark. Prev.* **2005**, *14*, 1847–1850. [CrossRef]
84. Bossuyt, P.M.M.; Reitsma, J.B.; Linnet, K.; Moons, K.G.M. Beyond Diagnostic Accuracy: The Clinical Utility of Diagnostic Tests. *Clin. Chem.* **2012**, *58*, 1636–1643. [CrossRef]
85. Cristofanilli, M.; Pierga, J.Y.; Reuben, J.; Rademaker, A.; Davis, A.A.; Peeters, D.J.; Fehm, T.; Nolé, F.; Gisbert-Criado, R.; Mavroudis, D.; et al. The clinical use of circulating tumor cells (CTCs) enumeration for staging of metastatic breast cancer (MBC): International expert consensus paper. *Crit. Rev. Oncol. Hematol.* **2019**, *134*, 39–45. [CrossRef]
86. Budd, G.T.; Cristofanilli, M.; Ellis, M.J.; Stopeck, A.; Borden, E.; Miller, M.C.; Matera, J.; Repollet, M.; Doyle, G.V.; Terstappen, L.W.M.M.; et al. Circulating Tumor Cells versus Imaging–Predicting Overall Survival in Metastatic Breast Cancer. *Clin. Cancer Res.* **2006**, *12*, 6403–6409. [CrossRef]
87. Alunni-Fabbroni, M.; Müller, V.; Fehm, T.; Janni, W.; Rack, B. Monitoring in Metastatic Breast Cancer: Is Imaging Outdated in the Era of Circulating Tumor Cells? *Breast Care* **2014**, *9*, 16–21. [CrossRef]
88. Bidard, F.-C.; Pierga, J.-Y. Clinical Utility of Circulating Tumor Cells in Metastatic Breast Cancer. *J. Clin. Oncol.* **2015**, *33*, 1622. [CrossRef]
89. Helissey, C.; Berger, F.; Cottu, P.; Diéras, V.; Mignot, L.; Servois, V.; Bouleuc, C.; Asselain, B.; Pelissier, S.; Vaucher, I.; et al. Circulating tumor cell thresholds and survival scores in advanced metastatic breast cancer: The observational step of the CirCe01 phase III trial. *Cancer Lett.* **2015**, *360*, 213–218. [CrossRef]
90. Van Poznak, C.; Somerfield, M.R.; Bast, R.C.; Cristofanilli, M.; Goetz, M.P.; Gonzalez-Angulo, A.M.; Hicks, D.G.; Hill, E.G.; Liu, M.C.; Lucas, W.; et al. Use of Biomarkers to Guide Decisions on Systemic Therapy for Women With Metastatic Breast Cancer: American Society of Clinical Oncology Clinical Practice Guideline. *J. Clin. Oncol.* **2015**, *33*, 2695–2704. [CrossRef]
91. Bidard, F.-C.; Proudhon, C.; Pierga, J.-Y. Circulating tumor cells in breast cancer. *Mol. Oncol.* **2016**, *10*, 418–430. [CrossRef]
92. Heller, G.; McCormack, R.; Kheoh, T.; Molina, A.; Smith, M.R.; Dreicer, R.; Saad, F.; de Wit, R.; Aftab, D.T.; Hirmand, M.; et al. Circulating Tumor Cell Number as a Response Measure of Prolonged Survival for Metastatic Castration-Resistant Prostate Cancer: A Comparison With Prostate-Specific Antigen Across Five Randomized Phase III Clinical Trials. *J. Clin. Oncol.* **2018**, *36*, 572–580. [CrossRef]
93. Scher, H.I.; Graf, R.P.; Schreiber, N.A.; McLaughlin, B.; Jendrisak, A.; Wang, Y.; Lee, J.; Greene, S.; Krupa, R.; Lu, D.; et al. Phenotypic Heterogeneity of Circulating Tumor Cells Informs Clinical Decisions between AR Signaling Inhibitors and Taxanes in Metastatic Prostate Cancer. *Cancer Res.* **2017**, *77*, 5687–5698. [CrossRef]
94. Sciarra, A.; Gentilucci, A.; Silvestri, I.; Salciccia, S.; Cattarino, S.; Scarpa, S.; Gatto, A.; Frantellizzi, V.; Von Heland, M.; Ricciuti, G.P.; et al. Androgen receptor variant 7 (AR-V7) in sequencing therapeutic agents for castratrion resistant prostate cancer: A critical review. *Medicine* **2019**, *98*, e15608. [CrossRef]
95. IJzerman, M.J.; Berghuis, A.M.S.; de Bono, J.S.; Terstappen, L.W.M.M. Health economic impact of liquid biopsies in cancer management. *Expert Rev. Pharmacoecon. Outcomes Res.* **2018**, *18*, 593–599. [CrossRef]
96. Sholl, L.M.; Aisner, D.L.; Allen, T.C.; Beasley, M.B.; Cagle, P.T.; Capelozzi, V.L.; Dacic, S.; Hariri, L.P.; Kerr, K.M.; Lantuejoul, S.; et al. Liquid Biopsy in Lung Cancer: A Perspective From Members of the Pulmonary Pathology Society. *Arch. Pathol. Lab. Med.* **2016**, *140*, 825–829. [CrossRef]
97. Vishnoi, M.; Haowen Liu, N.; Yin, W.; Boral, D.; Scamardo, A.; Hong, D.; Marchetti, D. The identification of a TNBC liver metastasis gene signature by sequential CTC-xenograft modelling. *Mol. Oncol.* **2019**. [CrossRef]
98. Gorges, T.M.; Penkalla, N.; Schalk, T.; Joosse, S.A.; Riethdorf, S.; Tucholski, J.; Lücke, K.; Wikman, H.; Jackson, S.; Brychta, N.; et al. Enumeration and Molecular Characterization of Tumor Cells in Lung Cancer Patients Using a Novel *In Vivo* Device for Capturing Circulating Tumor Cells. *Clin. Cancer Res.* **2016**, *22*, 2197–2206. [CrossRef]
99. Kuske, A.; Gorges, T.M.; Tennstedt, P.; Tiebel, A.-K.; Pompe, R.; Preißer, F.; Prues, S.; Mazel, M.; Markou, A.; Lianidou, E.; et al. Improved detection of circulating tumor cells in non-metastatic high-risk prostate cancer patients. *Sci. Rep.* **2016**, *6*, 39736. [CrossRef]
100. Mandair, D.; Vesely, C.; Ensell, L.; Lowe, H.; Spanswick, V.; Hartley, J.A.; Caplin, M.E.; Meyer, T. A comparison of CellCollector with CellSearch in patients with neuroendocrine tumours. *Endocr. Relat. Cancer* **2016**, *23*, L29–L32. [CrossRef]
101. Dizdar, L.; Fluegen, G.; van Dalum, G.; Honisch, E.; Neves, R.P.; Niederacher, D.; Neubauer, H.; Fehm, T.; Rehders, A.;

Krieg, A.; et al. Detection of circulating tumor cells in colorectal cancer patients using the GILUPI CellCollector: Results from a prospective, single-center study. *Mol. Oncol.* **2019**, *13*, 1548–1558. [CrossRef]

102. Gallerani, G.; Cocchi, C.; Bocchini, M.; Piccinini, F.; Fabbri, F. Characterization of tumor cells using a medical wire for capturing circulating tumor cells: A 3D approach based on immunofluorescence and DNA FISH. *J. Vis. Exp.* **2017**, *130*, e56936. [CrossRef]
103. Galanzha, E.I.; Menyaev, Y.A.; Yadem, A.C.; Sarimollaoglu, M.; Juratli, M.A.; Nedosekin, D.A.; Foster, S.R.; Jamshidi-Parsian, A.; Siegel, E.R.; Makhoul, I.; et al. In vivo liquid biopsy using Cytophone platform for photoacoustic detection of circulating tumor cells in patients with melanoma. *Sci. Transl. Med.* **2019**, *11*, eaat5857. [CrossRef]
104. Andree, K.C.; Mentink, A.; Zeune, L.L.; Terstappen, L.W.M.M.; Stoecklein, N.H.; Neves, R.P.; Driemel, C.; Lampignano, R.; Yang, L.; Neubauer, H.; et al. Toward a real liquid biopsy in metastatic breast and prostate cancer: Diagnostic LeukApheresis increases CTC yields in a European prospective multicenter study (CTCTrap). *Int. J. Cancer* **2018**, *143*, 2584–2591. [CrossRef]
105. Fischer, J.C.; Niederacher, D.; Topp, S.A.; Honisch, E.; Schumacher, S.; Schmitz, N.; Zacarias Föhrding, L.; Vay, C.; Hoffmann, I.; Kasprowicz, N.S.; et al. Diagnostic leukapheresis enables reliable detection of circulating tumor cells of nonmetastatic cancer patients. *Proc. Natl. Acad. Sci. USA* **2013**, *110*, 16580–16585. [CrossRef]
106. Franken, A.; Driemel, C.; Behrens, B.; Meier-Stiegen, F.; Endris, V.; Stenzinger, A.; Niederacher, D.; Fischer, J.C.; Stoecklein, N.H.; Ruckhaeberle, E.; et al. Label-Free Enrichment and Molecular Characterization of Viable Circulating Tumor Cells from Diagnostic Leukapheresis Products. *Clin. Chem.* **2019**, *65*, 549–558. [CrossRef]
107. Lambros, M.B.; Seed, G.; Sumanasuriya, S.; Gil, V.; Crespo, M.; Fontes, M.; Chandler, R.; Mehra, N.; Fowler, G.; Ebbs, B.; et al. Single-Cell Analyses of Prostate Cancer Liquid Biopsies Acquired by Apheresis. *Clin. Cancer Res.* **2018**, *24*, 5635–5644. [CrossRef]
108. Cayrefourcq, L.; Mazard, T.; Joosse, S.; Solassol, J.; Ramos, J.; Assenat, E.; Schumacher, U.; Costes, V.; Maudelonde, T.; Pantel, K.; et al. Establishment and characterization of a cell line from human circulating colon cancer cells. *Cancer Res.* **2015**, *75*, 892–901. [CrossRef]
109. Coumans, F.A.W.; Ligthart, S.T.; Uhr, J.W.; Terstappen, L.W.M.M. Challenges in the enumeration and phenotyping of CTC. *Clin. Cancer Res.* **2012**, *18*, 5711–5718. [CrossRef]
110. Sharma, R.; Sharma, S. *Physiology, Blood Volume*; StatPearls Publishing LLC: Tampa/St. Petersburg, FL, USA, 2019.
111. Coumans, F.A.; Siesling, S.; Terstappen, L.W. Detection of cancer before distant metastasis. *BMC Cancer* **2013**, *13*, 283. [CrossRef]
112. Stoecklein, N.H.; Fischer, J.C.; Niederacher, D.; Terstappen, L.W.M.M. Challenges for CTC based liquid biopsies: Low CTC frequency and diagnostic leukapheresis as a potential solution. *Expert Rev. Mol. Diagn.* **2016**, *16*, 147–164. [CrossRef]
113. Fehm, T.N.; Meier-Stiegen, F.; Driemel, C.; Jäger, B.; Reinhardt, F.; Naskou, J.; Franken, A.; Neubauer, H.; Neves, R.P.L.; van Dalum, G.; et al. Diagnostic leukapheresis for CTC analysis in breast cancer patients: CTC frequency, clinical experiences and recommendations for standardized reporting. *Cytom. Part. A* **2018**, *93*, 1213–1219. [CrossRef]
114. Steeg, P.S. Targeting metastasis. *Nat. Rev. Cancer* **2016**, *16*, 201–218. [CrossRef] [PubMed]
115. Ghajar, C.M. Metastasis prevention by targeting the. *Nat. Rev. Cancer* **2015**, *15*, 238–247. [CrossRef] [PubMed]

8

Microfluidic Device for On-Chip Immunophenotyping and Cytogenetic Analysis of Rare Biological Cells

Kumuditha M. Weerakoon-Ratnayake [1,2], Swarnagowri Vaidyanathan [2,3], Nicholas Larkey [2,4], Kavya Dathathreya [1,2], Mengjia Hu [2,4], Jilsha Jose [2], Shalee Mog [1,2], Keith August [5], Andrew K. Godwin [4], Mateusz L. Hupert [6,*], Malgorzata A. Witek [1,2,*] and Steven A. Soper [1,2,4,6,7,*]

1 Department of Chemistry, The University of Kansas, Lawrence, KS 66047, USA; kratnayake@ku.edu (K.M.W.-R.); kavya87.atreya@gmail.com (K.D.); shaleemog@ku.edu (S.M.)
2 Center of BioModular Multiscale Systems for Precision Medicine, Lawrence, KS 66045, USA; swarna_vaidy@ku.edu (S.V.); larkeyn@ku.edu (N.L.); m367h922@kumc.edu (M.H.); jiljose97@gmail.com (J.J.)
3 Bioengineering, The University of Kansas, Lawrence, KS 66045, USA
4 Department of Pathology & Laboratory Medicine, University of Kansas Medical Center, Kansas City, KS 66160, USA; agodwin@kumc.edu
5 Children's Mercy Hospital, Kansas City, MO 64108, USA; kjaugust@cmh.edu
6 Biofluidica Inc., BioFluidica Research Laboratory, Lawrence, KS 66047, USA
7 Department of Mechanical Engineering, The University of Kansas, Lawrence, KS 66045, USA
* Correspondence: matt@biofluidica.com (M.L.H.); mwitek@ku.edu (M.A.W.); ssoper@ku.edu (S.A.S.).

Abstract: The role of circulating plasma cells (CPCs) and circulating leukemic cells (CLCs) as biomarkers for several blood cancers, such as multiple myeloma and leukemia, respectively, have recently been reported. These markers can be attractive due to the minimally invasive nature of their acquisition through a blood draw (i.e., liquid biopsy), negating the need for painful bone marrow biopsies. CPCs or CLCs can be used for cellular/molecular analyses as well, such as immunophenotyping or fluorescence in situ hybridization (FISH). FISH, which is typically carried out on slides involving complex workflows, becomes problematic when operating on CLCs or CPCs due to their relatively modest numbers. Here, we present a microfluidic device for characterizing CPCs and CLCs using immunofluorescence or FISH that have been enriched from peripheral blood using a different microfluidic device. The microfluidic possessed an array of cross-channels (2–4 µm in depth and width) that interconnected a series of input and output fluidic channels. Placing a cover plate over the device formed microtraps, the size of which was defined by the width and depth of the cross-channels. This microfluidic chip allowed for automation of immunofluorescence and FISH, requiring the use of small volumes of reagents, such as antibodies and probes, as compared to slide-based immunophenotyping and FISH. In addition, the device could secure FISH results in <4 h compared to 2–3 days for conventional FISH.

Keywords: microfluidics; immunophenotyping; fish; liquid biopsy; circulating leukemia cells; circulating plasma cells

---

1. Introduction

Molecular diagnostics are growing immensely due in part to the Precision Medicine Initiative (www.whitehouse.gov/precision-medicine), which seeks to match appropriate therapies to the molecular characteristics of a patient's disease. Unfortunately, the majority of molecular diagnostic tests are expensive, involve slow turnaround times from centralized laboratories, and require highly specialized equipment with seasoned technicians to carry out the assay. In addition, acquisition of the molecular biomarkers requires a solid tissue or bone marrow biopsy, which can be an invasive procedure, especially for anatomically inaccessible organs. For example, bone marrow biopsies are typically required to monitor leukemia or multiple myeloma status, which not only complicates sample acquisition but limits the frequency of testing.

Liquid biopsies are generating a significant amount of interest in the medical community owing to the minimally invasive nature of acquiring biomarkers and the fact that they can enable precision decisions on managing a variety of diseases [1,2]. Liquid biopsy markers include but are not limited to circulating tumor cells (CTCs), cell-free DNA (cfDNA), and extracellular vesicles (EVs). As an example of the utility of liquid biopsy analysis for some blood cancers, we have shown that circulating plasma cells (CPCs) can be used to stage patients diagnosed with multiple myeloma [3]. Circulating leukemia cells (CLCs) can be used to determine relapse from minimum residual disease (MRD) in patients with acute myeloid leukemia (AML) or acute lymphoblastic leukemia (ALL) [4], all of which typically require a highly painful and invasive bone marrow biopsy. The challenge with using CLCs or CPCs is that their abundance in blood is lower than what is found in the bone marrow, requiring highly sensitive assays to analyze their molecular content following enrichment to remove interfering white and red blood cells.

A cytogenetic method called fluorescence in-situ hybridization (FISH), which was discovered in the early 1980s [5], can be used to detect chromosomal modifications [6–8]. FISH identifies abnormalities in chromosomes using fluorescent DNA probes that hybridize to a specific gene region. When properly hybridized to its complementary sequence, FISH allows for the visualization of chromosomal aberrations, such as deletions, fusions, balanced translocations, etc. [9]. For example, cytogenetic abnormalities are found in most cases of multiple myeloma, in which IGH translocations initiate events associated with tumorigenesis and disease progression [10]. The progression of multiple myeloma was discovered in clinical studies where investigators found frequent chromosomal aberrations, such as 13q14 deletions (del13q14), 1q21 gains (amp1q21), and monosomy 13 and 17p13 deletions (del17p13) [10,11]. B-type acute lymphoblastic leukemia (B-ALL), which is a common childhood malignancy, is prognosed by BCR/ABL [t(9;22)], MLL [t(4;1)], and TEL/AML1 [t12;21] [12] aberrations.

While FISH assays are widely used in clinical settings, the workflow requires labor intensive and time-consuming protocols. Conventional slide-based methods for FISH utilize workflows that necessitate the need for highly trained professionals and relatively high volumes of costly FISH probes; the full assay may require 2-3 days of processing. Therefore, it is critical to develop alternative methods and platforms to undertake FISH that address the aforementioned limitations [9,13].

Microfluidic FISH assays can address many of the limitations associated with slide-based FISH, such as providing process automation and reducing reagent requirements and processing time [9]. Even though there are microfluidic assays that have been developed over the past years for genetic techniques, such as PCR and DNA microarrays, less effort has been devoted toward realizing the implementation of FISH assays using microfluidic devices [14–16]. A summary of microfluidic devices for performing FISH are summarized in Table S1 along with their operational characteristics.

The first microfluidic for FISH was developed by Sieben and coworkers in 2007 [17]. The study demonstrated the ability to detect a FISH signal with 10-fold higher throughput and 1/10th reagent consumption compared to slide-based FISH. Modification of the microfluidic substrate to achieve cell adherence is a common protocol to allow for processing the cells and imaging. Therefore, in this report, TiO_2-modified glass slides were used for cell adherence to identify FISH signals, with the PDMS fluidic network used to introduce FISH reagents to the target cells. The PDMS was removed from the TiO_2-modified glass slide, and imaging was undertaken using a 100× objective [18]. Liu et al. [19] performed FISH on centromere-sized cell arrays modified with 3-aminopropyltriethoxysilane (APTES) or polyethylene glycol (PEG)-coated glass slides with overnight hybridization. Wang et al. [20] introduced an APTES-coated glass slide with a PDMS microfluidic for stretching chromosomal DNA from a single cell to perform FISH. In a recent study, a microfluidic consisting of a Pyrex-Si stack was generated and used for FISH to provide breast cancer prognosis [21–24]. This report reduced FISH reagent consumption by 70% and hybridization time to 2 h.

Mayer et al. [24] used a microfluidic device to investigate HER2 amplification and immunohistochemistry of breast cancer patients. In a further extension of this study, the authors reported short incubation microfluidic-assisted FISH [23]. The researchers were able to reduce FISH hybridization time to 15 min for cell lines and 35 min for human tissue slides. Microfluidic channels etched into a glass slide, called FISHing lines, allowed the processing of 10 samples on a glass slide with 0.2 μL of FISH probe. MRD screening using the BCR/ABL fusion gene for chronic myeloid leukemia was performed by Mughal et al. [25].

Researchers have also reported using plastics to make the microfluidic for FISH to reduce the fabrication cost. Kwasny et al. [26] described the use of cyclic olefin copolymer (COC) devices sealed with glass or COC cover plates that were surface modified to allow for stretching chromosomes. Perez-Torella and coworkers reported a COC chamber, which was capable of delivering FISH reagents to cells [27]. The chamber was modified with 2-hydroxyethyl cellulose (HEC) to allow for cell adherence to the chamber walls. Micro-FISH devices fabricated with CO_2 laser ablation in a plastic were reported, which resulted in a 20-fold reduction in the sample volume [28].

Our group has developed a highly sensitive microfluidic for rare cell isolation. The isolated cells were immunostained directly within the selection chip, released into a 96-well plate, and visualized/enumerated using fluorescence microscopy. In addition, following cell selection and release from the isolation chip, the cells could be subjected to FISH using conventional slide-based approaches [3,4]. Unfortunately, the workflow required extensive manual handling of cells.

To address the workflow challenge, we developed a microfluidic device named "microtrap" for both immunophenotyping and FISH analysis of biological cells enriched from clinical samples, which we report herein. Cells isolated on the isolation chip could be transferred to the microtrap device for immunophenotyping and/or FISH in an automated fashion with reduced processing time.

The microtrap device consisted of an array of 80,000 containment pores generated by cross-channels (2–4 μm in width and depth) connecting a network of interleaving fluidic channels. Unique to this device is that it was fabricated with containment pores or microtraps that arrayed single cells in a 2-D format and within a common imaging plane. The device does not rely on modification of the surface to retain cells; rather, it physically entraps the cells. Additionally, imaging of the cells in the microtrap device on different z-planes demonstrated better spatial resolution of the hybridization probes to designate the detection of a genetic abnormality. This feature cannot be achieved in Flow-FISH analysis [29,30]. The microtrap device provided reduced processing time (18 → 3 h) and lower amounts of FISH probes compared to slide-based processing. The microtrap device integration to a cell selection microfluidic allowed for automated cell processing as well as minimized the amount of operator handling of the enriched cells, simplifying the workflow and reducing cell loss.

We demonstrate the utility of the device for the processing of non-adherent cells, such as CPCs and CLCs, with the device being able to perform immunophenotyping and FISH from liquid biopsy markers. In addition, we demonstrate that we can affinity-enrich B-ALL CLCs from a pediatric patient blood sample using a cell enrichment microfluidic decorated with anti-CD19 antibodies and perform FISH and immunophenotyping on the enriched cells using the device reported herein.

2. Materials and Methods

2.1. Design and Fabrication of the Microtrap Device

The design of the microtrap device is shown in Figure 1. There are two basic renditions, with each differing in terms of the number of microtraps the device possessed. Design 1—single bed with 7200 microtraps; and Design 2—8-bed device with 80,000 microtraps. The microtrap size could vary depending on the width and depth of the cross-channels and was as small as 4 μm (width) × 2 μm (depth) to accommodate the containment of smaller cells, such as CPCs and CLCs (6–16 μm in diameter), compared to the larger-sized CTCs [31]. The 8-bed device (shown in Figure 1A,B) consisted of a significantly increased number of microtraps as compared to our previous design due to the high numbers of cells that are enriched for leukemia and multiple myeloma diseases compared to CTCs' affinity enriched from epithelial cancers [3,4,31].

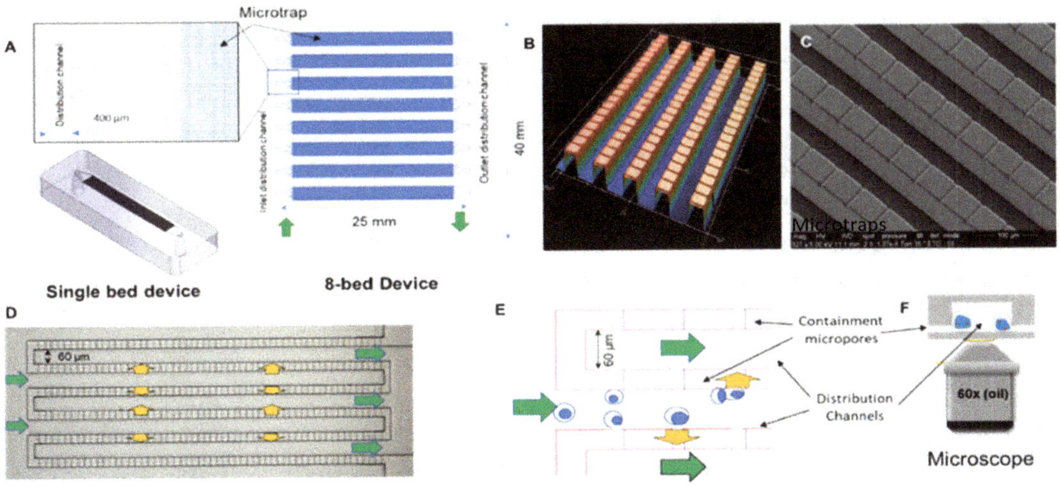

Figure 1. Microfluidic device for performing automated immunophenotyping and FISH. (**A**) Design of the microfluidic network composed of a single bed with 7200 microtraps and the 8-bed device containing 10,000 microtraps in each bed for a total of 80,000 traps per device. Microtrap size: 4 × 2 × 50 μm (w × d × l). (**B**) Profilometer scan of the microtrap chip replicated in PDMS from a 3-level SU-8 relief and a Si master showing microchannel depth varying between input/output distribution channels, interleaving channels, and cross-channels. (**C**) Cross-channels and the deeper interleaving channels are shown in the SEM image. (**D**) Optical microscope image of a lithographically patterned 2-level SU-8 relief for preparing a single bed microtrap device. The arrows show the fluid path. (**E**) Schematic showing operation of the microtrap chip. Cells in solution (green arrows) are contained at the entrances of the microtraps, letting the fluid pass (yellow arrows) into the outlet channels of the interleaving network. (**F**) Schematic showing the 3-dimensionality of cells captured in the microtrap chip and imaging using a high magnification (60× or 100×) objective through a thin cover plate.

The lithography and fabrication steps of both devices are discussed in detail in the Supplementary Materials. The architecture of the single bed device (2-step lithography to create two levels) consisted of two independent networks of interleaving channels that were 60 × 40 μm^2 (W × D) and interconnected using an array of cross-channels orthogonally placed in between the interleaving channels (see Figure S1). The volume of this device was ~1 μL. For the 8-bed device, 3-level lithography was used to allow generation of a deep layer for the distribution channels, interleaving channels, and cross-channels (see Figure S2). Profilometry scans of the 8-bed chip showed the 3 levels for the SU-8 relief (Figure 1B). Figure 1C shows an SEM of the lithographically patterned 2-level SU-8 relief of a single bed device, with a magnified section showing the replica PDMS device with interleaving channels (30 μm) and shallow cross-channels (2–4 μm). In the 8-bed device, the 3 rd lithography layer was used to fabricate a deeper distribution channel network to distribute fluid to all 8 beds equally [32]. The distribution channels of the 8-bed device were 400 x 150 μm (W × D). The internal volume of this device, including all fluidic channels, was ~10 μL.

The fluidic operation of the device is shown in Figure 1D,E, where the flow from the interleaving channel input flows across (90° angle) the interconnecting cross-channels and then to the output channels of the interleaving network. The cross-channel, when a cover plate was bonded to the chip substrate, generated a microtrap structure where cells were contained but still allowed fluid to flow through it.

The optimal device design was chosen from testing 14 different designs for which we determined the containment efficiency of CPCs and CLCs. The 14 devices with different pore widths (4, 6, and 8 μm) and different depths (2 and 4 μm) were evaluated in terms of their ability to trap CPCs and CLCs (data not shown). The schematic in Figure 1D shows how the cells were contained at the entrance of the microtraps and high magnification imaging of the trapped cell as shown in Figure 1E.

An SU-8 relief was used to prepare PDMS trapping devices that were bonded to glass cover plates to allow for high resolution fluorescence imaging. Reservoirs were formed in the PDMS device using a sharp biopsy puncture. PDMS devices and glass cover plates (No.1 coverslips with a thickness of 0.13–0.16 mm) were cleaned with IPA, then washed with water, and air dried. We used thin glass coverslips to facilitate the use of high NA objectives to accommodate their short working distances and high magnification necessary for FISH. PDMS devices were bonded to the glass coverslips by treating the surfaces with O_2 plasma (50 W for 1 min). After plasma treatment, slight pressure was applied starting from one edge to the other to avoid the trapping of air bubbles. Peak tubing was sealed with epoxy glue to the reservoirs and a syringe pump was used to provide continuous fluid flow through the device.

2.2. *Sample loading*

The microtrap device was flooded with a continuous flow of 25 μL/min of PBS buffer (pH 7.4, 100 mM) followed by flowing 0.5% BSA/PBS solution through the device using a PHD2000 syringe pump (Harvard Apparatus, Holliston, MA, USA). Forty μL/min flow rates were used to remove air bubbles from the device. Once the device was fully wetted with the 0.5% BSA/PBS solution, cells were introduced into the device for either immunophenotyping or FISH at a volume flow rate of 10 μL/min.

2.3. *On-Chip immunostaining*

For immunostaining, live cells (RPMI-8226 or SUP-B15) were loaded as described above onto the microtrap device. After loading target cells onto the device, 2% paraformaldehyde (PFA) was injected at 10 μL/min for 2 min and allowed to incubate for 15 min. After incubation, the device was washed with PBS buffer for 5 min. Fixed cells were then treated with human Fc blocker (IgG1) for 15 min to block any Fc receptors on the cell surface followed by incubation with monoclonal antibodies for 30 min. For RPMI-8226 cells, anti-CD138 FITC (MI15, 5.0 μg/mL), anti-CD56-PE (MEM-188 clone, 20 μg/mL), and anti-CD38-APC (HIT2 clone, 2.5 μg/mL) monoclonal antibodies were used. Cells were then washed with PBS and permeabilized with 0.1% Triton X-100 for 5 min followed by counter staining with DAPI for 2 min.

2.4. *Sample Preparation for FISH*

RPMI-8226 cells were cultured as a model CPC cell line for multiple myeloma and similarly, a SUP-B15 cell line was cultured as a model CLC for B-ALL (see Supplementary Materials for more information). Once the cells were removed from the culture media, they were washed with 1× PBS twice. Following the wash, the cells were re-suspended in 0.075 M KCl hypotonic solution plus 100 μL of Colcemid to swell the cells. Colcemid helps the chromosomes to stretch, thus enhancing the clarity and the resolution of the fluorescent probes used for FISH. Cells were fixed using Carnoy's fixative (methanol: glacial acetic acid = 3:1 (v/v)). Carnoy's solution is a light fixative (no crosslinking) as opposed to 2% PFA. During fixation, Carnoy's fixative was added dropwise to the cells and mixed with gentle agitation. The solution was then centrifuged at 1200 rpm for 10 min. This step was repeated 3 times and the fixed cells were stored at −20 °C until being used for experiments.

2.5. *On-Chip FISH*

Before using fixed cell samples prepared as described above, they were mixed with fresh Carnoy's solution, and introduced into the microtrap device at 10 μL/min. In the case of live cell samples, cells were fixed on-chip before processing for FISH. Once the live cells were injected and contained at the microtraps in the device, 0.056 M KCl hypotonic solution was injected and cells were incubated for 10 min. After KCl treatment, cells were fixed by injecting Carnoy's fixative and incubated for 30 min, replacing the solution with fresh Carnoy's every 10 min. Then, all of the Carnoy's solution was removed by washing with PBS.

Next, the chip was washed with 2X SSC for 2 min at room temperature followed by a series of ethanol washes (70%, 85%, and 100% EtOH) injected for 1–2 min each and dried with 100% EtOH for 5 min. After the EtOH wash, the chip was dried completely by heating and evaporating all of the EtOH. One µL and 10 µL of 5× diluted FISH probe mix was introduced into the single and 8-bed device, respectively, while applying light vacuum to the outlet tubing. Once the device was filled with FISH probes, the inlet and outlet were sealed with rubber cement. The chip was heated to 75 °C for 5 min and incubated in a hybridization oven (Bambino II™ Hybridization Oven) (Boekel Scientific, Feasterville, PA, USA) for 2 h at 37 °C. Afterwards, the rubber cement was removed, and the chip was kept at 72 °C (±1 °C) while washing for 2 min with 0.4X SSC (pH 7.0). The temperature of this step is critical as it will remove the remaining free probes and keep hybridized ones associated to the target DNA to improve imaging of the FISH signal. Next, the device was washed with 2 × SSC + 0.05%, and Tween-20 (pH 7.0) for 1 min at room temperature. Finally, the device was washed with EtOH for 5 min and dried completely before applying 2 µL of 4-diamidino-2-phenylindole (DAPI II) for nuclear staining. After these steps, the chip could be stored at 4 °C until imaging.

2.6. Imaging On-chip and Image Processing

All imaging was performed using a Keyence BZ-X710 microscope (Keyence Cooperation of America, Itasca, IL, USA) equipped with BZ-X filters; DAPI (Ex: 360/40 nm, Em: 460/50 nm, dichroic mirror wavelength (DMW): 400 nm), GFP (Ex: 470/40, Em: 525/50 nm, DMW: 495 nm), TRITC (Ex: 545/25 nm, Em: 605/70 nm, DMW: 565 nm), and Cy5 (Ex: 620/60 nm, Em: 700/75 nm, DMW: 660 nm). The microscope was equipped with a Nikon objectives (Nikon Instruments, Melville, NY, USA) CFI Plan Apo 4× (λ 4×, NA 0.20 WD 20.00 mm), 10× (λ 10×, NA 0.45 WD 4.00 mm), 20× (λ 20×, NA 0.75 WD 1.00 mm), 40× (λ 40×, NA 0.95 WD 0.21 mm), and 60× (Apo VC 60× NA 1.40 NA WD 0.13 mm) objectives. For immunophenotyping, exposure times of 50 ms for DAPI, 500 ms for FITC, and 1500 ms for TRITC/Cy3 and Cy5/APC were used at 10×, 20×, and 40× magnification. All images for FISH experiments were acquired in high-resolution mode. Cell nuclei were imaged with the 60× objective using DAPI (blue) filters (200 ms) and the FISH signals were acquired using FITC/GFP (green, 1500 ms) and (Cy5/APC)/Cy3 (red, 2500/3000 ms) channels. Due to the cell's 3-D profile while they were contained at the microtrap inlets, it was necessary to do imaging across the *z*-axis. For the imaging of slides, we used $\Delta z = 2$ µm for 5 different planes while for on-chip FISH imaging, cells were imaged at 10 different planes along the *z*-axis with $\Delta z = 1$ µm. All images were processed using BZ-X Analyzer (Keyence Cooperation) and FIJI software (NIH) [33].

2.7. Patient Sample Processing for FISH

CLCs from B-ALL patients were captured using the CTC isolation device described in our earlier publications (see Supplementary Materials for more information), which was also used for acute myeloid leukemia (AML) and multiple myeloma work [3,4]. Healthy donor blood sample was obtained from the University of Kansas Medical Center (KUMC) IRB-approved Biospecimen Repository Core Facility. Blood samples from a patient diagnosed with B-ALL was collected according to an approved Children's Mercy Hospital Institutional Review Board procedure. Written informed consent was obtained from the patient included in the study before enrollment. Peripheral blood samples (5 mL) were drawn by venipuncture into Vacuette® K3EDTA (Greiner Bio-one, Monroe, NC, USA) tubes. Following affinity enrichment using the microfluidic enrichment chip, cells were released from the capture device's surface using USER enzyme that cleaved a single-stranded bifunctional linker containing a uracil residue [34]. The released cells were collected into a microfuge tube and centrifuged to prepare for immunophenotyping or FISH. For FISH, once the cells were spun down, supernatant was removed and pre-heated (37 °C) in 0.056 M KCl and incubated for 10 min. The mixture was centrifuged again and after removing the supernatant, ice-cold Carnoy's fixative (if fixation was done before introduction to the microtrap device) was added to the cells and centrifuged again. This step was repeated 3 more times and cells were stored in Carnoy's fixative at 20 °C until further use. CLCs were spun down and resuspended in fresh Carnoy's solution before use. These samples were infused into the microtrap device and processed according to the on-chip FISH procedure as described above.

3. Results

3.1. Microtrap Device for Immunophenotyping and Cytogenetic Analysis

The large number of traps in the microtrap device needed to be contained for analysis of CLCs' and CPCs' affinity selected from a blood sample (see Table 1). Contrary to CTCs, which use, for example, EpCAM as the enrichment antigen, the number of non-diseased cells selected by antibodies targeting leukemic associated and multiple myeloma antigens is high because non-diseased cells can also express the enrichment antigen (i.e., CD19, CD34, CD117, CD33, or CD138). Following CLC or CPC affinity selection and release from the selection chips [3], cells were trapped without surface modifications at containment pores of the microtrap device, which were arranged in a 2-D format to make it easier for imaging single cells. This simplified imaging was compared to cells stochastically arranged on glass slides or wells possessing the appropriate surface chemistry. In addition, we will show that we can interface the microtrap device for immunophenotyping and FISH of CLCs and CPCs to a device used for their enrichment from the blood of patients.

Table 1. Number of CLCs and CPCs compared to CTCs detected in cancer patients.

Cancer Cells (Antigen Used for Selection)	Target Cells (mL^{-1})	Non-Aberrant Cells (mL^{-1})	Cell Diameter (μm)	Reference
CLCs (AML, CD33, CD34, CD117)	11–2684	10–2450	11–16	[4]
CLCs (B-ALL, CD19)	40–840	400–2050	6–12	this work
CPCs (CD138)	10–5900	43–1875	12–16	[3]
CTCs in epithelial tumors (EpCAM)	1–800	3–10	10–23	[35]

Note: The references used here were taken from a single type of cell selection chip (sinusoidal device) so that comparisons could be made as to the numbers of the CLCs, CPCs, and CTCs secured from liquid biopsies.

Considering the size of the CLCs and CPCs with an average diameter of 6–16 μm, we used 4×2 μm^2 microtraps to maximize the containment efficiency of CLCs and CPCs. Table 1 shows the number of CLCs for both AML and B-ALL as well as CPCs enriched using the appropriate marker(s). The numbers of CTCs enriched from different epithelial cancers (i.e., pancreatic, breast, prostate, colorectal, and ovarian) is shown as well in Table 1. Not only are the number of aberrant cells enriched higher for the leukemia and multiple myeloma diseases compared to epithelial cancers, but the number of non-diseased cells enriched is higher as well compared to epithelial cancers due to the fact that non-diseased blood cells can carry the same antigens as CLCs and CPCs, whereas for CTCs the blood cells do not express EpCAM. Due to the higher number of cells anticipated for the leukemic and multiple myeloma diseases, we had to build devices with high numbers of microtraps.

3.2. Microfluidic Containment Device Operation

The microtrap device (both single and 8-bed devices) used two independent networks of interleaving channels that were interconnected using smaller cross-channels positioned orthogonally to the interleaving channels. The cross-channel, when sealed with a cover plate, generated a pore structure whose dimensions were determined by the size of the cross-channels (Figure 2A) and allowed fluid flow between the input and output interleaving channels. During operation, CPCs/CLCs are released from the isolation chip and directed to the microtrap device, where they become physically trapped with the efficiency of trapping dependent on the size of the microtrap with respect to the cell of interest. Trapped cells could then be immunophenotyped or subjected to FISH followed by imaging with a fluorescence microscope to read out the appropriate signals. Microtrap beds were connected in parallel to provide sufficient numbers of equally accessible pores to retain a large number of cells (see Table 1). If the eight beds were connected in series, a majority, if not all, of the cells would be retained within the first bed, generating cell pileup and crowding, thus making imaging difficult. Conversely, placing the beds in parallel (see Figure 1A) allowed access to all beds from a common input, and thus, dispersed the cells with equal probability at the 80,000 containment pores associated with the 8-bed device.

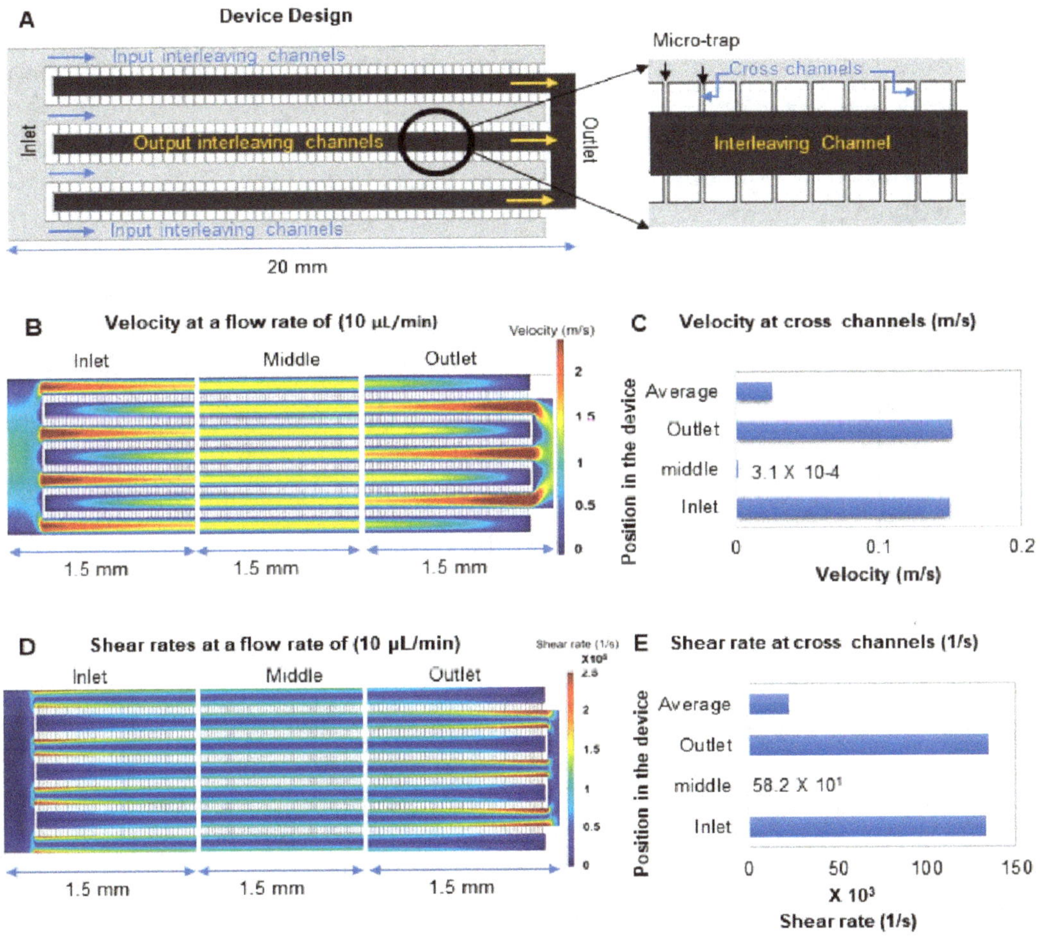

Figure 2. Simulations of the microtrap device. (**A**) 2-D CAD design of the microtrap device used for COMSOL simulations showing the interleaving network for the flow of fluid, and the cross-channels, which produce the microtraps when a cover plate is sealed to the device. The magnified image of the microtrap area is shown on the right with a single interleaving output channel (red) and two interleaving input channels (gray). (**B**) The simulated linear fluid velocity throughout the microtrap chip. The simulation shows three sections of the device: (i) input section; (ii) middle section; and (iii) outlet section. Flow was simulated across the interleaving input/output channels and the cross-channels. The dashed box shown here is the region of the device that was simulated in Figure S3 (see Supplementary Materials). (**C**) Bar graph representing the mean velocities expressed in m/s observed for the cross-channels at different sections of the device and at a 10 μL/min volume flow rate. The sections labeled here correspond to the sections of the device simulated in (**B**). (**D**) Simulated shear rate at three different sections of the device, inlet, middle, and outlet sections. (**E**) Bar graphs representing the mean shear rates across the cross-channels at different sections of the device at a volume flow rate of 10 μL/min. The sections of the device listed here correspond to those sections shown in (**D**).

Even though earlier studies from our group showed that live cells with a cell diameter of ~16 μm (CTCs) were successfully contained by microtraps with dimensions of 8 × 4 μm^2 [31], the current study required containment of smaller cells and cell numbers that were higher (see Table 1). RPMI-8226 cells (multiple myeloma model) have a size range of 6–16 μm with an average diameter of ~13 μm whereas SUP-B15 cells (B-ALL model cell line) have an average size of ~10 μm with a size range of 8–12 μm. Thus, when devices with microtraps of 8 × 4 μm^2 (width × depth) were used, we observed that the containment efficiency was <50% (data not shown). However, devices with microtraps of 4 × 2 μm^2 produced a containment efficiency >90% for these cells (Figures S1 and S3B).

3.3. Device Design and COMSOL Simulations

Finite element analysis (COMSOL) was performed on the microtrap device to deduce the projected linear velocity through the fluidic network and the corresponding shear rates to help determine the containment efficiency of the microfluidic device to physically trap live cells without damaging them. Laminar flow was validated across the entire device for the following flow rates: 1, 3, 5, and 10 μL/min with an aqueous fluid (Figure S3). The rationale behind choosing these flow rates was to test the optimal flow rate for effective containment of live cells without damage but having sufficient pressure to fill the device without generating air bubbles. The cells used for our studies were human cancer cell lines having a diffusion coefficient of 5×10^{-14} m^2/s [36].

The flow within the microtrap device was driven hydrodynamically, hence, a parabolic flow profile existed with higher velocity in the center of each individual channel as compared to the channel walls (no-slip condition; see Figure 2B) [37]. Figure 2A shows a CAD drawing of the device with 4-μm-wide microtraps and a depth of 2 μm. This device was found to provide 90% containment efficiency for unfixed RPMI-8226 cells as shown in Figure S3A. For larger epithelial cells (i.e., SKBR3), the containment efficiency was 96%. Cells were evenly distributed throughout the microtrap device as well (Figure 3; Figure 4). In the experiments evaluating trapping efficiency, the cells were DAPI stained and counted using a microscope to verify the number of cells captured. Cells that were not retained were collected at the outlet of the microtrap device into a flat-bottomed plate and enumerated and inspected for damage. The containment efficiency was determined from the ratio of cells in the microtrap to the total number of cells introduced (i.e., cells trapped and cells passing through the microtrap device).

The average velocity at the inlet and outlet interleaving channels toward the input end of the device was 2.5 m/s and in the interleaving channels in the center of the device it was 1.5 m/s at a volumetric flow rate of 10 μL/min. Even at these relatively high velocities, the flow was still laminar (see Figure S3B,C). Accounting for differences in the velocity between the interleaving and cross channels, the average velocity was calculated for the cross-channels at different sections of the device at a 10 μL/min volume flow rate and is plotted in Figure 2C.

The average linear velocity in the cross-channels was 0.02 m/s at the 10 μL/min volumetric flow rate. To measure the pressure drop across the device, the relative pressure at the outlet was defined in absolute terms ($p_A = p + p_{ref}$, where p_A is the absolute pressure, p is the relative pressure, and p_{ref} is the reference pressure, which was set to 1 atm (101 kPa) [3]. A gradual drop in pressure across the length of the device was noted, with this drop being ~14 kPa (16 and 2 kPa at the inlet and outlet, respectively, at 10 μL/min). The calculated shear rates at different volumetric flow rates were used to determine the shear stress in the microtrap device [38]. According to Newton's law, shear stress is the shear rate times the viscosity:

$$\text{Shear stress (dynes/(cm}^2\text{))} = \text{Shear rate (1/s)} \times \text{T}, \quad (1)$$

where T is the dynamic viscosity (T for water is 8.90×10^{-3} dynes*s/cm^2 at 25 °C).

We calculated the average shear stress on the cells experienced in the microtrap device through the entire device at different flow rates. At a flowrate of 1 μL/min, the shear rate calculated was 6042 s^{-1}, which corresponds to a shear stress of 54 dynes/cm^2 and is 10 times higher at 10 μL/min (Table 2). Moreover, higher shear rates were observed in the inlet and outlet of the device, where cells have potentially the highest probability of being damaged when flowing near the wall of the device as opposed to the center of the channel or the center area of the device where lower shear stress is observed (Figure 2D,E). Shear rate distributions across a section of the device can be found in Figure S3D,E.

Table 2. Average shear rate and calculated shear stress on cells at each microtrap for the flow rates listed.

Flow Rate (μL/min)	Shear Rate (1/s)	Shear Stress (dynes/cm^2)
1	6042	53.8
3	18,206	162.0
5	30,454	271.0
10	63,750	567.4

The shear stress experienced by cells in physiological conditions as they travel through capillaries and arterioles ranges between 40 and 55 dynes/cm^2 [39–42]. Interestingly, even though the shear stress at a flow rate of 10 μL/min was 10 times higher than the average shear stress of cells traveling through arterioles, we did not observe obvious damage of RPMI-8226 or SUP-B15 cells when contained at the entrance of any microtrap within the device. Cells tolerate transiently high shear stress of ~3000 dyn/cm^2 or higher [43,44]. In our microtrap device, the average transit time of the cell traveling through the chip is 210 ms if it is not retained by the microtrap and only 1 min when cells are retained within the microtrap device. Because the cells experience high shear stress briefly, they remain intact during the transport in high shear stress environments [45].

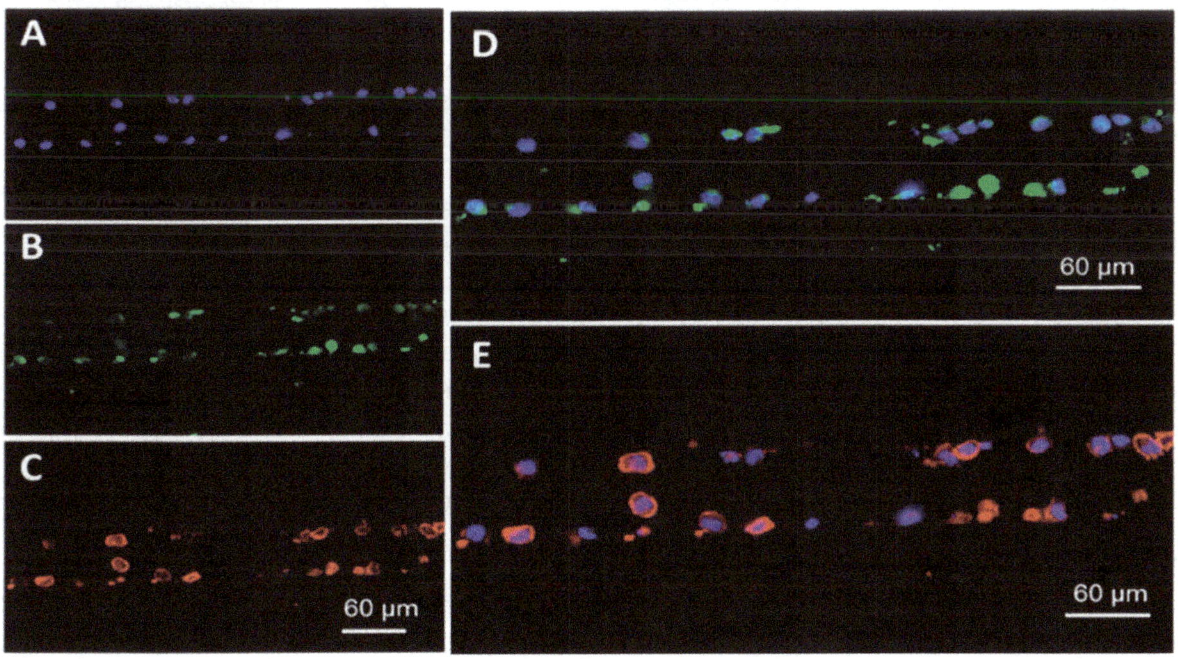

Figure 3. On-chip immunophenotyping of RPMI-8226 cells. (**A**) DAPI-labeled RPMI-8226 cell nucleus aligned at the entrance of the microtraps formed by the cross-channels and the cover plate assembled to the device. (**B**) CD138 expression of the RPMI-8226 cells and (**C**) CD38 expression for the same cells. (**D**) Composite image of CD138 expression (FITC channel) with the cell nucleus (DAPI channel of the microscope). (**E**) Composite image of CD38 expression (APC channel) with the cell nucleus that was DAPI stained. Exposure times were DAPI 50 ms, FITC 500 ms, and APC 1500 ms with 20× magnification. All images were collected using the Keyence fluorescence microscope. Shown in this fluorescence image are cells aligned along one interleaving input channel with cross-channels on either side of that channel.

As a control for on-chip immunophenotyping using the microtrap device, we performed immunophenotyping on a slide (see Figures S4 and S5) to compare with our on-chip results. RPMI-8226 cells were labeled with FITC-anti-CD138 antibodies and APC-anti-CD38 antibodies. RPMI-8226 cells express these markers [3,46,47]. See Figures S4 and S5 for the flow cytometry results for RPMI-8226 and on-slide immunophenotyping data, respectively.

Figure 3A shows DAPI-stained cells positioned at the microtraps in the single bed device. Here, we injected live cells followed by introduction of a fixative (2% PFA) to demonstrate the ability to fix cells on-chip. It was evident that our device could contain live cells without damaging their integrity. Trapped cells were stained with anti-CD138-FITC and anti-CD38-APC human antibodies (Figure 3B,C, respectively). Figure 3D,E show the composite images of the cell nucleus (DAPI channel) with the corresponding FITC and APC fluorescence emission signals.

Imaging of retained cells in the device for immunophenotyping was rapid. The microtrap device could be imaged at 20× magnification for all three colors (DAPI 50 ms, FITC 500 ms and APC 1500 ms) in <2 min. An advantage of using the microtrap device is the fact that single cells are positioned at the micropore entrance in a 2–D format, making them easy to locate. For immunophenotyping using the 8-bed device, >98% of cells were imaged in one plane without requiring z-stacking using a 20× microscope objective.

3.4. *Microchip Processing and Imaging of a Large Number of Single Cells*

Results for isolating CTCs, CLCs, and CPCs are summarized in Table 1, and show a high number of cells enriched from 1 mL of a patient blood sample using the sinusoidal microfluidic enrichment chip when analyzing CPCs and CLCs due to the fact that the enrichment antibody also selects non-diseased cells as opposed to CTCs, where the enriched fraction possesses only a few non-diseased cells. As such, thousands of cells may be required to be analyzed via immunophenotyping or FISH [3] to identify cancer cells (i.e., CLCs and CPCs). To facilitate the analysis of a vast number of cells, we designed the microtrap device with eight beds capable of entrapping enriched cells, subject them to staining, and present them in a 2-D array format for microscopic evaluation.

The 8-bed microtrap device possessed 80,000 containment pores patterned in PDMS from SU-8 reliefs. Three-level reliefs (i.e., three different heights of microstructures; see Supplementary Materials for fabrication description) were required to reduce the pressure in the chip and achieve well-balanced flow through the entire fluidic network (Figures 1 and 3). Because CLCs and CPCs have a diameter ranging from 6–16 μm (see Table 1), to ensure maximum containment efficiency by the microtraps, we used 4 × 2 μm^2 cross-sections for the containment microtraps (see Figure S3A). Fluorescence microscope images of cells contained at the microtraps and immunostained are presented in Figure 4.

Figure 4A shows a brightfield image of a single bed in the 8-bed device. Merged images of cells aligned at micropore entrances were imaged with DAPI (50 ms acquisition time) and anti-CD38-APC antibodies (1500 ms acquisition time) to identify CD38 on the RPMI-8226 cell surfaces; see Figure 4B. Figure 4C shows a single bed device containing DAPI-stained cells. DAPI-stained cells in two consecutive beds from the 8-bed device are imaged and presented in Figure 4D. As can be seen from Figure 4C,D, the contained cells are fairly well distributed throughout the microtrap 2-D array in spite of the decrease in the linear velocity seen down the length of the interleaving input channels (see Figure 2B). We noticed no loss of cell integrity at the microtraps even in the region of the input/output ends of the microtrap array where the shear stress was high (Figure 3D). Finally, when the flow was stopped, the cells remained at their trapped location.

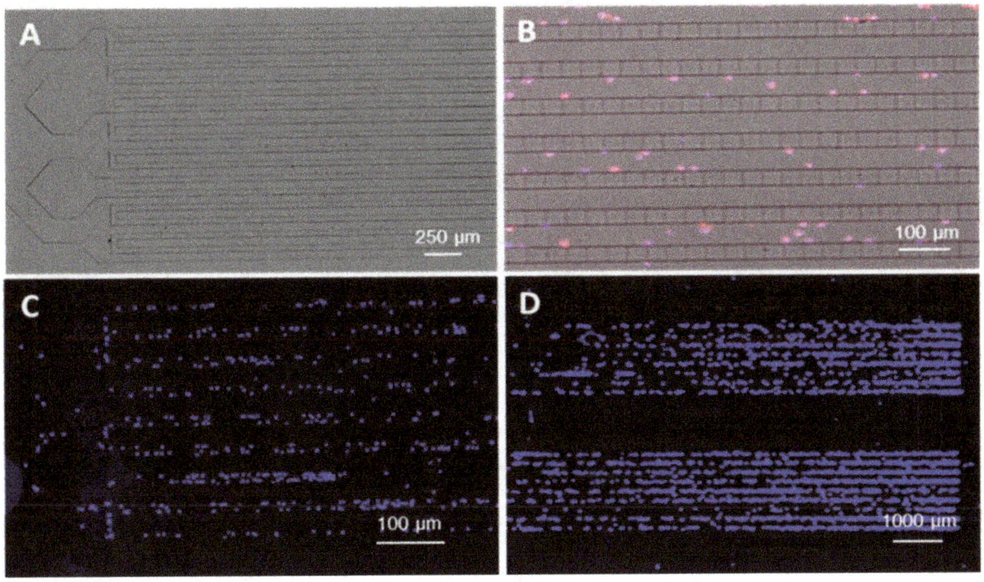

Figure 4. (**A**) Brightfield image of the bifurcated entrance channels of the microtrap device. RPMI-8226 cells were injected into the device at 10 μL/min and contained at the entrance of the microtraps. Cell images were processed according to the procedure listed in the materials and methods section of this manuscript and labeled with DAPI (nuclear stain) and CD38-APC markers. (**B**) Brightfield image merged with DAPI and APC channels showing the presence of the cell nucleus and CD38 on the cell surface aligned mainly at the microtrap entrances. (**C**) Entrance of the single bed device imaged using DAPI. RPMI-8226 cells were trapped inside the device at the entrance to the microtraps. (**D**) Two consecutive beds of the 8-bed device imaged with the DAPI channel of the microscope for stained RPMI-8226 cells.

3.5. On-Chip FISH

The microtrap device can be used for immunophenotyping and cytogenetic analysis, such as FISH. FISH determines aberrations in a metaphase chromosome or chromosomes buried in interphase nuclei from a fixed cytogenetic sample. The procedure for FISH processing on-chip is detailed in the materials and methods section as well as the Supplementary Materials. FISH experiments were carried out on RPMI-8226 and SUP-B15 cells with Cytocell FISH probes. The conventional workflow using microscope slides for FISH is a tedious and time-consuming process (see Figure S6), which requires 2–3 days including overnight hybridization of FISH probes. Figure 5 shows a step-by-step workflow for the on-chip FISH procedure, which required ~4 h of processing time and 2 μL of stock FISH probes for the 8-bed microtrap device, producing a 5-fold reduction in FISH probe volume compared to the slide-based FISH assay.

Figure 6A (i–iv) shows FISH signals from RPMI 8226-cells processed on-chip. FISH signals present in both the red and green channels of the fluorescence microscope in all of the cells were seen except for the image shown in Figure 6A (ii), where only one green signal was present due to deletion of the target gene region. In some of the cells, only one set (1 red and 1 green) of signal was present. One reason for losing some FISH signals is that the cells possess a 3-D structure (Figure 1E) even after entrapment at the microtrap and the fact that a high numerical aperture (NA) microscope was used with a short focal length; the images shown in Figure 6A were processed using only a single imaging plane (z-axis). This issue was addressed by using z-stacking of the imaging planes over a range equal to the average cell diameter.

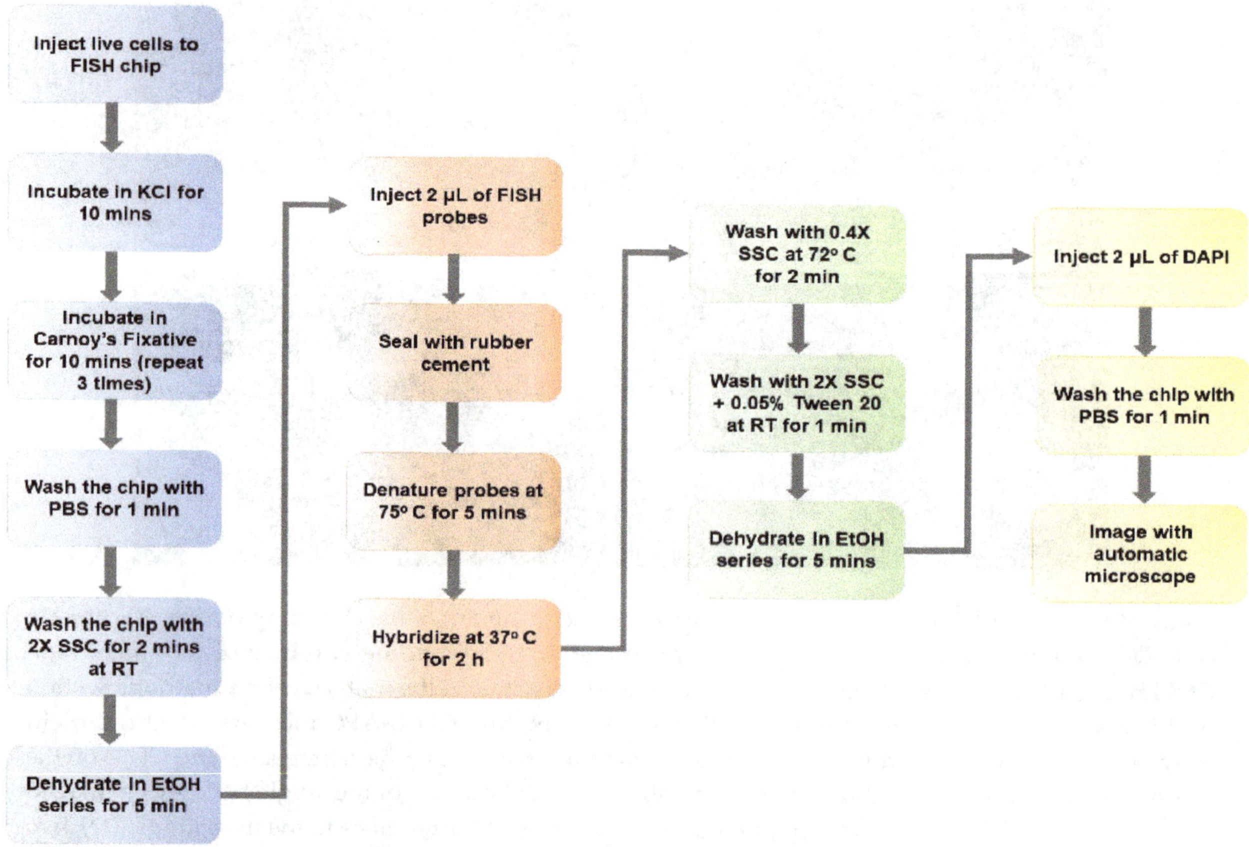

Figure 5. Workflow of FISH using the microtrap device. The workload was reduced from 2 days (slide method) to 4 h using the microtrap device primarily due to the hybridization time reduced from overnight to 2 h. The probe volume required for the assay was also reduced from 10 to 2 µL as well as using the microtrap device. Live cells were injected into the microtrap device at a flow rate of 10 µL/min and the washing steps were done at 5 µL/min to reduce the shear stress on the fixed cells contained within the microfluidic device.

Figure 6B shows a set of SUP-B15 cells processed using the microtrap chip for the TEL/AML1 FISH probes imaged in one image plane (i.e., no z-stacking). For TEL/AML1, probe TEL (ETV6—Erythroblastosis Variant Gene 6 translocation, ETS) refers to a region in chromosome 12 p-arm (12p13.2), and AML1 (or RUNX1—Runt-Related Transcription Factor 1) refers to the region in the q-arm of chromosome 21 (21q22.12). In a normal cell, there should be two red and two green signals and in a diseased cell, two yellow fusion signals are expected due to translocation of the TEL and AML1 genes. All of the images (Figure 6B) showed distinct red and green signals, with no clear indication of a yellow fusion signal to identify any cell as positive for the t(12;21) translocation.

Figure 6C,D shows two examples of SUP-B15 cells processed with BCR/ABL1 FISH probes for Ph t(9;22) (q34.12; q11.23) imaged with z-stacking (1-µm increments along the *z*-axis over ~15 µm). Ph t(9;22) (q34.12; q11.23) consists of two gene regions, with one from chromosome 9 corresponding to ABL1 gene (red labeled) and the other for the BCR (breakpoint cluster region) gene in chromosome 22 (green labeled). In a cell without a chromosomal fusion aberration, there are two green and two red signals. If there are yellow fusion signals detected, the cell can be identified as Ph(+). Imaging of 1-µm z-planes over a 15-µm range covered the entire cell as noted in Figure 6C,D. Those images showed two or more signals present within the cells entrapped by the microtrap device. Figure 6C shows two distinct red and green signals (confirmation as a cell not possessing the fusion product) and Figure 6D shows one yellow fusion signal (second signal not visible or merged with the first one) and one red and green signal, confirming it as a B-ALL cell that is Ph(+).

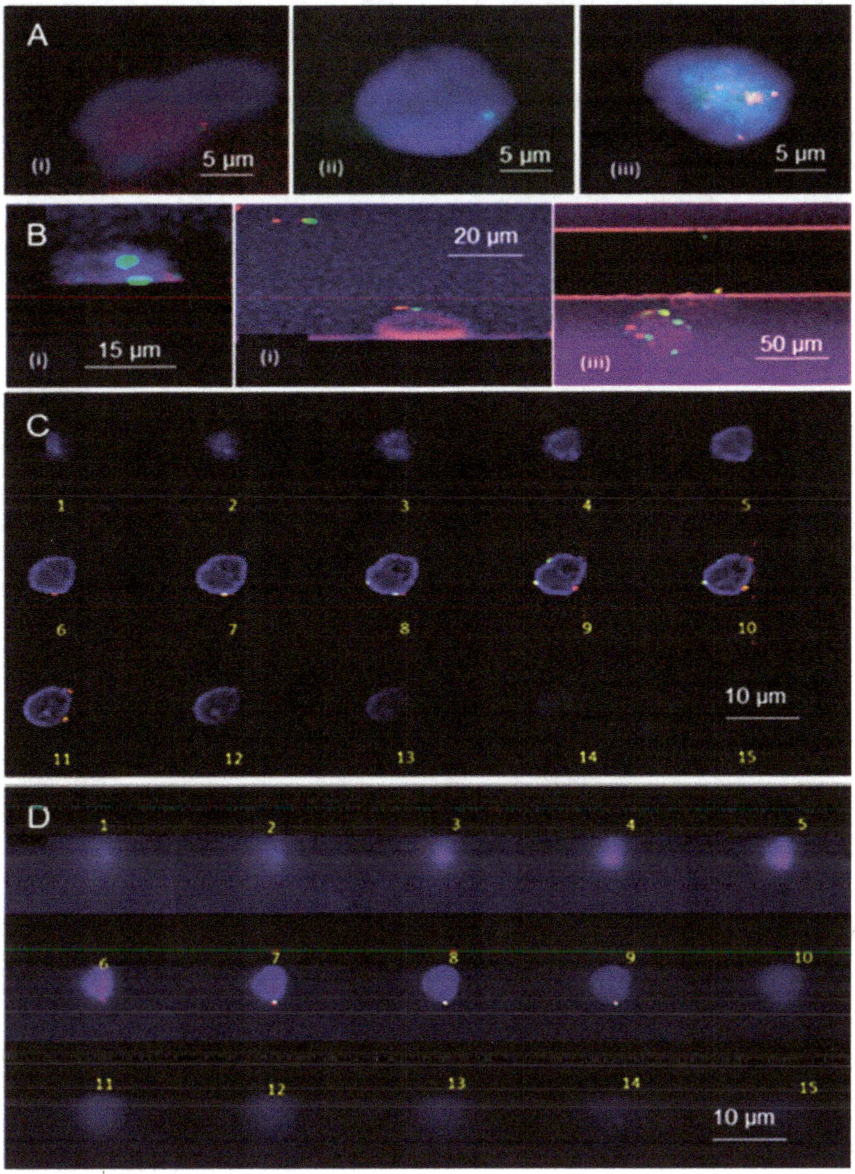

Figure 6. FISH-on-chip analysis of RPMI-8226 and SUP-B15 cells. (**A**) RPMI-8226 cells after FISH processing using the microtrap chip with the D13S319 plus deletion probe; (i) a cell that shows one green and one red FISH signal; (ii) a cell with only one green signal; (iii) a cell with 2 green and 2 red signals. (**B**) FISH analysis of SUP-B15 cells processed with the TEL/AML1 translocation, dual fusion probes showing the TEL (ETV6, 12p13.2) region in red, and AML1 (RUNX1, 21q22.12) region in green. (i) Two cells contained at the entrance of two different microtraps, but the FISH probes were visible in only one cell with two green signals; (ii) shows one red and two green signals with no clear yellow signals present; (iii) two cells that show distinct red and green signals, one cell captured at the entrance of microtrap shows one red, and one green signal with a possible yellow fusion signal. Both (**A**,**B**) were imaged in one single z-plane without z-stacking. (**C**,**D**) show z-stacking planes of 15 different image planes for FISH images from SUP-B15 cells captured at the microtrap and FISH processed with BCR/ABL plus translocation, dual fusion probe. (**C**) SUP-B15 cell with two green and two red signals. (**D**) SUP-B15 cell with one yellow fusion signal (second yellow signal not visible) and one red and green signal. Each image shows 15 separate images through the 15 µm distance range taken at 1-µm imaging intervals. FISH probes were specific to the BCR/ABL gene region, Philadelphia (Ph) chromosome tagging. All images were acquired using a Nikon 60× oil objective with DAPI—200 ms, FITC—1500 ms, TRITC—2500 ms integration times. The average SNR was 59 for the green probe and 68 for the red probe.

3.6. Measurement of MRD Status in Pediatric B-ALL Patients Using CLCs

CLCs and normal B-cells were affinity selected from a patient's peripheral blood with anti-human CD19 antibodies attached to the surface of a CLC enrichment (sinusoidal) microchip via a single-stranded oligonucleotide cleavable linker containing a dU residue [34]. Blood was collected from a pediatric patient (1–18 years) diagnosed with B-ALL undergoing induction and consolidation chemotherapy. Released cells following enrichment were immunophenotyped to distinguish CLCs from normal B-cells using the microtrap device; cells that demonstrated the expression of terminal deoxynucleotidyl transferase (TdT) in the nucleus were classified as CLCs. Additionally, the staining cocktail contained CD19/CD34/CD10 fluorescently labeled monoclonal antibodies to provide additional phenotypic data (Figure 7A).

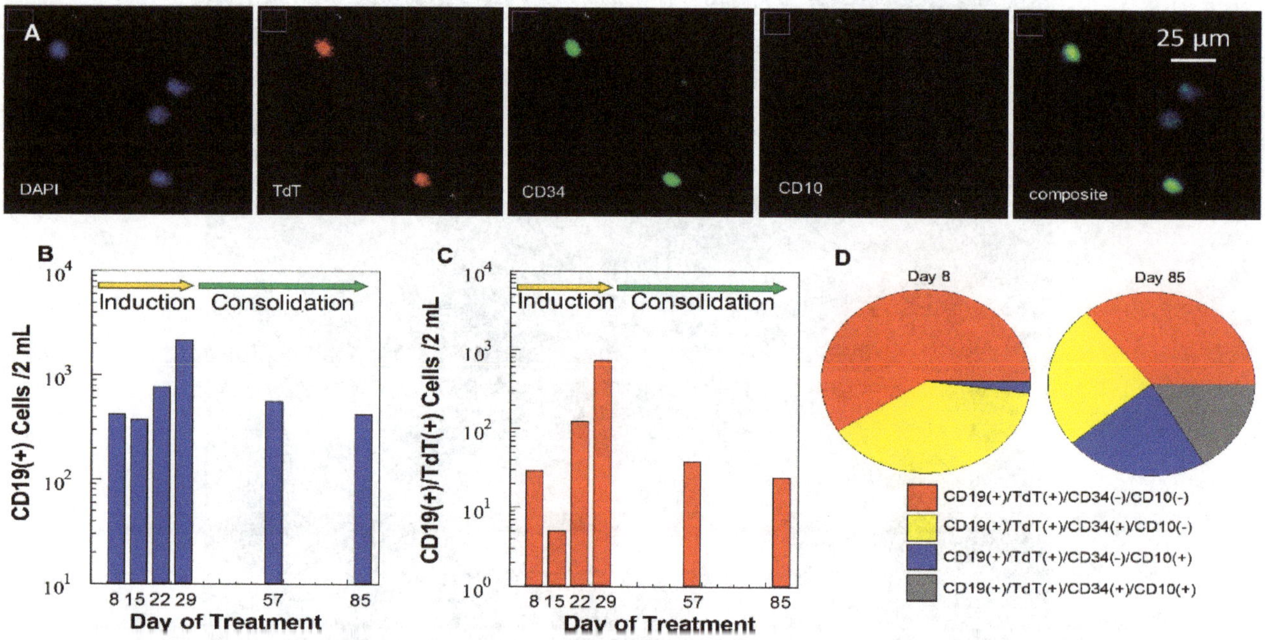

Figure 7. (**A**) Immunophenotyping of cells enriched from peripheral blood of a B-ALL patient by targeting cells with that express the CD19 antigen. The cells were stained using DAPI (nucleus), and monoclonal antibodies directed against TdT (FITC), CD34 (Cy3), and CD10 (Cy5). The images were acquired using a 40× microscope objective. The CLCs shown were DAPI(+)/CD34(+) and TdT(+), but CD10(−). (**B**) Microfluidic monitoring of a B-ALL patient from day 8 to 85 of chemotherapy. Total cell count represents all DAPI(+)/CD19(+) cells selected. (**C**) Number of CLCs identified as DAPI(+)/CD19(+)/TdT(+)/CD34(±)/CD10(±). (**D**) Change in phenotype among CLCs for this patient for days 8 and 85.

We analyzed the blood of a pediatric B-ALL patient to determine MRD status during chemotherapy on days 8, 15, 22, 29, 57, and 85 (Figure 7B–D). The clinical specificity was determined based on a threshold value established from the analysis of healthy donors as negative controls (average CD19 expressing cells was 68 cells/mL of peripheral blood). Grounded on that, we classified this patient as MRD(−) upon completion of induction and consolidation therapy on day 85. In this particular patient, on day 85, we observed a new phenotypic population of cells (Figure 7D) not observed during the first two analyses, which were secured on days 8 and 15 during induction therapy. Cells with the CD19(+)/TdT(+)/CD34(+)/CD10(+) phenotype began to appear in the blood on day 22. Although leukemic cell phenotype changes are common in B-ALL due to the effects of steroids as part of chemotherapy (i.e., loss of CD34), it is likely that the aforementioned cells represent normal immature lymphoid precursors whose morphology and immunophenotype are similar to the CLCs found in B-ALL. MRD status of this patient was determined to be positive only once, which was on day 29 of treatment when the level of enriched cells classified as CLCs (361 cells/mL) were above the threshold value.

To confirm the chromosomal status of the CD19-expressing cells on day 29 of treatment, we tested the enriched cells for chromosomal aberrations via FISH. Figure 8 shows a stitched image of cells contained on the microtrap device from a B-ALL patient. The sample was processed using TEL/AML1 FISH probes, which were able to identify the t(12;21) translocation. The TEL/AML1 fusion FISH probes identifies the most common rearrangements in childhood B-ALL, which is seen in around 17% of patients [48]. Figure 8a shows a single cell with two green FISH signals and Figure 8b shows two green FISH signals, with one signal showing a yellow signal (see arrow). Figure 8c shows a single cell with discrete red, green, and yellow signals. There was no visible evidence of a second yellow signal for confirmation of both t(12;21) translocations. Figure 8d shows a cell with the same observation of one red, one green, and one yellow signal, which were closely packed together. In Figure 8e, there are two green and one red signal visible. Figure 8a,b,e shows a lack of a red signal from the TEL gene. It has been observed that in B-ALL patients, there is the possibility for deletion of one TEL allele [49–51].

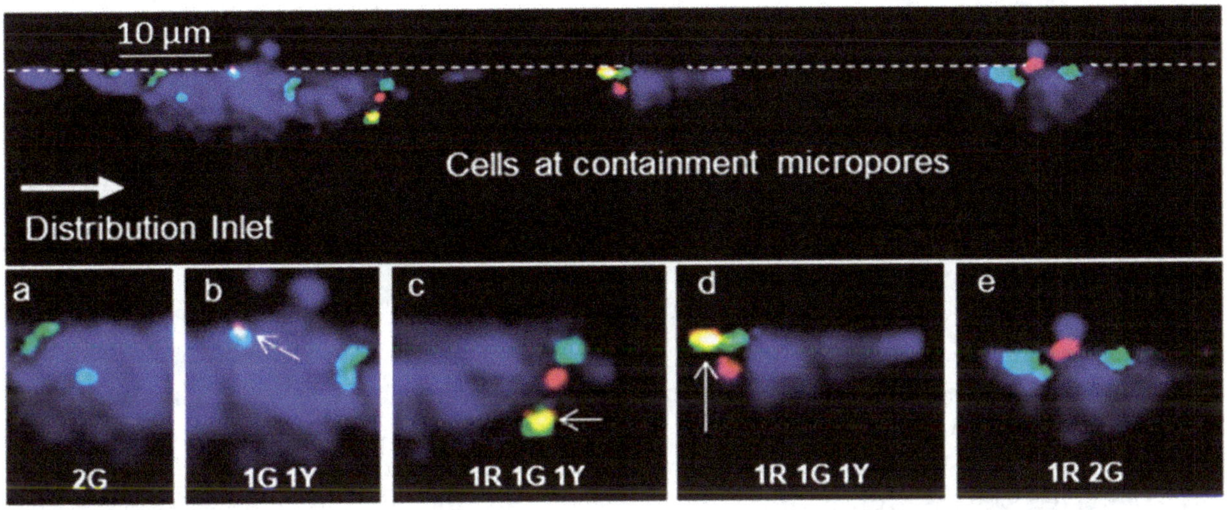

Figure 8. On-chip FISH processed B-cells isolated from a diagnosed B-ALL patient. TEL/AML1 FISH probes were used for the chromosomal aberration of t (12;21) translocation. Cells were imaged at the microtraps. Zoomed images show (**a**) single cell with 2 green FISH signals; (**b**) single cell with one green and one yellow signal; (**c**) a single cell with one red, one green, and one yellow FISH signal; (**d**) single cell with one red, one green, and one yellow FISH signals close to each other in the cell; and (**e**) single cell with one red and two green signals. In all cases, the images were collected using a 60× objective with z-stacking.

4. Discussion

FISH testing constitutes important and independent prognostic factors and is considered obligatory for analyzing patient outcome [52]. Of the current ~117 human genetic tests approved by the US Food and Drug Administration, 18 of these are FISH-based assays and most are directed toward hematological diseases, such as AML, multiple myeloma, and ALL.

AML arises from mutations occurring in progenitor cells of the myeloid lineage, which results in the inability of these cells to differentiate into functional blood cells. AML is the most common adult leukemia, with >21,000 new cases in the US in 2018, with a 5-year survival rate of 25%. The primary cause of death for AML patients is due to disease relapse [53]. The WHO currently categorizes patients into four groups [54]. For example, one category is patients with recurrent genetic abnormalities, which can consist of seven different chromosomal aberrations (typically balanced translocations or inversions, inv). Some of these aberrations are t(8;21)(q22;q22) associated with the *RUNX1/RUNX1T1* genes in chromosome 8, inv(16)(p13.1q22)/t(16;16)(p13.1;q22) occurring in the *CBFB/MYH11* genes of chromosome 16, and inv(3)(q21q26)/t(3;3)(q21;q26) of the *RPN1/EVI1* genes in chromosome 3. While AML MRD is typically managed using bone marrow biopsies, we have shown that CLCs can be used to determine recurrence from MRD in AML. The CLCs were enriched from blood samples using three sinusoidal microfluidic devices, with each one targeting a specific AML-associated antigen, CD117, CD34, and CD33 [54].

Multiple myeloma is associated with the abnormal expansion of terminally differentiated B clonal plasma cells in the bone marrow that produces an abnormal monoclonal paraprotein [55,56]. Multiple myeloma has three clinically defined stages: (i) MGUS (monoclonal gammopathy of undetermined significance), which is the asymptomatic stage; (ii) SMM (smoldering multiple myeloma) an intermediate phase; and (iii) the symptomatic stage referred to as active multiple myeloma [57]. In most cases, bone marrow biopsies are used to manage multiple myeloma. However, we and others have shown that CPCs can be used to manage this disease, which used a minimally invasive liquid biopsy [3,4,31]. In our study, we used a microfluidic device containing an array of sinusoidal microchannels with anti-CD138 monoclonal antibodies used to enrich CPCs from multiple myeloma patients [3]. It has been reported that in 16–50% of all multiple myeloma cases, chromosome 13q aberrations are present [58,59]. More than 90% of reported cases show the chromosomal aberration specifically in the 13q14 region [60]. We were able to perform FISH in the CPCs to detect the presence of chromosome 13q deletions using a slide-based FISH method (see Figures S6 and S7).

The FISH probes used for the RPMI-8226 cells, a model of multiple myeloma, identifies the DLEU region of chromosome 13 covering the 13q14 gene and used a red (APC channel) fluorescent probe. The control gene, 13qter located at the end of chromosome 13, was labeled with a green fluorescent probe (FITC channel). In a normal cell, there are two green signals and two red signals. However, due to the polyploidy nature in some cells, there may be multiple chromosomes (>2). In CPCs, it is expected that one or both DLEU regions (DLEU1 and DLEU2) may be deleted [61].

Figures 3 and 6 show immunophenotyping and FISH processing of RPMI-8226 cells using our microtrap device. As expected, the data seen in Figures S4 and S5 in the Supplementary Materials and our previous studies [3] confirmed the expression of CD138 and CD38 proteins for RPMI-8226 cells. We detected the presence of chromosome 13 as a green FISH signal corresponding to the 13qter gene (100 kb), which was present in all images, as shown in Figure 6A. Figure 6A (i) and (ii) shows deletion of the red signal corresponding to gene regions covering DLEU1, DLEU2, D13S319, D13S272, and RH47934 (156 kb) as expected for the RPMI-8226 cell line, as well. Most of the RPMI-8226 cells contained both red and green signals (Figure S7 in Supplementary Materials) lacking deletion, which is consistent with the karyotype data for this cell line.

B-ALL is the most common cancer diagnosed in children, representing ~30% of cancer diagnoses [62]. Despite significant improvements in the overall survival of children with B-ALL, there is a group of patients that experience relapse and ultimately die from their disease [63]. In fact, the likelihood of relapse is 80% for patients who have MRD at the end of induction therapy, indicative of active disease [64]. Monitoring of MRD, therefore, is considered a powerful predictor of outcome in B-ALL.

Cytogenetic abnormalities detected at diagnosis or generated during chemotherapy constitute important prognostic factors [52]. In B-ALL patients, 25–30% of patients have hyperdiploidy, 25% have t(12;21), 3–5% have t(9;22), 10% have MLL translocations, and 2% have iAMP21 chromosomal abnormalities. Once an aberration is detected, it can aid in the determination of the treatment regimen [65,66]. As another example, the detection of specific chromosome aberrations, such as t(9;22)(q34;q11.2) for *BCR-ABL1*, which results in the formation of the Philadelphia (Ph) chromosome, or t(12;21) aberrations of TEL/AML1 gene translocations are used to assign B-ALL patients to specific targeted therapies [67].

For the SUP-B15 cell line, which is a model for B-ALL [68], there are a few targeted gene variations that are typically evaluated using FISH [50,69,70]. MLL break-apart probes are used to detect the breakage of the MLL gene, which is frequently found in infant B-ALL [71–73]. BCR/ABL1 probes are used to detect the formation of the Philadelphia (Ph) chromosome produced by the fusion of two genes from chromosome 9 and 22, which is one the most important prognostic indicators for several hematological disorders, including B-ALL (see Figures S7, S8, and S9 for on-slide FISH analysis for some of these chromosomal abnormalities) [74–76].

In this study, the SUP-B15 cells were tested for TEL/AML1 translocations and BCR/ABL1 (Ph chromosome) using the microtrap device for FISH. For the SUP-B15 cell line, it is expected to see two distinct red and two green signals with the TEL/AML1 FISH probes. The TEL (ETV6) gene region marked in red corresponds to the 12p13.2 in chromosome 12 covering 168 kb of D12S1898 region and the green marker covers AML1 (RUNX1) in chromosome 21q22.12, a 167 kb gene region, including the CLIC6 gene [50]. On-chip FISH results (Figure 6B) showed distinct red and green signals corresponding to the presence of chromosome 12 and 21. Lack of a yellow signal confirmed t(12;21)(p13.2;q22.12) translocations were not present in the SUP-B15 cell line, which agrees with the karyotype as noted in the literature. The Philadelphia chromosome results from translocations of the ABL1 gene (9q34.11-q34.12, red) in chromosome 9 and the BCR gene (22q11.22-q11.23, green) in chromosome 22. The BCR probe region covers the GNAZ and RAB36 genes in a 169 kb region plus 148 kb region in telemetric BCR. The ABL1 probe covers a 346 kb region in the middle of the FUBP3 gene. As for the karyotype data of SUP-B15, we expected to see >90% of cells possessing the Ph chromosome (yellow fusion signal present). No yellow signal would be considered a cell with no Ph chromosome.

Figure 6C showed that SUP-B15 cells processed on-chip with BCR/ABL1 genes expressed two distinct red and green signals. This confirmed that there was no Ph chromosome present while in Figure 6D it showed the presence of one yellow fusion signal, confirming Ph(+) in that cell. In Figure 6D, we did not identify the fusion signal in this cell. However, the patient sample processed for FISH on-chip showed improved FISH signals as seen in Figure 8.

The microtrap FISH assay resulted in an SNR of 59 for the green signal, and 68 for the red signal. In the case of the slide-based assays, the SNR for the green and red signals were 64 and 63, respectively (see Figure S9), indicating that the ability to detect single molecules associated with the fluorescent reporter attached to each FISH probe was clearly visible using the microtrap device. The challenge is that in the FISH experiments, we used a high numerical objective with a small focal distance and as seen in Figure 1E, z-stacking was necessary to cover the genetic material housed within the nucleus. This may have been the reason that some signals were missed. This can be obviated by using a high numerical objective with a larger focal distance to better cover the entire nuclear region when the cells are located at the pore entrance.

Most FISH-based assays are predicated on the use of bone marrow, which is enriched in diseased cells compared to blood. For example, in the case of multiple myeloma, CPCs in peripheral blood are reported to be >100-fold lower than in bone marrow [77,78]. If disease relapse and chromosomal defects could be detected from peripheral blood, painful bone marrow biopsies could be avoided, and physicians could obtain information in near real time and potentially implement changes in treatment to affect better outcomes for patients with hematological diseases.

To obviate the need for a bone marrow biopsy, we used a liquid biopsy secured from a B-ALL patient using an affinity microfluidic chip and performed immunophenotyping and FISH on those enriched cells using our microtrap device. Similar to our previous work on the isolation of leukemic cells from blood of patients diagnosed with AML, a sinusoidal microfluidic chip with positive affinity selection was used [3,4], but in this case the affinity selection used a different antibody (anti-CD19 monoclonal antibodies) to enrich B-cells. While the enrichment of the CLCs and CPCs in our previous work was accomplished using a microfluidic chip, the downstream analysis was done off-chip, including immunophenotyping and FISH. Standard FISH workflow demands highly trained and experienced personnel (see Figure S6), making it difficult to implement in clinical laboratories not possessing the specialized facilities and trained personnel. Additionally, blood cells collected from a bone marrow biopsy are used for cytogenetic analysis, requiring an invasive procedure [64].

Using our microfluidic assay, a blood sample was subjected to affinity enrichment with high efficiency in terms of recovery and purity, and thus, the entire leukocyte population of peripheral blood did not require cytogenetic interrogation. The only cell population interrogated was those that expressed the target antigen, which in the case of B-ALL was CD19-expressing cells. Additionally, enriched cells were distributed in an array-like format as determined by the position of the microtraps of the microtrap device (see Figure 4C,D), which made them easier to image as opposed to stochastically distributed on a properly functionalized surface to induce cell adhesion to the surface. In addition, the 8-bed version of the microtrap device possessed 80,000 pores for retaining cells. While the data displayed in Table 1 show the total number of cells affinity selected (aberrant and non-aberrant) were 5314 for AML, 2840 for ALL, and 7775 for multiple myeloma, these numbers were based on a per mL sample volume. In some cases, it may be necessary to use larger input volumes, such as 10 mL, to search for rare CLCs and CPCs to find cells in the correct phase to elicit proper FISH signals. In these cases, the full advantage of the large dynamic range of the 8-bed device can be realized.

To reduce the workflow for FISH, microfluidics has been suggested by several groups, with the processing time reduced from several days using conventional slide-based FISH to several hours using FISH-on-chip platforms [14,26–28]. In addition, microfluidics has also resulted in a reduction in the use of expensive FISH probes. However, the reported platforms (see Table S1) required the use of special surface coatings to allow for cells to adhere to the surface of the device. Because of the stochastic nature of the attachment to the surface of the chip, this can create cell aggregates that made it difficult to image single cells under high magnification to determine the chromosomal status of the cells. Our device obviated the need of surface coatings and ordered the cells in a 2-D format to reduce device preparation steps and simplify single cell imaging, respectively. The microfluidic was comprised of an array of microtraps that were easily formed via a replication step in PDMS from a relief prepared by lithography. The relief also possessed the fluidic network.

We showed in this work that the microtrap device could be coupled to a rare cell enrichment chip to allow processing of circulating cells, such as CPCs or CLCs, with the ability to perform immunophenotyping and FISH of the enriched cells directly from blood samples. Moreover, experiments showed that our microtrap device was capable of containing live cells with >90% efficiency with sufficient traps to process CLCs and CPCs enriched from blood. The microtrap device was operated at a 10 μL/min volume flow rate to facilitate proper filling of the device without air bubbles. At this flow rate, even though cells experienced ~570 dynes/cm^2 shear stress, no obvious cell damage was observed for cells contained by the microtraps. Cell physical survival was attributed to the cell membrane's ability to handle relatively high shear stress for a brief time.

Using our 8-bed device with 80,000 microtraps, we could process thousands of cells for molecular profiling following enrichment. FISH results were achieved in <4 h by reducing the hybridization time from overnight to 2 h. Also, automated imaging was demonstrated for phenotyping in 2 min and FISH imaging in < 5 min, reducing the workflow for FISH compared to conventional slide-based assays, which requires 2–3 days of processing time, most of which is done manually. The FISH-on-chip provides full process automation.

When the cells were physically retained at the microtraps, they did possess a 3-D structure, requiring z-stacking to cover all of the FISH probes present inside the cell nucleus. Unlike immunophenotyping, where it was possible to use a single focusing plane because a lower magnification and associated longer focal length was required, FISH required the use of a high numerical aperture objective with a smaller focal length to capture high resolution images along several focal planes to image all of the FISH probes present in the cell nucleus. Even though z-stacking was necessary for FISH imaging, it was possible to automate this process. We set a common upper and lower threshold point in the device and selected the points where the cells were present at the arrayed microtraps and proceeded with automated imaging with pre-set exposure times for different filters (DAPI 200 ms, FITC 1500 ms, Cy3/Cy5 2500 ms). Processing of the captured images using BZ-X Analyzer and FIJI is detailed in the Supplementary Materials.

As opposed to our previous version of this device, which possessed larger microtraps [31], this device was designed to accommodate smaller CLCs and CPCs and the higher number of cells to analyze (see Table 1). For example, the CLCs (i.e., B-ALL cells) are smaller than CTCs and, as such, required the use of a smaller containment pore ($4 \times 2\ \mu m^2$ compared to $8 \times 6\ \mu m^2$ in our previous report) [31]. Because the affinity selection process for CLCs and CPCs results in the enrichment of a much larger number of cells due to the fact that the even non-diseased cells express the capture antigen, a larger number of containment pores were required (80,000 herein compared to 5000 in our previous device) [31]. In addition, our previous report only performed immunophenotyping and did not carry out FISH on the chip, as was demonstrated here.

In the current rendition, the microtrap device was made from PDMS by casting it against a relief. However, the same device architecture can be made from a thermoplastic, such as cyclic olefin copolymer (COC), which has some decisive advantages compared to PDMS [79]. For example, because COC is a thermoplastic, it can be injection molded to allow production of devices at high rates and at significantly lower chip cost compared to PDMS [79]. In fact, the entire device can be injection molded in a single cycle, with the only requirement being cover plate bonding as a finishing step. Also, COC has excellent optical properties that allow for high sensitivity imaging using the spectral range typically employed for FISH [27]. COC can be UV/O_3 activated to change its wettability to allow efficient filling with aqueous solutions without creating air bubbles and does not show the typical rapid hydrophobic recovery as seen with PDMS [80]. This will allow for the generation of a low-cost disposable appropriate for in vitro diagnostics. When the microtrap device is physically integrated to the cell enrichment device via a fluidic motherboard, fully automated processing of liquid biopsy samples can be envisioned to enable clinical use.

5. Conclusions

The ability of our microtrap device was demonstrated using CPCs and CLCs for immunophenotyping and FISH analyses of relatively small cells (D_{avg} ~12 μm) and in high numbers. The same device could be used to identify expression patterns of proteins and detect targeted chromosomal aberrations in single cells. Using the 8-bed device with 80,000 containment microtraps, we could process thousands of live or fixed cells enriched using a cell isolation microchip. FISH results were achieved in <4 h by primarily reducing the hybridization time from overnight to 2 h, and lowering the volume of the FISH probes required for analysis. The microtrap device was used for automated imaging for phenotypic identification of cells in 2 min and for FISH in <5 min for all fluorescence channels without the need to scan a relatively large area. Moreover, we were able to enrich B-cells from an ALL patient and process those cells to identify chromosomal aberrations. In future work, the use of a thermoplastic, such as COC, instead of PDMS will be undertaken to produce a low-cost, disposable device appropriate for use in clinical applications.

Supplementary Materials:
Table S1. Summary of FISH-based microfluidic technologies. Figure S1: metrology of microtrap device, Figure S2: fabrication steps for 8-bed microtrap device, Figure S3: COMSOL simulations of microtrap device, Figure S4: flow cytometry results of RPMI-8226 cells, Figure S5: immunophenotyping of RPMI-8226 cells, Figure S6: chart of FISH workflow using slide-based method, Figure S7: slide-based FISH results for RPMI-8226 cells, Figure S8: slide-based FISH results for SUP-B15 cells, Figure S9: slide-based FISH results for SUB-B15 cells.

Author Contributions: Conceptualization, K.A., M.L.H., M.A.W., S.A.S.; methodology, K.A., A.K.G., M.L.H., M.A.W., S.A.S.; formal analysis, K.A., M.A.W., S.A.S.; investigation, K.M.W.-R., S.V., N.L., K.D., M.H., J.J., S.M., M.L.H., M.A.W.; writing—original draft preparation, K.M.W.-R., S.V.; writing—review and editing, K.A., M.A.W., S.A.S.; supervision, M.A.W., S.A.S.; project administration, K.A., S.A.S.; funding acquisition, S.A.S., M.A.W. All authors have read and agreed to the published version of the manuscript.

Acknowledgments: The authors also acknowledge the KU Nanofabrication Facility, and the University of Kansas Cancer Center's Biospecimen Repository Core (NCI: P30-CA168524) for sample collection. We thank Lindsey Roe for help in editing the manuscript.

References

1. Macías, M.; Alegre, E.; Díaz-Lagares, A.; Patiño-García, A.; Pérez-Gracia, J.L.; Sanmamed, M.F.; López-López, R.; Varo, N.; Gonzalez, A. Liquid Biopsy: From Basic Research to Clinical Practice. In *Advances in Virus Research*; Elsevier BV: Bethesda, MD, USA, 2018; pp. 73–119.
2. Jeffrey, S.S.; Toner, M. Liquid biopsy: A perspective for probing blood for cancer. *Lab Chip* **2019**, *19*, 548–549. [CrossRef]
3. Kamande, J.W.; Lindell, M.A.M.; Witek, M.A.; Voorhees, P.M.; Soper, S.A. Isolation of circulating plasma cells from blood of patients diagnosed with clonal plasma cell disorders using cell selection microfluidics. *Integr. Biol.* **2018**, *10*, 82–91. [CrossRef]
4. Jackson, J.M.; Taylor, J.B.; Witek, M.A.; Hunsucker, S.; Waugh, J.P.; Fedoriw, Y.; Shea, T.C.; Soper, S.A.; Armistead, P.M. Microfluidics for the detection of minimal residual disease in acute myeloid leukemia patients using circulating leukemic cells selected from blood. *Analyst* **2016**, *141*, 640–651. [CrossRef]
5. Bauman, J.; Wiegant, J.; Borst, P.; Van Duijn, P. A new method for fluorescence microscopical localization of specific DNA sequences by in situ hybridization of fluorochrome-labelled RNA. *Exp. Cell Res.* **1980**, *128*, 485–490. [CrossRef]
6. Langer-Safer, P.R.; Levine, M.; Ward, D.C. Immunological method for mapping genes on Drosophila polytene chromosomes. *Proc. Natl. Acad. Sci. USA* **1982**, *79*, 4381–4385. [CrossRef]
7. Andreeff, M.; Pinkel, D. *Introduction to Fluorescence in Situ Hybridization: Principles and Clinical Applications*; Wiley-Liss: New York, NY, USA, 1999.
8. Ye, C.; Stevens, J.; Liu, G.; Ye, K.; Yang, F.; Bremer, S.; Heng, H. Combined multicolor-FISH and immunostaining. *Cytogenet. Genome Res.* **2006**, *114*, 227–234. [CrossRef]
9. Sato, K. Microdevice in Cellular Pathology: Microfluidic Platforms for Fluorescence in situ Hybridization and Analysis of Circulating Tumor Cells. *Anal. Sci.* **2015**, *31*, 867–873. [CrossRef]
10. Dowd, A.; Homeida, S.; Elkarem, H.A. Detection of chromosome 13 (13q14) deletion among Sudanese patients with multiple myeloma using a molecular genetics fluorescent in situ hybridization technique (FISH). *Malays. J. Pathol.* **2015**, *37*, 95–100.
11. Hu, Y.; Chen, L.; Sun, C.; She, X.-M.; Ai, L.; Qin, Y. Clinical significance of chromosomal abnormalities detected by interphase fluorescence in situ hybridization in newly diagnosed multiple myeloma patients. *Chin. Med. J.* **2011**, *124*, 2981–2985.
12. Conter, V.; Bartram, C.R.; Valsecchi, M.G.; Schrauder, A.; Panzer-Grümayer, R.; Moricke, A.; Aricò, M.; Zimmermann, M.; Mann, G.; De Rossi, G.; et al. Molecular response to treatment redefines all prognostic factors in children and adolescents with B-cell precursor acute lymphoblastic leukemia: Results in 3184 patients of the AIEOP-BFM ALL 2000 study. *Blood* **2010**, *115*, 3206–3214. [CrossRef]
13. Tai, C.-H.; Ho, C.-L.; Chen, Y.-L.; Chen, W.L.; Lee, G.-B. A novel integrated microfluidic platform to perform fluorescence in situ hybridization for chromosomal analysis. *Microfluid. Nanofluidics* **2013**, *15*, 745–752. [CrossRef]
14. Sieben, V.J.; Debes-Marun, C.S.; Pilarski, L.M.; Backhouse, C.J. An integrated microfluidic chip for chromosome enumeration using fluorescence in situ hybridization. *Lab Chip* **2008**, *8*, 2151. [CrossRef]
15. Chen, L.; Manz, A.; Day, P.J.R. Total nucleic acid analysis integrated on microfluidic devices. *Lab Chip* **2007**, *7*, 1413. [CrossRef]
16. Packard, M.M.; Shusteff, M.; Alocilja, E. Microfluidic-Based Amplification-Free Bacterial DNA Detection by Dielectrophoretic Concentration and Fluorescent Resonance Energy Transfer Assisted in Situ Hybridization (FRET-ISH). *Biosensors* **2012**, *2*, 405–416. [CrossRef]
17. Sieben, V.; Marun, C.D.; Pilarski, P.; Kaigala, G.; Pilarski, L.; Backhouse, C. FISH and chips: Chromosomal analysis on microfluidic platforms. *IET Nanobiotechnol.* **2007**, *1*, 27. [CrossRef]
18. Zanardi, A.; Bandiera, D.; Bertolini, F.; Corsini, C.; Gregato, G.; Milani, P.; Barborini, E.; Carbone, R. Miniaturized FISH for screening of onco-hematological malignancies. *Biotechniques* **2010**, *49*, 497–504. [CrossRef]
19. Liu, Y.; Kirkland, B.; Shirley, J.; Wang, Z.; Zhang, P.; Stembridge, J.; Wong, W.; Takebayashi, S.-I.; Gilbert, D.M.; Lenhert, S.; et al. Development of a single-cell array for large-scale DNA fluorescence in situ hybridization. *Lab Chip* **2013**, *13*, 1316–1324. [CrossRef]

20. Wang, X.; Takebayashi, S.-I.; Bernardin, E.; Gilbert, D.M.; Chella, R.; Guan, J. Microfluidic extraction and stretching of chromosomal DNA from single cell nuclei for DNA fluorescence in situ hybridization. *Biomed. Microdevices* **2012**, *14*, 443–451. [CrossRef]
21. Ciftlik, A.T.; Lehr, H.-A.; Gijs, M.A.M. Microfluidic processor allows rapid HER2 immunohistochemistry of breast carcinomas and significantly reduces ambiguous (2+) read-outs. *Proc. Natl. Acad. Sci. USA* **2013**, *110*, 5363–5368. [CrossRef]
22. Nguyen, H.T.; Trouillon, R.; Matsuoka, S.; Fiche, M.; De Leval, L.; Bisig, B.; Gijs, M.A. Microfluidics-assisted fluorescence in situ hybridization for advantageous human epidermal growth factor receptor 2 assessment in breast cancer. *Lab. Investig.* **2016**, *97*, 93–103. [CrossRef]
23. Nguyen, H.T.; Dupont, L.N.; Cuttaz, E.A.; Jean, A.M.; Trouillon, R.; Gijs, M.A.M. Breast cancer HER2 analysis by extra-short incubation microfluidics-assisted fluorescence in situ hybridization (ESIMA FISH). *Microelectron. Eng.* **2018**, *189*, 33–38. [CrossRef]
24. Mayer, J.A.; Pham, T.; Wong, K.L.; Scoggin, J.; Sales, E.V.; Clarin, T.; Pircher, T.J.; Mikolajczyk, S.D.; Cotter, P.D.; Bischoff, F.Z. Fish-Based Determination of Her2 Status in Circulating Tumor Cells Isolated with the Microfluidic Cee™ Platform. *Cancer Genet.* **2011**, *204*, 589–595. [CrossRef]
25. Mughal, F.; Baldock, S.J.; Karimiani, E.G.; Telford, N.; Goddard, N.J.; Day, P.J. Microfluidic channel-assisted screening of hematopoietic malignancies. *Genes Chromosom. Cancer* **2013**, *53*, 255–263. [CrossRef]
26. Kwasny, D.; Mednova, O.; Vedarethinam, I.; Dimaki, M.; Silahtaroglu, A.; Tümer, Z.; Almdal, K.; Svendsen, W.E. A Semi-Closed Device for Chromosome Spreading for Cytogenetic Analysis. *Micromachines* **2014**, *5*, 158–170. [CrossRef]
27. Perez-Toralla, K.; Guneri, E.T.; Champ, J.; Mottet, G.; Bidard, F.-C.; Pierga, J.-Y.; Klijanienko, J.; Draskovic, I.; Malaquin, L.; Viovy, J.-L.; et al. FISH in chips: Turning microfluidic fluorescence in situ hybridization into a quantitative and clinically reliable molecular diagnosis tool. *Lab Chip* **2015**, *15*, 811–822. [CrossRef]
28. Vedarethinam, I.; Shah, P.; Dimaki, M.; Tümer, Z.; Tommerup, N.; Svendsen, W.E. Metaphase FISH on a Chip: Miniaturized Microfluidic Device for Fluorescence in situ Hybridization. *Sensors* **2010**, *10*, 9831–9846. [CrossRef]
29. Baerlocher, G.M.; Vulto, I.; De Jong, G.; Lansdorp, P.M. Flow cytometry and FISH to measure the average length of telomeres (flow FISH). *Nat. Protoc.* **2006**, *1*, 2365–2376. [CrossRef]
30. Arrigucci, R.; Bushkin, Y.; Radford, F.; Lakehal, K.; Vir, P.; Pine, R.; Martin, D.; Sugarman, J.; Zhao, Y.; Yap, G.S.; et al. FISH-Flow, a protocol for the concurrent detection of mRNA and protein in single cells using fluorescence in situ hybridization and flow cytometry. *Nat. Protoc.* **2017**, *12*, 1245–1260. [CrossRef]
31. Kamande, J.W.; Hupert, M.L.; Witek, M.A.; Wang, H.; Torphy, R.J.; Dharmasiri, U.; Njoroge, S.K.; Jackson, J.M.; Aufforth, R.D.; Snavely, A.; et al. Modular Microsystem for the Isolation, Enumeration, and Phenotyping of Circulating Tumor Cells in Patients with Pancreatic Cancer. *Anal. Chem.* **2013**, *85*, 9092–9100. [CrossRef]
32. Hupert, M.L.; Jackson, J.M.; Wang, H.; Witek, M.A.; Kamande, J.; Milowsky, M.I.; Whang, Y.E.; Soper, S.A. Arrays of High-Aspect Ratio Microchannels for High-Throughput Isolation of Circulating Tumor Cells (CTCs). *Microsyst. Technol.* **2013**, *20*, 1815–1825. [CrossRef]
33. Schindelin, J.; Arganda-Carreras, I.; Frise, E.; Kaynig, V.; Longair, M.; Pietzsch, T.; Preibisch, S.; Rueden, C.; Saalfeld, S.; Schmid, B.; et al. Fiji: An Open-Source Platform for Biological-Image Analysis. *Nat. Methods* **2012**, *9*, 676–682. [CrossRef]
34. Nair, S.V.; Witek, M.A.; Jackson, J.M.; Lindell, M.A.M.; Hunsucker, S.; Sapp, T.; Perry, C.E.; Hupert, M.L.; Bae-Jump, V.; Gehrig, P.A.; et al. Enzymatic cleavage of uracil-containing single-stranded DNA linkers for the efficient release of affinity-selected circulating tumor cells. *Chem. Commun.* **2015**, *51*, 3266–3269. [CrossRef]
35. Witek, M.A.; Aufforth, R.D.; Wang, H.; Kamande, J.W.; Jackson, J.M.; Pullagurla, S.R.; Hupert, M.L.; Usary, J.; Wysham, W.Z.; Hilliard, D.; et al. Discrete microfluidics for the isolation of circulating tumor cell subpopulations targeting fibroblast activation protein alpha and epithelial cell adhesion molecule. *NPJ Precis. Oncol.* **2017**, *1*, 24. [CrossRef]
36. Bray, D. *Cell Movements: From Molecules to Motility*, 2nd ed.; Garland Science: New York, NY, USA, 1992; pp. 1–386. [CrossRef]
37. Lauga, E.; Brenner, M.P.; Stone, H. Microfluidics: The No-Slip Boundary Condition. In *Springer Handbook of Experimental Fluid Mechanics*; Springer Science and Business Media LLC: New York, NY, USA, 2007; pp. 1219–1240.

38. Rosser, J.; Thomas, D. Bioreactor processes for maturation of 3D bioprinted tissue. In *3D Bioprinting for Reconstructive Surgery*; Elsevier BV: Duxford, UK, 2018; pp. 191–215.
39. Lipowsky, H.H.; Kovalcheck, S.; Zweifach, B.W. The distribution of blood rheological parameters in the microvasculature of cat mesentery. *Circ. Res.* **1978**, *43*, 738–749. [CrossRef]
40. Lipowsky, H.H.; Usami, S.; Chien, S. In vivo measurements of "apparent viscosity" and microvessel hematocrit in the mesentery of the cat. *Microvasc. Res.* **1980**, *19*, 297–319. [CrossRef]
41. Varma, S.; Voldman, J. A cell-based sensor of fluid shear stress for microfluidics. *Lab Chip* **2015**, *15*, 1563–1573. [CrossRef]
42. Papaioannou, T.G.; Stefanadis, C. Vascular wall shear stress: Basic principles and methods. *Hell. J. Cardiol.* **2005**, *46*, 9–15.
43. Malek, A.M.; Alper, S.L.; Izumo, S. Hemodynamic shear stress and its role in atherosclerosis. *JAMA* **1999**, *282*, 2035–2042. [CrossRef]
44. Strony, J.; Beaudoin, A.; Brands, D.; Adelman, B. Analysis of shear stress and hemodynamic factors in a model of coronary artery stenosis and thrombosis. *Am. J. Physiol. Circ. Physiol.* **1993**, *265*, H1787–H1796. [CrossRef]
45. Barnes, J.M.; Nauseef, J.T.; Henry, M. Resistance to Fluid Shear Stress Is a Conserved Biophysical Property of Malignant Cells. *PLoS ONE* **2012**, *7*, e50973. [CrossRef]
46. Pellat-Deceunynck, C.; Barillé, S.; Puthier, D.; Rapp, M.J.; Harousseau, J.L.; Bataille, R.; Amiot, M. Adhesion molecules on human myeloma cells: Significant changes in expression related to malignancy, tumor spreading, and immortalization. *Cancer Res.* **1995**, *55*, 3647–3653. [PubMed]
47. Sahara, N.; Takeshita, A.; Shigeno, K.; Fujisawa, S.; Takeshita, K.; Naito, K.; Ihara, M.; Ono, T.; Tamashima, S.; Nara, K.; et al. Clinicopathological and prognostic characteristics of CD56-negative multiple myeloma. *Br. J. Haematol.* **2002**, *117*, 882–885. [CrossRef] [PubMed]
48. Jamil, A.; Theil, K.S.; Kahwash, S.; Ruymann, F.B.; Klopfenstein, K.J. TEL/AML-1 fusion gene. Its frequency and prognostic significance in childhood acute lymphoblastic leukemia. *Cancer Genet. Cytogenet.* **2000**, *122*, 73–78. [CrossRef]
49. Mikhail, F.M.; Serry, K.A.; Hatem, N.; Mourad, Z.I.; Farawela, H.M.; El Kaffash, D.M.; Coignet, L.; Nucifora, G. AML1 gene over-expression in childhood acute lymphoblastic leukemia. *Leukemia* **2002**, *16*, 658–668. [CrossRef]
50. Raynaud, S.; Cave, H.; Baens, M.; Bastard, C.; Cacheux, V.; Grosgeorge, J.; Guidal-Giroux, C.; Guo, C.; Vilmer, E.; Marynen, P.; et al. The 12;21 translocation involving TEL and deletion of the other TEL allele: Two frequently associated alterations found in childhood acute lymphoblastic leukemia. *Blood* **1996**, *87*, 2891–2899. [CrossRef]
51. Chung, H.Y.; Kim, K.-H.; Jun, K.R.; Jang, S.; Park, C.-J.; Chi, H.-S.; Im, H.J.; Seo, J.J.; Seo, E.-J. Prognostic Significance ofTEL/AML1Rearrangement and Its Additional Genetic Changes in Korean Childhood Precursor B-Acute Lymphoblastic Leukemia. *Korean J. Lab. Med.* **2010**, *30*, 1. [CrossRef]
52. Mrózek, K.; Harper, D.P.; Aplan, P.D. Cytogenetics and Molecular Genetics of Acute Lymphoblastic Leukemia. *Hematol. Clin. N. Am.* **2009**, *23*, 991–1010. [CrossRef]
53. Jaso, J.M.; Wang, S.; Jorgensen, J.L.; Lin, P. Multi-color flow cytometric immunophenotyping for detection of minimal residual disease in AML: Past, present and future. *Bone Marrow Transplant.* **2014**, *49*, 1129–1138. [CrossRef]
54. Morrissette, J.J.; Bagg, A. Acute Myeloid Leukemia: Conventional Cytogenetics, FISH, and Moleculocentric Methodologies. *Clin. Lab. Med.* **2011**, *31*, 659–686. [CrossRef]
55. Tschumper, R.C.; Asmann, Y.W.; Hossain, A.; Huddleston, P.; Wu, X.; Dispenzieri, A.; Eckloff, B.W.; Jelinek, D.F. Comprehensive Assessment of Potential Multiple Myeloma Immunoglobulin Heavy Chain V-D-J Intraclonal Variation Using Massively Parallel Pyrosequencing. *Oncotarget* **2012**, *3*, 502–513. [CrossRef]
56. Hideshima, T.; Anderson, K.C. Molecular mechanisms of novel therapeutic approaches for multiple myeloma. *Nat. Rev. Cancer* **2002**, *2*, 927–937. [CrossRef] [PubMed]
57. Lee, H.C.; Wang, H.; Baladandayuthapani, V.; Lin, H.; He, J.; Jones, R.J.; Kuiatse, I.; Gu, N.; Wang, Z.; Ma, W.; et al. RNA Polymerase I Inhibition with CX-5461 as a Novel Therapeutic Strategy to Target MYC in Multiple Myeloma. *Br. J. Haematol.* **2017**, *177*, 80–94. [CrossRef] [PubMed]

58. Fonseca, R.; Oken, M.M.; Harrington, D.; Bailey, R.J.; A Van Wier, S.; Henderson, K.J.; E Kay, N.; Van Ness, B.; Greipp, P.R.; Dewald, G.W. Deletions of chromosome 13 in multiple myeloma identified by interphase FISH usually denote large deletions of the q arm or monosomy. *Leukemia* **2001**, *15*, 981–986. [CrossRef] [PubMed]
59. Bullrich, F.; Fujii, H.; Calin, G.A.; Mabuchi, H.; Negrini, M.; Pekarsky, Y.; Rassenti, L.; Alder, H.; Reed, J.C.; Keating, M.J.; et al. Characterization of the 13q14 tumor suppressor locus in CLL: Identification of ALT1, an alternative splice variant of the LEU2 gene. *Cancer Res.* **2001**, *61*, 6640. [PubMed]
60. Shaughnessy, J.; Erming, T.; Jeffrey, S.; Klaus, B.; Reid, L.; Ashraf, B.; Christopher, M.; Guido, T.; Joshua, E.; Bart, B. High Incidence of Chromosome 13 Deletion in Multiple Myeloma Detected by Multiprobe Interphase Fish. *Blood* **2000**, *4*, 1505–1511. [CrossRef]
61. Bullrich, F.; Veronese, M.; Kitada, S.; Jurlander, J.; Caligiuri, M.; Reed, J.; Croce, C. Minimal region of loss at 13q14 in B-cell chronic lymphocytic leukemia. *Blood* **1996**, *88*, 3109–3115. [CrossRef]
62. Childhood Acute Lymphoblastic Leukemia Treatment (PDQ®): Health Professional Version. Available online: https://www.ncbi.nlm.nih.gov/pubmed/26389206 (accessed on 21 February 2020).
63. Seibel, N.L. Treatment of Acute Lymphoblastic Leukemia in Children and Adolescents: Peaks and Pitfalls. *Hematology* **2008**, *2008*, 374–380. [CrossRef]
64. Coustan-Smith, E.; Sancho, J.; Hancock, M.L.; Razzouk, B.I.; Ribeiro, R.C.; Rivera, G.K.; Rubnitz, J.; Sandlund, J.T.; Pui, C.-H.; Campana, D. Use of peripheral blood instead of bone marrow to monitor residual disease in children with acute lymphoblastic leukemia. *Blood* **2002**, *100*, 2399–2402. [CrossRef]
65. Blunck, C.B.; Terra-Granado, E.; Noronha, E.P.; Wajnberg, G.; Passetti, F.; Pombo-De-Oliveira, M.S.; Emerenciano, M. CD9 predicts ETV6-RUNX1 in childhood B-cell precursor acute lymphoblastic leukemia. *Hematol. Transfus. Cell Ther.* **2019**, *41*, 205–211. [CrossRef]
66. Van Delft, F.W.; Horsley, S.; Colman, S.; Anderson, K.; Bateman, C.; Kempski, H.; Zuna, J.; Eckert, C.; Saha, V.; Kearney, L.; et al. Clonal origins of relapse in ETV6-RUNX1 acute lymphoblastic leukemia. *Blood* **2011**, *117*, 6247–6254. [CrossRef]
67. Marshall, G.M.; Pozza, L.D.; Sutton, R.; Ng, A.; A De Groot-Kruseman, H.; Van Der Velden, V.H.; Venn, N.C.; Berg, H.V.D.; De Bont, E.S.J.M.; Egeler, R.M.; et al. High-risk childhood acute lymphoblastic leukemia in first remission treated with novel intensive chemotherapy and allogeneic transplantation. *Leukemia* **2013**, *27*, 1497–1503. [CrossRef] [PubMed]
68. Thomas, E.; Young, M.; Wimalendra, M.; Tunstall, O. PO-0168 Leukaemia Cutis: An Unusual Paediatric Presentation Of Acute Lymphoblastic Leukaemia. *Arch. Dis. Child.* **2014**, *99*, 1030–1043. [CrossRef]
69. Borkhardt, A.; Cazzaniga, G.; Viehmann, S.; Valsecchi, M.G.; Ludwig, W.D.; Mangioni, S.; Schrappe, M.; Riehm, H.; Lampert, F.; Basso, G.; et al. Incidence and Clinical Relevance of Tel/Aml1 Fusion Genes in Children with Acute Lymphoblastic Leukemia Enrolled in the German and Italian Multicenter Therapy Trials. *Blood* **1997**, *2*, 571–577. [CrossRef]
70. Mosad, E.; Hamed, H.; Bakry, R.; Ezz-Eldin, A.M.; Khalifa, N.M. Persistence of TEL-AML1 fusion gene as minimal residual disease has no additive prognostic value in CD 10 positive B-acute lymphoblastic leukemia: A FISH study. *J. Hematol. Oncol.* **2008**, *1*, 17. [CrossRef]
71. Rubnitz, J.E.; Link, M.P.; Shuster, J.J.; Carroll, A.J.; Hakami, N.; Frankel, L.S.; Pullen, D.J.; Cleary, M.L. Frequency and Prognostic Significance of Hrx Rearrangements in Infant Acute Lymphoblastic Leukemia: A Pediatric Oncology Group Study. *Blood* **1994**, *2*, 570–573. [CrossRef]
72. Secker-Walker, L.M.; Moorman, A.V.; Bain, B.J.; Mehta, A.B. Secondary acute leukemia and myelodysplastic syndrome with 11q23 abnormalities. *Leukemia* **1998**, *12*, 840–844. [CrossRef]
73. Rowley, J.D. The Critical Role of Chromosome Translocation in Human Leukemias. *Annu. Rev. Genet.* **1998**, *32*, 495–519. [CrossRef]
74. Naumovski, L.; Morgan, R.; Hecht, F.; Link, M.P.; E Glader, B.; Smith, S.D. Philadelphia chromosome-positive acute lymphoblastic leukemia cell lines without classical breakpoint cluster region rearrangement. *Cancer Res.* **1988**, *48*, 2876.
75. Harata, M.; Soda, Y.; Tani, K.; Ooi, J.; Takizawa, T.; Chen, M.; Bai, Y.; Izawa, K.; Kobayashi, S.; Tomonari, A.; et al. CD19-targeting liposomes containing imatinib efficiently kill Philadelphia chromosome-positive acute lymphoblastic leukemia cells. *Blood* **2004**, *104*, 1442–1449. [CrossRef]
76. Groffen, J.; Stephenson, J.R.; Heisterkamp, N.; De Klein, A.; Bartram, C.R.; Grosveld, G. Philadelphia chromosomal breakpoints are clustered within a limited region, bcr, on chromosome 22. *Cell* **1984**, *36*, 93–99. [CrossRef]

77. Vij, R.; Mazumder, A.; Klinger, M.; O'Dea, D.; Paasch, J.; Martin, T.; Weng, L.; Park, J.; Fiala, M.; Faham, M.; et al. Deep Sequencing Reveals Myeloma Cells in Peripheral Blood in Majority of Multiple Myeloma Patients. *Clin. Lymphoma Myeloma Leuk.* **2014**, *14*, 131–1390. [CrossRef]
78. Rajkumar, S.V.; Rajkumar, S.V.; Greipp, P.R.; Witzig, T.E. Cell proliferation of myeloma plasma cells: Comparison of the blood and marrow compartments. *Am. J. Hematol.* **2004**, *77*, 7–11.
79. Aghvami, S.A.; Opathalage, A.; Zhang, Z.; Ludwig, M.; Heymann, M.; Norton, M.; Wilkins, N.; Fraden, S. Rapid prototyping of cyclic olefin copolymer (COC) microfluidic devices. *Sens. Actuators B Chem.* **2017**, *247*, 940–949. [CrossRef]
80. Jackson, J.M.; Witek, M.A.; Hupert, M.L.; Brady, C.; Pullagurla, S.; Kamande, J.; Aufforth, R.D.; Tignanelli, C.; Torphy, R.J.; Yeh, J.J.; et al. UV activation of polymeric high aspect ratio microstructures: Ramifications in antibody surface loading for circulating tumor cell selection. *Lab Chip* **2014**, *14*, 106–117. [CrossRef]

9

S100-EPISPOT: A New Tool to Detect Viable Circulating Melanoma Cells

Laure Cayrefourcq [1], Aurélie De Roeck [2], Caroline Garcia [2], Pierre-Emmanuel Stoebner [2], Fanny Fichel [2], Françoise Garima [1], Françoise Perriard [3], Jean-Pierre Daures [3], Laurent Meunier [2,†] and Catherine Alix-Panabières [1,*,†]

1. Laboratory of Rare Human Circulating Cells (LCCRH), University Medical Centre of Montpellier, UPRES EA2415, 34093 Montpellier, France
2. Department of Dermatology, Nîmes University Hospital, University of Montpellier, 30029 Nîmes, France
3. UPRES EA2415, University Institute of Clinical Research (IURC), Montpellier University, 34093 Montpellier, France

* Correspondence: c-panabieres@chu-montpellier.fr

† These authors have contributed equally to the work.

Abstract: Metastatic melanoma is one of the most aggressive and drug-resistant cancers with very poor overall survival. Circulating melanoma cells (CMCs) were first described in 1991. However, there is no general consensus on the clinical utility of CMC detection, largely due to conflicting results linked to the use of heterogeneous patient populations and different detection methods. Here, we developed a new EPithelial ImmunoSPOT (EPISPOT) assay to detect viable CMCs based on their secretion of the S100 protein (S100-EPISPOT). Then, we compared the results obtained with the S100-EPISPOT assay and the CellSearch® CMC kit using blood samples from a homogeneous population of patients with metastatic melanoma. We found that S100-EPISPOT sensitivity was significantly higher than that of CellSearch®. Specifically, the percentage of patients with ≥2 CMCs was significantly higher using S100-EPISPOT than CellSearch® (48% and 21%, respectively; $p = 0.0114$). Concerning CMC prognostic value, only the CellSearch® results showed a significant association with overall survival ($p = 0.006$). However, due to the higher sensitivity of the new S100-EPISPOT assay, it would be interesting to determine whether this functional test could be used in patients with non-metastatic melanoma for the early detection of tumor relapse and for monitoring the treatment response.

Keywords: circulating tumor cells; melanoma; liquid biopsy; EPISPOT; CellSearch®

---

1. Introduction

Melanoma is the most malignant skin cancer, and its incidence rate is increasing worldwide. Early stage and localized melanoma can be cured by surgical resection. Conversely, metastatic melanoma is one of the most aggressive and drug-resistant cancers with very poor overall survival (OS) (six to nine months). Melanoma management has recently undergone revolutionary changes with the discovery of predictive tumor biomarkers (BRAF mutations and immune checkpoint inhibitors such as programmed cell death protein 1 (PD-1), its ligand (PD-L1), and cytotoxic T-lymphocyte antigen 4 (CTLA-4)) and the development of the associated treatments. These new treatments, alone or in combination, have dramatically improved the outcome of patients with metastatic melanoma. For example, the anti-PD-1 drug pembrolizumab has demonstrated benefits to progression-free survival (34–38%) and objective responses (21–25%) at six months compared with chemotherapy (16% and 4%; $p < 0.0001$) [1]. However, despite the good response rates, immunotherapy results in systemic toxicity, and it is not effective in all patients.

Circulating tumor cells (CTCs) are cancer cells that are shed from the primary and metastatic tumor(s). They can be detected in peripheral blood samples using different technologies, but their identification and characterization require extremely sensitive and specific analytical methods [2–6]. Their analysis is considered as a real-time liquid biopsy for patients with cancer [7–10]. In 2011, the U.S. Food and Drug Administration (FDA) cleared the CellSearch® system (Menarini Silicon Biosystems) for CTC analysis to monitor patients with metastatic breast, colorectal and prostate cancer [11–13]. The CellSearch® epithelial cell-based assay has clearly demonstrated its clinical significance and is now used as the gold standard in clinical studies evaluating different cancer types. Even though a very limited number of studies have evaluated melanoma CTCs using the CellSearch® Circulating Melanoma Cell Kit, they all provided similar results, reflecting the robustness and reproducibility of this assay. The detection of circulating melanoma cells (CMCs) was described for the first time in 1991. Since then, the many studies on CMCs from patients with melanoma at different stages and using different detection approaches have reported conflicting results [14]. Indeed, metastatic melanoma is a highly heterogeneous tumor and CMCs may display different phenotypes and functional states.

Moreover, CMC analysis with the CellSearch® detection kit does not allow discriminating between dead and viable CMCs, the only CMCs involved in metastatic development [15]. The functional EPithelial ImmunoSPOT (EPISPOT) assay was described in 2005 and allows the identification of viable CTCs in peripheral blood samples of patients with cancer (e.g., breast, prostate, and colon cancer) [16–20] by detecting proteins secreted/released/shed by single viable epithelial cancer cells [21].

The aim of this study was to compare CMC detection using the CellSearch® system and a new EPISPOT assay (S100-EPISPOT assay) designed to identify viable CMCs that secrete S100, a protein expressed and secreted by melanoma cells [22], in blood samples from patients with metastatic melanoma.

2. Materials and Methods

2.1. Patient Cohort

A prospective controlled observational comparative study (Circulating Tumor Cells and Melanoma: Comparing the EPISPOT and CellSearch Techniques; NCT01558349) was conducted at the Nîmes University Hospital, Nîmes, France, between June 2013 and June 2017. The main objective was to determine if we can observe more positive patients with the EPISPOT assay than the CellSearch® system. All patients with melanoma signed a written informed consent before enrolment in the CELLCIRC study and treatment initiation. The study was carried out in accordance with the World Medical Association Declaration of Helsinki. The experimental protocol was approved by the French bioethical committee "Sud Méditerranée III" (Approval reference No. 2012.06.10). Blood samples from healthy volunteers (n = 38) and patients with metastatic malignant melanoma (n = 50; before any treatment) were collected in the morning and processed within 24 h.

2.2. Melanoma Cell Lines

The melanoma cancer cell lines WM-266-4 (ATCC® CRL-1676™) and MV3 (kindly provided by Klaus Pantel, University of Tumor Biology, Hamburg, Germany) were used for optimizing the S100-EPISPOT assay. WM-266-4 cells were maintained in αMEM medium (22571, Gibco, Grand Island, USA) supplemented with 10% fetal calf serum (FCS), and MV3 cells in RPMI 1640 medium (L0501, Dominique Dutscher, Brumath, France), supplemented with 5mM L-glutamine (25030, Gibco, Grand Island, USA) and 10% FCS.

2.3. Flow Cytometry Experiments

Intracellular expression of the S100 protein in WM-266-4 and MV3 cells was determined by flow cytometry using a Cyan cytometer (Beckman-Coulter, Villepinte, France) and a fixation/permeabilization kit (Beckman Coulter, Brea, USA). The two anti-S100 antibodies (clones 8B10 and 6G1) used in the EPISPOT assay were tested to confirm S100 expression in these melanoma cell lines.

2.4. *Immunofluorescence Assay*

Melanoma cell lines were immunostained with the two anti-S100 antibodies (8B10 and 6G1), as described for the flow cytometry experiments. Then, cells were seeded on glass slides using a Cytospin 4 centrifuge (Shandon, Runcorn, England) and mounted with ProLong Gold Antifade reagent with 4′,6-diamidino-2-phenylindole (DAPI) (Invitrogen). S100 expression was analyzed with a fluorescent microscope (Axio Imager M1, Carl Zeiss Vision, Halbermoos, Germany).

2.5. *S100-EPISPOT Assay*

As CMCs are rare in peripheral blood, a pre-enrichment step was performed before the EPISPOT assay. To separate erythrocytes and leukocytes from CMCs, the RosettSep™ reagent (20 µL/mL) was added to 13–15 mL of blood collected in EDTA tubes, and enrichment was performed following the manufacturer's instructions (RosetteSep™ CTC Enrichment Cocktail containing Anti-CD36, STEMCELL Technologies). During the S100-EPISPOT assay, enriched CMCs were cultured on a membrane coated with the anti-S100 8B10 antibody (10 ng/µL; Abcam) for 2 days. Then, secreted S100 captured by the 8B10 antibody was detected by incubation with another antibody against S100 (6G1: 3 ng/µL; Abcam) conjugated to AlexaFluor 488. Single fluorescent S100 immunospots were counted under a fluorescent microscope equipped with a camera and computer-assisted analysis (KS ELISPOT, Carl Zeiss Vision). The detailed procedure of the EPISPOT assay has been described by Soler et al. [21]. Results were corrected as "number of cells per 7.5 mL of blood" to be comparable with those obtained with the CellSearch system.

2.6. *Cell Search® System*

Patient blood samples were collected in special CellSave® tubes and the analysis was performed using the Circulating Melanoma Cell Kit (9594V, Menarini, Bologna, Italy), according to the manufacturer's instructions, as previously described [23]. Reagents consisted of ferrofluids coated with anti-CD146 antibodies to enrich melanoma cells and endothelial cells, a phycoerythrin-conjugated antibody that binds to high molecular weight melanoma-associated antigen (HMW-MAA) to identify melanoma cells, a mixture of two allophycocyanin-conjugated antibodies against CD45 to identify leukocytes and against CD34 to identify endothelial cells, and the nuclear dye DAPI to identify nucleated cells. The criteria to identify an object as a melanoma cell included: round to oval morphology, visible (DAPI-positive) nucleus, positive staining for HMW-MAA, and negative staining for CD45 and CD34. Results were expressed as number of cells per 7.5 mL of blood.

2.7. *Statistical Analysis*

The sensitivity and specificity of each technology (S100-EPISPOT and CellSearch®) were assessed by comparing the true positive/negative and false positive/negative results for healthy donors and patients with melanoma using the McNemar's test and the McNemar's test with continuity correction.

Patients' characteristics were described using medians and ranges (quantitative variables). The Fisher's test was used to evaluate the association between categorical clinical characteristics and CMC detection. The association between lactate dehydrogenase (LDH) value and CMC number was assessed using Spearman and Kendall rank correlation coefficients.

The median OS was analyzed using the Kaplan–Meier method. The end point was death by any cause. Survival curves were compared with the non-parametric log rank test ($p \leq 0.05$ considered as significant). The univariate Cox proportional hazard regression model was used to estimate the hazard ratio and 95% confidence intervals (CI) for the CellSearch® assay, and the risk of death in function of the CMC number (<2 vs. ≥2) after verifying the hazard proportional test.

Statistical analyses were performed using SAS, version 9.4 (SAS Institute).

3. Results

3.1. S100 Expression in Melanoma Cell Lines

To develop a new EPISPOT assay for CMC detection, two different melanoma cell lines (WM-266-4 and MV3), derived from metastatic sites, were used. Analysis of S100 protein expression by flow cytometry (FC) showed that both cell lines expressed S100, but at different levels (one log difference between cell lines) (Figure 1a). These results were confirmed by immunofluorescence (IF) experiments (Figure 1b). Both cell lines expressed S100, but the signal intensity was higher in WM-266-4 than in MV3 cells (same exposure time). Similar results were obtained with the two anti-S100 antibodies (8B10 and 6G1) used in the S100-EPISPOT assay. These findings indicated that both melanoma cell lines could be used to optimize the S100-EPISPOT assay.

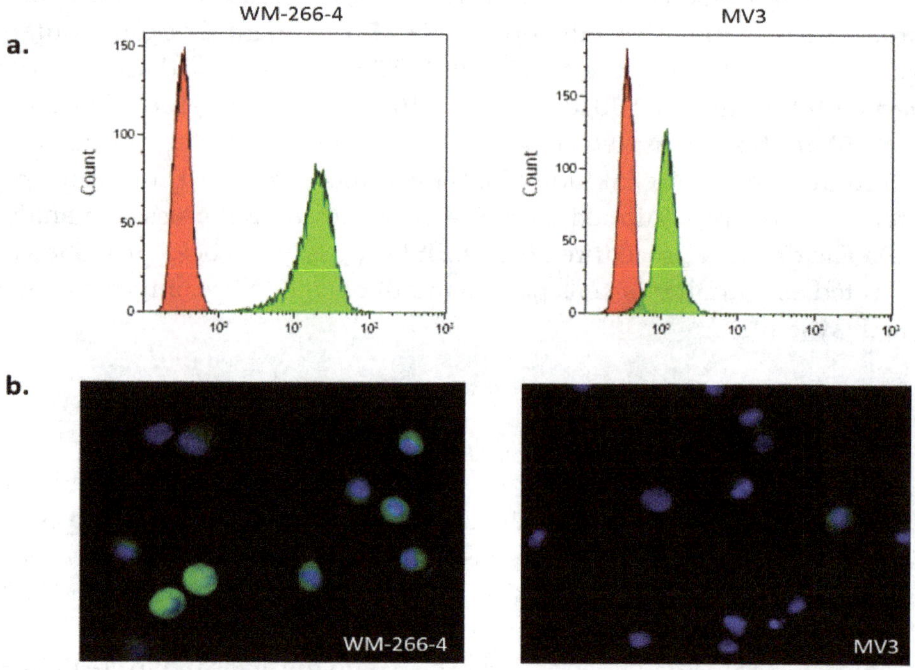

Figure 1. S100 protein expression in WM-266-4 and MV3 cells by (**a**) flow cytometry and (**b**) immunofluorescence analysis using the anti-S100 antibody 6G1 conjugated to AlexaFluor 488.

3.2. S100-EPISPOT Assay

Then, the two melanoma cell lines were used to evaluate the feasibility and detection threshold of the S100-EPISPOT assay. No immunospots (indicative of S100 secretion) could be detected when using MV3 cells, possibly due to their very weak S100 expression, as observed by FC and IF. Conversely, on average, 20–47% of WM-266-4 cells secreted S100 (Figure 2). This detection rate is considered normal for cell lines analyzed with the EPISPOT assay because of variations in cell cycle and protein productivity. In parallel, enriched peripheral blood mononuclear cells (PBMCs) from healthy donors were tested (n = 3) to confirm that S100 was not secreted by PBMCs (data not show) and validate this assay for CMC detection after blood enrichment.

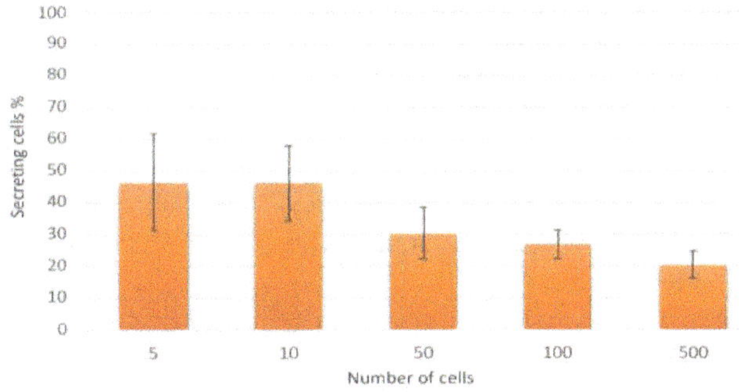

Number of cells	Secreting cells % (mean)	SD	Range
5	47	15	[0-5]
10	46	12	[1.3-8.9]
50	30	8	[8.7-23.8]
100	27	5	[19.6-66.7]
500	20	4	[83.3-141.5]

Figure 2. Melanoma cell detection using the S100-EPISPOT assay. Representative images of immunospots using WM-266-4 cells (upper panels) and percentage (n = 3) of S100-secreting cells recovered from serial dilutions of WM-266-4 cells (lower panels). SD = standard deviation.

Surprisingly, the detection rate was higher with smaller numbers of tested cells (Figure 2). These data demonstrate that the new S100-EPISPOT assay could be used to detect rare CMCs in peripheral blood. Previous studies also reported a better recovery rate for small number of cancer cells spiked in samples of healthy donors (1–20 cancer cells/10 mL) [24,25].

3.3. *Specificity of the S100-EPISPOT Assay*

Among the 38 healthy donors enrolled, six and four donors could not be evaluated by EPISPOT and CellSearch, respectively, for reasons listed in Figure 3a. The specificity of the S100-EPISPOT and CellSearch® assays were compared using blood samples from 38 healthy donors and two detection thresholds (≥1 and ≥2 CMCs) (Table 1a). The CellSearch® system gave similar results with both thresholds: 94% and 97%. The S100-EPISPOT assay specificity was lower for the ≥1 CMC threshold (78%), but reached 97%, as for CellSearch®, for the ≥2 CMC threshold. These results are consistent with previous reports and allowed for choosing a cut-off of 2 CMCs for the positive CMC detection [23,26].

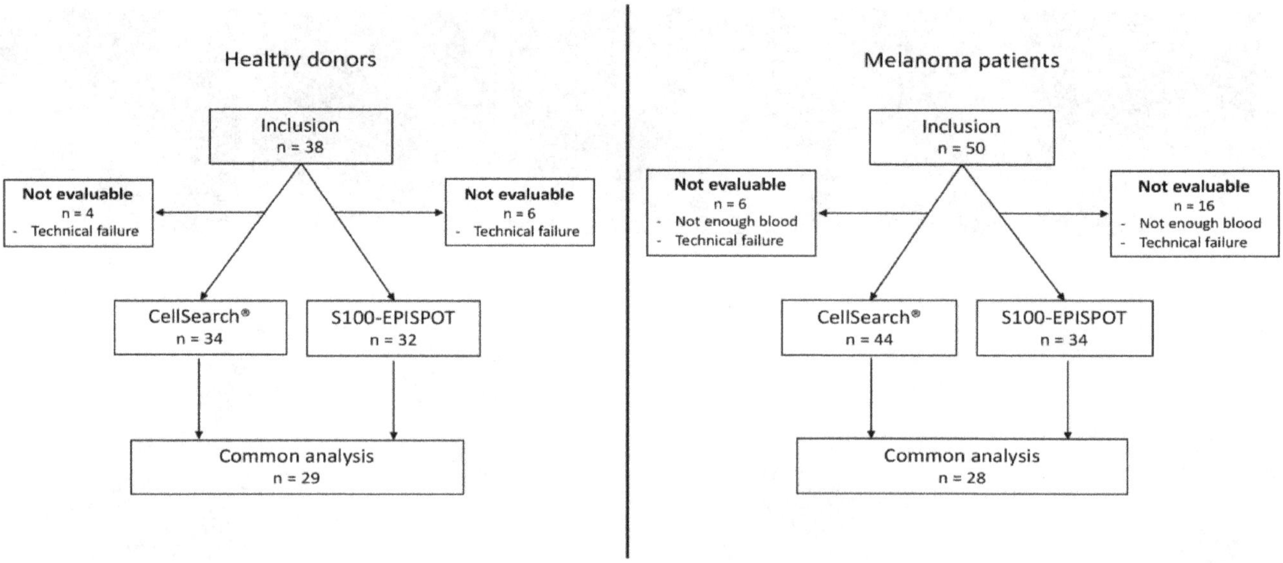

Figure 3. Study flowchart showing the number of included healthy donors and patients and the number of patients and healthy donors who underwent circulating melanoma cell (CMC) analysis using the S100- EPISPOT assay and CellSearch® system.

Table 1. (**a**) Specificity of the S100-EPISPOT and CellSearch CMC assays in healthy donors using two different CMC cut-offs. (**b**) Sensitivity and specificity of the two assays with the cut-off ≥2 CMCs (healthy donors: $n = 28$; patients with metastatic melanoma: $n = 29$).

a.

	S100-EPISPOT (n = 32)		**CellSearch CMC (n = 34)**	
<1	25	(78%)	32	(94%)
≥1	7	(22%)	2	(6%)
<2	31	(97%)	33	(97%)
≥2	1	(3%)	1	

b.

	S100 EPISPOT	**CellSearch CMC**	*p value* (Mac Nemar)	*p value* (corrected Mac Nemar)
Sensitivity (n = 29)	48%	21%	**0.0114**	**0.0269**
Specificity (n = 28)	100%	96%	0.3173	1

3.4. Circulating Melanoma Cell Detection in Patients with Metastatic Melanoma

Among the 50 patients enrolled, 16 and 6 could not be evaluated by EPISPOT and CellSearch, respectively, for reasons listed in Figure 3b. The patients included in this study (26 men and 24 women with a mean age of 64, range, 29–89) had metastatic melanoma, with a median survival of 6.72 months (95% CI = 4.24–12.16). CMCs were detected using the CellSearch® system and the S100-EPISPOT assay in blood samples from 44 and 34 patients, respectively. Using the cut-off of ≥2 CMCs per 7.5 mL of blood, 10/44 (23%) patients were positive for CMCs with the CellSearch® system, and 15/34 (44%) with the S100-EPISPOT assay (Table 2). The S100-EPISPOT assay gave a higher number of positive patients than the CellSearch® system, but with a smaller range: EPISPOT (0–450) and CellSearch® (0–4937). Analysis of the specificity and sensitivity using the results obtained for samples from healthy donors ($n = 28$) and patients with metastatic melanoma ($n = 29$) tested with both assays, showed that sensitivity was 48% for S100-EPISPOT and 21% for CellSearch® ($p = 0.0114$, corrected $p = 0.0269$), and specificity was 100% for S100-EPISPOT and 96% for CellSearch® (Table 1b). The correlation between both technologies was assessed using the Spearman ($r_s = 0.49$, $p = 0.0070$) and Kendall ($\tau = 0.44$, $p = 0.0055$) tests.

Table 2. Description of CMC detection for the S100-EPISPOT and CellSearch® assays in (**a**) all patients included in the study and (**b**) the subpopulation of patients tested with both assays.

a.

Nb Patients	Assay	Total	Failed	Mean	Standard Deviation	Median	Min	Max	Lower quartile	Upper quartile
50	S100-EPISPOT	34	16	21.59	80.35	1.00	0.00	450.00	0.00	3.00
	CellSearch®	44	6	142.57	752.01	0.00	0.00	4937.00	0.00	1.00

b.

Nb Patients	Assay	Total	Failed	Mean	Standard Deviation	Median	Min	Max	Lower quartile	Upper quartile
29	S100-EPISPOT	29	0	25.24	86.68	1.00	0.00	450.00	0.00	4.00
	CellSearch®	29	0	206.10	924.77	0.00	0.00	4937.00	0.00	1.00

3.5. Clinical Relevance of CMC Detection

No significant association between the patients' clinical characteristics and CMC detection with EPISPOT and CellSearch® was observed (Table 3), except for high LDH level (2-fold higher than normal) and CMC detection using CellSearch® ($p = 0.0315$). The correlation between LDH level and CTC detection by CellSearch® was significant using the Spearman ($r_s = 0.39, p = 0.0138$) and Kendall ($\tau = 0.30, p = 0.0132$) tests.

Table 3. Clinical characteristic of patients with metastatic melanoma in function of CTC detection with the S100-EPISPOT and CellSearch® assays.

	EPISPOT			CellSearch®		
	< 2 n = 19	≥ 2 n = 15	*p* value (Fisher)	< 2 n = 34	≥ 2 n = 10	*p* value (Fisher)
Sex						
Men	8 (42.11%)	9 (60.00%)	0.4905	18 (52.94%)	4 (40.00%)	0.7205
Women	11 (57.89%)	6 (40.00%)		16 (47.06%)	6 (60.00%)	
BRAF mutation						
No	8 (47.06%)	6 (54.55%)	1.0000	15 (48.39%)	2 (28.57%)	0.4267
Yes	9 (52.94%)	5 (45.45%)		16 (51.61%)	5 (71.43%)	
Ulceration						
Absence	4 (36.36%)	6 (85.71%)	0.0656	11 (52.38%)	4 (66.67%)	0.6618
Presence	7 (63.64%)	1 (14.29%)		10 (47.62%)	2 (33.33%)	
Metastatic sites						
Nb ≤ 2	10 (52.63%)	7 (46.67%)	1.0000	22 (64.71%)	3 (30.00%)	0.0738
Nb > 2	9 (47.37%)	8 (53.33%)		12 (35.29%)	7 (70.00%)	
LDH value						
Normal	6 (42.86%)	5 (35.71%)	0.2909	10 (34.48%)	1 (10.00%)	**0.0315**
> Normal	7 (50.00%)	4 (28.57%)		16 (55.17%)	4 (40.00%)	
2x > Normal	1 (7.14%)	5 (35.71%)		3 (10.34%)	5 (50.00%)	

Table 3. *Cont.*

	EPISPOT			CellSearch®		
	< 2 n = 19	≥ 2 n = 15	*p* value (Fisher)	< 2 n = 34	≥ 2 n = 10	*p* value (Fisher)
Primary tumor localization						
Other	1 (2.94%)	6 (17.65%)	**0.0106**	6 (13.95%)	3 (6.98%)	0.2017
Face	1 (2.94%)	0 (0.00%)		0 (0.00%)	1 (2.33%)	
Lower limb	8 (23.53%)	2 (5.88%)		12 (27.91%)	1 (2.33%)	
Upper limb	6 (17.65%)	1 (2.94%)		7 (16.28%)	1 (2.33%)	
Torso	3 (8.82%)	6 (17.65%)		9 (20.93%)	3 (6.98%)	

In univariate Kaplan–Meier analyses, OS was associated only with CMC detection by CellSearch® (CS) ($p = 0.0006$) (Figure 4a). Hazard ratio for death was 3.57 (95% CI 1.64–7.77; Cox regression analysis with CMC <2 vs. ≥2). No association was detected between OS and the classical clinical variables (age, sex, Breslow depth, Clark stage, LDH, BRAF mutation, and number of metastatic sites) (Table S1). Moreover, no significant association was observed for the S100-EPISPOT assay (EPI) alone (Figure S1). Analysis of the potential synergistic effect of combining both assays indicated that the "double negative" (EPI and CS < 2 CMCs) group was not associated with OS ($p = 0.3419$) compared with the "double and/or simple positive" (EPI and/or CS ≥2) group. Conversely, analysis of the "double positive" (EPI and CS ≥ 2 CMCs), "simple positive" (EPI or CS ≥2 CMCs) and "double negative" (EPI and CS < 2 CMCs) groups highlighted the association of the "double positive" group with OS ($p = 0.0081$) (Figure 4b).

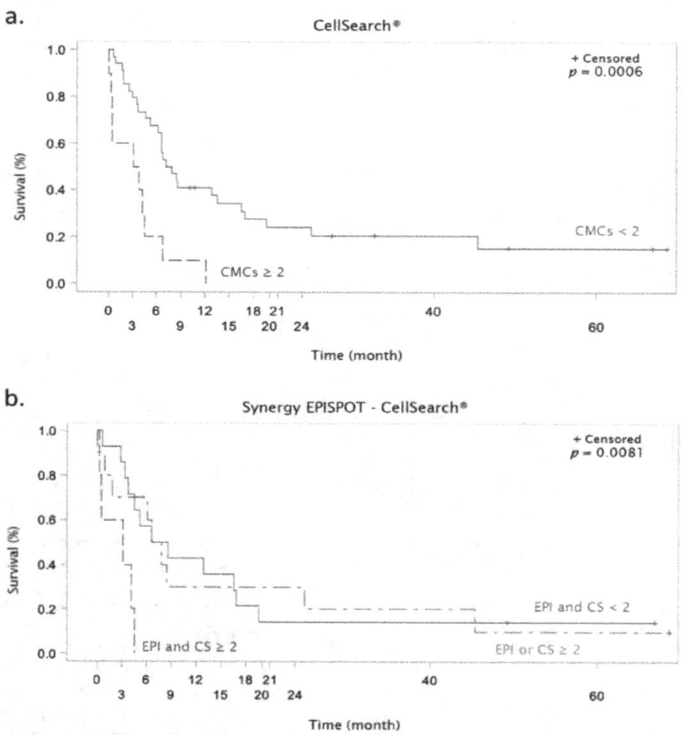

Figure 4. Kaplan–Meier survival curves of patients with metastatic melanoma according to the detection cut-off of 2 CMCs using the (**a**) CellSearch® system, and (**b**) when combining the S100-EPISPOT (EPI) and CellSearch® (CS) assay results.

4. Discussion

The cancer metastatic cascade is a complex process characterized by several events, including cell migration, local invasion, intravasation of tumor cells into the circulation and extravasation at distant sites to form detectable metastases [27]. The mechanisms involved in this multistep process are largely unknown, but recent studies suggested that some CTCs could be at the origin of distant metastases [28,29]. For this reason, CTC detection has become a great challenge for the personalized treatment of patients with cancer.

In the literature, the number of techniques used for CTC detection in patients with melanoma is almost equal to the number of studies published on this topic. These methods are based on the physical or phenotypical properties of melanoma cells [3,30,31]. Up to now, none of them has been validated because of the lack of reproducible results. Nestin, CD133 [32], receptor activator of NF-k B (RANK) [33], ATP-binding cassette sub-family B member 5 (ABCB5) [34], CD20 [35], and CD271 [36] have been identified as potential candidates for the identification of melanoma-initiating cells.

Compared with CTC detection techniques based only on the expression of surface markers, assays that identify viable/functional cells could be more interesting and more relevant for monitoring the response to treatment. Functional CTC analysis also offers the possibility to determine the biological properties of metastatic cells, including the identification of metastasis-initiating cells [37]. Currently, the EPISPOT assay is the only functional assay to detect viable, prognostically relevant CTCs at the single-cell level after enrichment by leukocyte depletion. This system has been used to test blood samples from hundreds of patients with different tumor types [17–19]. As this technique could be considered to be a protein secretion-profiling assay, the analyzed proteins should be specific and significantly produced and released by tumor cells. Among the proteins expressed and released by melanoma cells, the S100 family is the most studied [22,38]. S100B expression is increased in melanoma cells compared with melanocytes and can be used for the diagnosis of metastatic malignant melanoma by immunohistochemistry. Moreover, serum S100B level increases in patients with melanoma, independently of the cancer stage. Its expression is clearly correlated with the presence of metastases, tumor burden, prognosis, and survival [39,40]. Recent data demonstrated that isolated CMCs obtained from patients with metastatic melanoma uniformly express S100 [41].

In this study, we first described the new S100 EPISPOT assay to detect viable CMCs, and then compared its results with those obtained with the CellSearch® system (Circulating Melanoma Cell Kit). In two melanoma cell lines, we found that S100 expression and secretion were heterogeneous. Consequently, only the cell line that strongly expresses S100 (WM-266-4 cells) secreted enough protein to be detected by the S100-EPIPSOT assay at the single-cell level. The subpopulation of cancer cells that weakly express S100 are missed using this assay. Moreover, WM-266-4 cell detection by the S100-EPISPOT system was more efficient when only a few cells were present in the sample. These experimental observations were confirmed by data obtained from blood samples from a homogeneous group of patients with metastatic melanoma, indicating that the S100-EPISPOT assay sensitivity is significantly higher than that of the CellSearch®system. Specifically, the percentage of patients with ≥2 CMCs was 48% with the S100-EPISPOT assay and 21% with the CellSearch® system (corrected $p = 0.0269$).

CMC detection (both methods) was not associated with Breslow depth and BRAF mutation status. Only LDH value was significantly correlated with CMC detection (CellSearch® method) in Spearman ($r_s = 0.39$, $p = 0.0138$) and Kendall ($k = 0.30$, $p = 0.0132$) tests.

Analysis of the prognostic value of CMC detection (both assays) indicated that only CMC ≥ 2 with CellSearch® was significantly associated with the OS ($p = 0.006$), as previously reported for the CellSearch® CMC detection system. By using this CellSearch® system, Rao et al. [23] detected at least two CMCs in 23% (18/79) of blood samples from 44 patients with metastatic melanoma. They also found that OS was shorter in patients with ≥2 CMCs per 7.5 mL of whole blood compared with the <2 CMC group. Similarly, Khoja et al. [26] showed that 26% of patients with metastatic skin melanoma had more than two CMCs per 7.5 mL of blood, and that the median OS was significantly shorter in this group.

In our patient cohort, no other clinical features, such as number of metastatic sites, Breslow depth, BRAF mutational status and even LDH level (routinely measured in the clinical practice) provided prognostic information. The S100-EPISPOT assay results alone did not predict the patients' clinical outcome, although this test was more sensitive than the CellSearch® system. Conversely, by combining the data obtained by both methods, OS was significantly associated ($p = 0.0005$) with "double positive results" (CellSearch® and S100-EPISPOT) compared with "simple positive result" (CellSearch® or EPISPOT) and "double negative result" (CellSearch® and S100-EPISPOT). This suggests that OS is shorter in patients in whom CMCs are detected with both methods.

However, in our patients with advanced disease and short survival probability (0.38, 95% CI 0.25–0.51 after 12 months and 0.16, 95% CI 0.07–0.27 after 24 months), measuring OS is not of clinical interest. The crucial aim of liquid biopsy is to obtain reliable and "real-time" information before, during, and after treatment to monitor the patient response. This is especially important for melanoma because currently no melanoma-specific blood-based biomarker is available for routine use in clinical practice. Another circulating biomarker of interest in advanced melanoma is ctDNA. The majority of studies have been applied to patients treated with BRAF inhibitors, via monitoring of the singular BRAFV600 mutation, predicting response to therapy and prognosis in metastatic melanoma [42]. Nevertheless, clinical trials that look at patient outcome as a result of ctDNA-guided clinical decisions are required before ctDNA can be successfully established as a melanoma-specific biomarker in clinical practice.

Finally, the main interest of our study is the finding that the new S100-EPISPOT assay has a good sensitivity (48%) for CMC detection and an acceptable specificity. It would be interesting to determine whether the S100-EPISPOT functional assay could be used for the early detection of tumor relapse or for monitoring therapy response in patients with non-metastatic melanoma.

Moreover, a completely new optimized EPISPOT assay, named EPIDROP (EPISPOT in a DROP) [43], is currently under development. This innovative micro-droplet technology allows not only the detection of viable S100-secreting CMCs at the single-cell level, but also the immunostaining of all CMCs before their encapsulation, for instance for the identification of tumor cells that could be targeted by immunotherapy (e.g., PD-L1-positive). This technology represents a new combination of standard CMC detection by IF, like the CellSearch® system, with a functional assay to identify the subset of functional and potentially metastasis-competent CMCs.

Author Contributions: Conceptualization, L.C., C.A.-P. and L.M.; Methodology, J-P.D.; Validation, L.C., C.A.-P. and L.M.; Formal analysis F.G., L.C. and F.P.; Investigation, L.C and L.M.; Resources, A.D.R., C.G., P-E.S., F.F. and L.M.; Writing—original draft preparation, L.C.; Writing—review and editing, F.G., C.A.-P., L.M. and F.P.; Supervision, C.A.-P. and L.M.; Funding acquisition, C.A.-P. and L.M.

Acknowledgments: We thank Nathalie Bedos, Sabrina Nicolas and Lyamin Bendjeddou, research nurses at the Nîmes University Hospital, involved in this study. We thank Elisabetta Andermarcher for assistance with her comments and proof-reading that greatly improved the manuscript.

References

1. Dummer, R.; Daud, A.; Puzanov, I.; Hamid, O.; Schadendorf, D.; Robert, C.; Schachter, J.; Pavlick, A.; Gonzalez, R.; Hodi, F.S.; et al. A randomized controlled comparison of pembrolizumab and chemotherapy in patients with ipilimumab-refractory melanoma. *J. Transl. Med.* **2015**, *13*. [CrossRef]
2. Alix-Panabieres, C.; Pantel, K. Technologies for detection of circulating tumor cells: Facts and vision. *Lab Chip* **2014**, *14*, 57–62. [CrossRef] [PubMed]
3. Nezos, A.; Msaouel, P.; Pissimissis, N.; Lembessis, P.; Sourla, A.; Armakolas, A.; Gogas, H.; Stratigos, A.J.; Katsambas, A.D.; Koutsilieris, M. Methods of detection of circulating melanoma cells: A comparative overview. *Cancer Treat. Rev.* **2011**, *37*, 284–290. [CrossRef] [PubMed]

4. van der Toom, E.E.; Verdone, J.E.; Gorin, M.A.; Pienta, K.J. Technical challenges in the isolation and analysis of circulating tumor cells. *Oncotarget* **2016**. [CrossRef] [PubMed]
5. Joosse, S.A.; Gorges, T.M.; Pantel, K. Biology, detection, and clinical implications of circulating tumor cells. *EMBO Mol. Med.* **2015**, *7*, 1–11. [CrossRef] [PubMed]
6. Alix-Panabieres, C.; Schwarzenbach, H.; Pantel, K. Circulating tumor cells and circulating tumor DNA. *Annu. Rev. Med.* **2012**, *63*, 199–215. [CrossRef] [PubMed]
7. Pantel, K.; Alix-Panabieres, C. Liquid biopsy: Potential and challenges. *Mol. Oncol.* **2016**, *10*, 371–373. [CrossRef] [PubMed]
8. Alix-Panabieres, C.; Pantel, K. Clinical Applications of Circulating Tumor Cells and Circulating Tumor DNA as Liquid Biopsy. *Cancer Discov.* **2016**. [CrossRef]
9. Khoo, B.L.; Grenci, G.; Jing, T.; Lim, Y.B.; Lee, S.C.; Thiery, J.P.; Han, J.; Lim, C.T. Liquid biopsy and therapeutic response: Circulating tumor cell cultures for evaluation of anticancer treatment. *Sci. Adv.* **2016**, *2*, e1600274. [CrossRef]
10. Alix-Panabieres, C.; Pantel, K. Circulating tumor cells: Liquid biopsy of cancer. *Clin. Chem.* **2013**, *59*, 110–118. [CrossRef]
11. de Bono, J.S.; Scher, H.I.; Montgomery, R.B.; Parker, C.; Miller, M.C.; Tissing, H.; Doyle, G.V.; Terstappen, L.W.; Pienta, K.J.; Raghavan, D. Circulating tumor cells predict survival benefit from treatment in metastatic castration-resistant prostate cancer. *Clin. Cancer Res.* **2008**, *14*, 6302–6309. [CrossRef]
12. Cohen, S.J.; Punt, C.J.; Iannotti, N.; Saidman, B.H.; Sabbath, K.D.; Gabrail, N.Y.; Picus, J.; Morse, M.; Mitchell, E.; Miller, M.C.; et al. Relationship of circulating tumor cells to tumor response, progression-free survival, and overall survival in patients with metastatic colorectal cancer. *J. Clin. Oncol.* **2008**, *26*, 3213–3221. [CrossRef]
13. Cristofanilli, M.; Budd, G.T.; Ellis, M.J.; Stopeck, A.; Matera, J.; Miller, M.C.; Reuben, J.M.; Doyle, G.V.; Allard, W.J.; Terstappen, L.W.M.M.; et al. Circulating Tumor Cells, Disease Progression, and Survival in Metastatic Breast Cancer. *N. Engl. J. Med.* **2004**, *351*, 781–791. [CrossRef]
14. Khoja, L.; Lorigan, P.; Dive, C.; Keilholz, U.; Fusi, A. Circulating tumour cells as tumour biomarkers in melanoma: Detection methods and clinical relevance. *Ann. Oncol. Off. J. Eur. Soc. Med Oncol. ESMO* **2015**, *26*, 33–39. [CrossRef]
15. Pantel, K.; Alix-Panabieres, C. Functional Studies on Viable Circulating Tumor Cells. *Clin. Chem.* **2016**, *62*, 328–334. [CrossRef]
16. Alix-Panabieres, C. EPISPOT assay: Detection of viable DTCs/CTCs in solid tumor patients. *Recent Results Cancer Res.* **2012**, *195*, 69–76. [CrossRef]
17. Deneve, E.; Riethdorf, S.; Ramos, J.; Nocca, D.; Coffy, A.; Daures, J.P.; Maudelonde, T.; Fabre, J.M.; Pantel, K.; Alix-Panabieres, C. Capture of viable circulating tumor cells in the liver of colorectal cancer patients. *Clin. Chem.* **2013**, *59*, 1384–1392. [CrossRef]
18. Ramirez, J.M.; Fehm, T.; Orsini, M.; Cayrefourcq, L.; Maudelonde, T.; Pantel, K.; Alix-Panabieres, C. Prognostic relevance of viable circulating tumor cells detected by EPISPOT in metastatic breast cancer patients. *Clin. Chem.* **2014**, *60*, 214–221. [CrossRef]
19. Alix-Panabieres, C.; Pantel, K. Liquid biopsy in cancer patients: Advances in capturing viable CTCs for functional studies using the EPISPOT assay. *Expert Rev. Mol. Diagn.* **2015**, *15*, 1411–1417. [CrossRef]
20. Kuske, A.; Gorges, T.M.; Tennstedt, P.; Tiebel, A.K.; Pompe, R.; Preisser, F.; Prues, S.; Mazel, M.; Markou, A.; Lianidou, E.; et al. Improved detection of circulating tumor cells in non-metastatic high-risk prostate cancer patients. *Sci. Rep.* **2016**, *6*, 39736. [CrossRef]
21. Soler, A.; Cayrefourcq, L.; Mazel, M.; Alix-Panabieres, C. EpCAM-Independent Enrichment and Detection of Viable Circulating Tumor Cells Using the EPISPOT Assay. *Methods Mol. Biol.* **2017**, *1634*, 263–276. [CrossRef]
22. Alegre, E.; Sammamed, M.; Fernández-Landázuri, S.; Zubiri, L.; González, Á. Chapter Two-Circulating Biomarkers in Malignant Melanoma. In *Advances in Clinical Chemistry*; Makowski, G.S., Ed.; Elsevier: Amsterdam, The Netherlands, 2015; Volume 69, pp. 47–89.
23. Rao, C.; Bui, T.; Connelly, M.; Doyle, G.; Karydis, I.; Middleton, M.R.; Clack, G.; Malone, M.; Coumans, F.A.; Terstappen, L.W. Circulating melanoma cells and survival in metastatic melanoma. *Int. J. Oncol.* **2011**, *38*, 755–760. [CrossRef]
24. Alix-Panabieres, C.; Brouillet, J.P.; Fabbro, M.; Yssel, H.; Rousset, T.; Maudelonde, T.; Choquet-Kastylevsky, G.;

Vendrell, J.P. Characterization and enumeration of cells secreting tumor markers in the peripheral blood of breast cancer patients. *J. Immunol. Methods* **2005**, *299*, 177–188. [CrossRef]

25. Alix-Panabières, C.; Rebillard, X.; Brouillet, J.-P.; Barbotte, E.; Iborra, F.; Segui, B.; Maudelonde, T.; Jolivet-Reynaud, C.; Vendrell, J.-P. Detection of Circulating Prostate-Specific Antigen–Secreting Cells in Prostate Cancer Patients. *Clin. Chem.* **2005**, *51*, 1538–1541. [CrossRef]
26. Khoja, L.; Lorigan, P.; Zhou, C.; Lancashire, M.; Booth, J.; Cummings, J.; Califano, R.; Clack, G.; Hughes, A.; Dive, C. Biomarker utility of circulating tumor cells in metastatic cutaneous melanoma. *J. Investig. Dermatol.* **2013**, *133*, 1582–1590. [CrossRef]
27. Pantel, K.; Speicher, M.R. The biology of circulating tumor cells. *Oncogene* **2015**. [CrossRef]
28. Baccelli, I.; Schneeweiss, A.; Riethdorf, S.; Stenzinger, A.; Schillert, A.; Vogel, V.; Klein, C.; Saini, M.; Bauerle, T.; Wallwiener, M.; et al. Identification of a population of blood circulating tumor cells from breast cancer patients that initiates metastasis in a xenograft assay. *Nat. Biotechnol.* **2013**, *31*, 539–544. [CrossRef]
29. Hodgkinson, C.L.; Morrow, C.J.; Li, Y.; Metcalf, R.L.; Rothwell, D.G.; Trapani, F.; Polanski, R.; Burt, D.J.; Simpson, K.L.; Morris, K.; et al. Tumorigenicity and genetic profiling of circulating tumor cells in small-cell lung cancer. *Nat. Med.* **2014**, *20*, 897–903. [CrossRef]
30. Rodic, S.; Mihalcioiu, C.; Saleh, R.R. Detection methods of circulating tumor cells in cutaneous melanoma: A systematic review. *Crit. Rev. Oncol. Hematol.* **2014**, *91*, 74–92. [CrossRef]
31. Xu, M.J.; Dorsey, J.F.; Amaravadi, R.; Karakousis, G.; Simone, C.B., 2nd; Xu, X.; Xu, W.; Carpenter, E.L.; Schuchter, L.; Kao, G.D. Circulating Tumor Cells, DNA, and mRNA: Potential for Clinical Utility in Patients With Melanoma. *Oncology* **2016**, *21*, 84–94. [CrossRef]
32. Fusi, A.; Reichelt, U.; Busse, A.; Ochsenreither, S.; Rietz, A.; Maisel, M.; Keilholz, U. Expression of the stem cell markers nestin and CD133 on circulating melanoma cells. *J. Investig. Dermatol.* **2011**, *131*, 487–494. [CrossRef]
33. Kupas, V.; Weishaupt, C.; Siepmann, D.; Kaserer, M.L.; Eickelmann, M.; Metze, D.; Luger, T.A.; Beissert, S.; Loser, K. RANK is expressed in metastatic melanoma and highly upregulated on melanoma-initiating cells. *J. Investig. Dermatol.* **2011**, *131*, 944–955. [CrossRef]
34. Schatton, T.; Murphy, G.F.; Frank, N.Y.; Yamaura, K.; Waaga-Gasser, A.M.; Gasser, M.; Zhan, Q.; Jordan, S.; Duncan, L.M.; Weishaupt, C.; et al. Identification of cells initiating human melanomas. *Nature* **2008**, *451*, 345–349. [CrossRef]
35. Fang, D.; Nguyen, T.K.; Leishear, K.; Finko, R.; Kulp, A.N.; Hotz, S.; Van Belle, P.A.; Xu, X.; Elder, D.E.; Herlyn, M. A tumorigenic subpopulation with stem cell properties in melanomas. *Cancer Res.* **2005**, *65*, 9328–9337. [CrossRef]
36. Boiko, A.D.; Razorenova, O.V.; van de Rijn, M.; Swetter, S.M.; Johnson, D.L.; Ly, D.P.; Butler, P.D.; Yang, G.P.; Joshua, B.; Kaplan, M.J.; et al. Human melanoma-initiating cells express neural crest nerve growth factor receptor CD271. *Nature* **2010**, *466*, 133–137. [CrossRef]
37. Alix-Panabieres, C.; Bartkowiak, K.; Pantel, K. Functional studies on circulating and disseminated tumor cells in carcinoma patients. *Mol. Oncol.* **2016**, *10*, 443–449. [CrossRef]
38. Palmer, S.R.; Erickson, L.A.; Ichetovkin, I.; Knauer, D.J.; Markovic, S.N. Circulating serologic and molecular biomarkers in malignant melanoma. *Mayo Clin. Proc.* **2011**, *86*, 981–990. [CrossRef]
39. Zarogoulidis, P.; Tsakiridis, K.; Karapantzou, C.; Lampaki, S.; Kioumis, I.; Pitsiou, G.; Papaiwannou, A.; Hohenforst-Schmidt, W.; Huang, H.; Kesisis, G.; et al. Use of proteins as biomarkers and their role in carcinogenesis. *J. Cancer* **2015**, *6*, 9–18. [CrossRef]
40. Kruijff, S.; Hoekstra, H.J. The current status of S-100B as a biomarker in melanoma. *Eur. J. Surg. Oncol. EJSO* **2012**, *38*, 281–285. [CrossRef]
41. Long, E.; Ilie, M.; Bence, C.; Butori, C.; Selva, E.; Lalvee, S.; Bonnetaud, C.; Poissonnet, G.; Lacour, J.P.; Bahadoran, P.; et al. High expression of TRF2, SOX10, and CD10 in circulating tumor microemboli detected in metastatic melanoma patients. A potential impact for the assessment of disease aggressiveness. *Cancer Med.* **2016**. [CrossRef]
42. Calapre, L.; Warburton, L.; Millward, M.; Ziman, M.; Gray, E.S. Circulating tumour DNA (ctDNA) as a liquid biopsy for melanoma. *Cancer Lett.* **2017**, *404*, 62–69. [CrossRef]
43. Pantel, K.; Alix-Panabieres, C. Liquid biopsy and minimal residual disease—Latest advances and implications for cure. *Nat. Rev. Clin. Oncol.* **2019**. [CrossRef]

10

New Frontiers in Diagnosis and Therapy of Circulating Tumor Markers in Cerebrospinal Fluid In Vitro and In Vivo

Olga A. Sindeeva [1], Roman A. Verkhovskii [1], Mustafa Sarimollaoglu [2], Galina A. Afanaseva [1,3], Alexander S. Fedonnikov [1,3], Evgeny Yu. Osintsev [1,3], Elena N. Kurochkina [1,3], Dmitry A. Gorin [4], Sergey M. Deyev [5], Vladimir P. Zharov [1,2] and Ekaterina I. Galanzha [1,6,*]

1 Laboratory of Biomedical Photoacoustics, Saratov State University, 83 Astrakhanskaya St, 410012 Saratov, Russia; mouse-oa@rambler.ru (O.A.S.); r.a.verhovskiy@mail.ru (R.A.V.); gafanaseva@yandex.ru (G.A.A.); fedonnikov@mail.ru (A.S.F.); dr_osintsev@mail.ru (E.Y.O.); e.katamadze@yandex.ru (E.N.K.); zharovvladimirp@uams.edu (V.P.Z.)

2 Arkansas Nanomedicine Center & Winthrop P. Rockefeller Cancer Institute, University of Arkansas for Medical Sciences, Little Rock, AR 72205, USA; msarimollaoglu@uams.edu

3 Saratov State Medical University, 112 Bolshaya Kazachia St., 410012 Saratov, Russia

4 Laboratory of Biophotonics, Skolkovo Institute of Science and Technology, 3 Nobelya Str., 121205 Moscow, Russia; gorinda@mail.ru

5 Shemyakin-Ovchinnikov Institute of Bioorganic Chemistry, Russian Academy of Sciences, Miklukho-Maklaya St., 16/10, Moscow 117997, Russia; biomem@mail.ru

6 Laboratory of Lymphatic Research, Diagnosis and Therapy (LDT), University of Arkansas for Medical Sciences, Little Rock, AR 72205, USA

* Correspondence: egalanzha@uams.edu

Abstract: One of the greatest challenges in neuro-oncology is diagnosis and therapy (theranostics) of leptomeningeal metastasis (LM), brain metastasis (BM) and brain tumors (BT), which are associated with poor prognosis in patients. Retrospective analyses suggest that cerebrospinal fluid (CSF) is one of the promising diagnostic targets because CSF passes through central nervous system, harvests tumor-related markers from brain tissue and, then, delivers them into peripheral parts of the human body where CSF can be sampled using minimally invasive and routine clinical procedure. However, limited sensitivity of the established clinical diagnostic cytology in vitro and MRI in vivo together with minimal therapeutic options do not provide patient care at early, potentially treatable, stages of LM, BM and BT. Novel technologies are in demand. This review outlines the advantages, limitations and clinical utility of emerging liquid biopsy in vitro and photoacoustic flow cytometry (PAFC) in vivo for assessment of CSF markers including circulating tumor cells (CTCs), circulating tumor DNA (ctDNA), microRNA (miRNA), proteins, exosomes and emboli. The integration of in vitro and in vivo methods, PAFC-guided theranostics of single CTCs and targeted drug delivery are discussed as future perspectives.

Keywords: cerebrospinal liquid biopsy; in vivo flow cytometry; tumor biomarkers; circulating tumor cells; ctDNA; miRNA; exosomes; emboli; targeted therapy

---

1. Introduction

Leptomeningeal and brain metastasis (LM and BM) as a result of metastatic dissemination of solid tumors (e.g., melanoma, breast cancer, lung cancer and colorectal cancer) and hematological neoplasms as well as primary brain tumors (BTs, e.g., glioma) are commonly fatal with minimum treatment options [1–11]. Relatively high number of underdiagnosed LM, BM and BT and often ineffective therapy are the major challenges. For example, autopsy data demonstrate that BM contribute to death in ~75% of melanoma patients but they are clinically diagnosed in only 37% cases [8].

Among other parts of central nervous system (CNS), cerebrospinal fluid (CSF) is the easiest accessible medium that can directly uptake tumor markers from different parts of CNS [12–17]. Normally, CSF is a colorless liquid (a total volume of 130–150 mL for human) that contains up to 5 cells/μL, mainly leukocytes (white blood cells [WBCs]) [18–20]. CSF is produced by the choroidal plexus of the ventricular system and ependymal brain cells from blood [18,20,21].

In tumor patients with CNS involvement, CSF contains various markers associated with disease progression and responses to therapy [2–4,13–17,22–30]. Among others, circulating tumor cells (CTCs) are direct seeds of metastasis and, therefore, their diagnostic significance encourages high attention of researchers and clinicians. Furthermore, multiple recent reports suggested that detection of tumor-derived markers such as exosomes, circulating tumor DNA (ctDNA), micro-RNA (miRNA) and proteins is relevant to LM, BM and BT. The diagnostic significance of these markers seems especially important for BT because some BTs are not metastatic and do not typically shed CTCs but may release tumor-derived markers in CSF. CTC aggregates (so-called clusters or emboli) in CSF may also have diagnostic value. This speculation is based on: (1) finding CTC emboli in CSF samples of patients with lung cancer and LM [30]; (2) detection of CTC clusters in blood of patients with BT (e.g., glioblastoma) assuming their leaving CNS through the compromised blood-brain barrier (BBB) [31]; and (3) experimental and clinical evidences that multicellular CTC aggregates in peripheral blood represent the aggressive cell subset responsible for initiating and promoting metastasis [31–40].

Based on the physiology of CNS and mechanisms of tumor development (e.g., compromising BBB to penetrate tumor cells [41]), CTCs, their aggregates and other tumor-derived markers may invade CSF through different mechanisms that include (1) crossing the compromised BBB by blood and lymphatic CTCs and/or (2) shedding tumor cells by existing BM and BT. The latter mechanism provides a solid basis for using CSF tumor markers to diagnose progression of BM and BT, and to estimate responses to therapy. The first mechanism likely works for LM and BM and suggests the origin of CSF tumor biomarkers from blood or/and lymph and their possible entry to CSF before colonization of brain tissue and meninges.

Thus, testing CSF might predict deadly LM, BM and BT; and advanced methods to assess CSF tumor markers in CSF are urgently needed to prolong life of patients suffering from CNS tumor lesions.

2. In Vitro Detection of CSF Tumor Markers

The gold standard for routine clinical examination of CSF is cytology after lumbar puncture [9–11, 24,42,43]. The detection approach is based on cytomorphology of tumor cells after staining samples with Wright-Giemsa or Papanicolaou dyes. However, the sensitivity of CSF cytology is estimated as low as 50% [9]. Furthermore, cytology is a relatively subjective method since its results depend on the ability of a laboratory technician to correctly identify types of cells, for example, to distinguish tumor cells from normal leukocytes [24–26]. This may lead to delaying of therapeutic interventions until other diagnostic criteria (e.g., abnormal magnetic resonance imaging [MRI]) and/or strong clinical symptoms emerge. As a result, involvement of CNS in some patients is found at autopsy only.

The limitations of cytology and deadly nature of LM, BM and BT encouraged researchers and clinicians to develop more sensitive and accurate markers using modern technologies. During the past decade, substantial efforts have been made to assess CSF samples using new concept of liquid biopsy (Figure 1) [2,15–17,23,26–30,44–48].

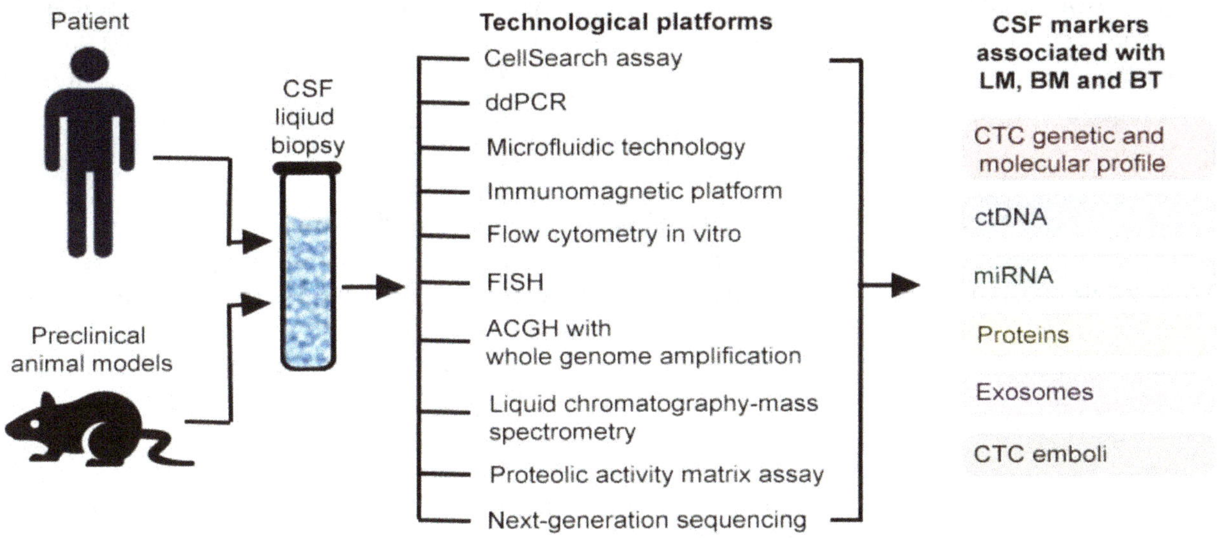

Figure 1. Cerebrospinal fluid (CSF) liquid biopsy detection of tumor markers in vitro.

CSF Liquid Biopsy

Several years ago, Patel et al. showed that FDA-approved CellSearch method can be used to identify CTCs in 7.5 mL CSF samples of breast cancer patients [22]. Compared to traditional cytology, the CellSearch assay has been demonstrated significantly higher number of CTCs [22,28,30,49,50]. Despite promise, this technological platform is limited in detection of only a few tumor markers, typically EpCam, for patients with epithelial cancers (e.g., breast cancer) and CD 146 and HMW-MMA for patients with melanoma [21,49]. Thus, CellSearch obviously cannot identify a highly heterogeneous population of CTCs and not suitable for diagnosis of many tumors such as glioblastoma. These limitations somewhat reduced enthusiasm to recommend this method in routine clinical practice.

Using real-time polymerase chain reaction (RT-PCR) for examination of patients with BM and LM has demonstrated higher sensitivity than conventional cytology [51]. However, relatively high rate of false-negatives during RT-PCR analysis make it a suboptimal method for CSF testing [9]. To solve this problem, cancer researchers and clinical oncologists recently explored the use of high-sensitive droplet-digital PCR (ddPCR) [52–56]. It was shown that ddPCR provides accurate and reliable CSF analysis. It can work with poor DNA quality and measure multiple parameters including absolute allele quantification, rare mutation, copy number variations, DNA methylation and gene rearrangements [52]. In a few clinical studies, ddPCR of CSF was able to detect ctDNA in patients with melanoma and CNS metastasis; and the obtained results were strongly correlated with cytology results and detection of abnormalities in MRI [52,56]. It is interesting that some patients with high level of ctDNA showed negative cytological results [56]. The small volume of CSF fluid required for testing ctDNA is definitely an additional advantage but high level of false results is a challenge. Overall, to date, it is too early to make conclusions on diagnostic value of ctDNA.

Another promising emerging data of CSF liquid biopsy have been obtained using immunofluorescence in situ hybridization (FISH) technology [24,57–60]. The published results hold promise to provide more accurate diagnosis of CSF CTCs than cytology. The main advantage of FISH is phenotypic and karyotypic identification and characterization of the highly heterogeneous CTCs, which can be assessed by both chromosome ploidy and the expression of various tumor markers [57]. However, FISH is not currently standardized for liquid biopsy and requires future development and research to clarify whether or not this method is reliable for identification of CTCs.

Integration of array comparative genomic hybridization (ACGH) analysis and whole genome amplification provided achieving the genomic characterization of rare CSF CTCs [61,62]. The clonal similarity between CSF CTCs and primary tumor genomic profiles with more copy number alterations in CTCs was demonstrated using samples of CSF and primary tumor from breast cancer patients with LM [62].

Analysis of CSF samples with conventional flow cytometry in vitro has been reported to diagnose CTCs in CSF [63–65]. Flow cytometry immunophenotypic testing of bulk breast cancer receptors, cancer stem cell markers and various WBC subpopulations looks interesting and suggests interplay of CSF and lymph fluid during CTC migration [63]. However, well-known limitation of flow cytometry to detect rare events might reduce enthusiasm for its use of assessment of CTCs which is supposed to be rare (up to 1-5 CTCs per sample) at early stage of CNS involvement.

In the past few years, the clinical potential of some other technological platforms including microfluidic technology, immunomagnetic platform, high performance liquid chromatography-mass spectrometry, next generation sequencing (NGS) and proteolytic activity matrix assay (PrAMA) has also been demonstrated [25,50,55,66–69]. Despite interest and promises, the singularity of these reports does not allow yet making conclusions on suitability of these methods to improve prognosis in patients.Overall, despite CSF liquid biopsy is expected to yield clinically significant biomarkers and assays, the main drawback to all aforementioned approaches in vitro is that their sensitivity is substantially limited by the volume of the sample [70,71]. Typically, up to 10 mL of CSF is used for examination, which is estimated to be less than 6–7% of the total 130–150 mL volume of human CSF. It means that in vitro testing misses up to 93–94% of CTCs [71]. A simple recalculation of the results in vitro, which detected minimum 1–2 CTCs per CSF patient sample (5–10 mL) with the existing LM and BM [21,49], shows that the real number of CTCs at the time of diagnosis was more than 15–20 cells in the total CSF volume. Serial analysis of multiple samples from repeated punctures increases sensitivity [28]. However, repeated punctures are a challenge because it can be performed over several days and may lead to delaying of therapies. In addition, the existing methods in vitro are burdened with: (1) low throughput, which may require many hours (if not, days) to assess a typical CSF sample and (2) multiple time-consuming sample-processing steps including staining, immunomagnetic capture, isolation and washing, which result in loss of many CTCs [21,23,30,49,51]. As a result, CTCs in small quantities may escape detection, which also contributes to late diagnosis and poor outcomes.

Based on this, liquid biopsy in vitro can provide advanced molecular and genetic analysis of tumor associated markers in CSF but it cannot detect rare CTCs at early stage of LM and BM and possibly, before LM and BM initiation (Table 1). The rarity of CSF CTCs definitely demands a new strategy. An attractive solution to these problems is to monitor almost entire CSF volume in vivo (Table 1).

Table 1. New and emerging technologies for detection of tumor biomarkers in CSF.

Detection Method	Biomarker	Disease	Approach		Significant Advantages	Main Limitations	Application		Refs
			In Vitro	In Vivo			Research	Clinical	
CellSearch	CTCs, Cell emboli	BM, LM	+		FDA-approved technology, Higher sensitivity than cytology	Small sample volume, Processing delay, Limited number of detected markers	+	+	[22,28,30,49,50]
Microfluidic technologies	CTCs	BM, BT	+		Single CTC capture in sub-nanoliter trap, Relatively quick (~1 h) analysis	Early stage of research using cell lines	+		[69]
Immuno-magnetic platform	CTCs	LM	+		Capable to detect and separate rare cells	Low sensitivity due to small sample volume, Limited number of detected markers		+	[25]
FC in vitro	CTCs	LM	+		Standardized technology, Higher sensitivity than cytology	Impossibility to detect rare cells	+	+	[63–65]
ddPCR	ctDNA, miRNA, CTCs	BT, BM, LM	+		High specificity	False-positivity, Not standardized	+	+	[52–56]
FISH	CTCs	LM, BT	+		Analysis poor DNA, Relatively high resolution	Early stage of research, Not standardized	+	+	[24,57–60]
ACGH	CTCs		+		Whole genome sequencing, High resolution compared to conventional CGH	Inability to detect aberrations that do not result in copy number changes		+	[61,62]
NGS	ctDNA, miRNA	LM	+		High-throughput whole genome sequencing, High specificity	High price, Complex data analysis		+	[50,56]
PrAMA	Proteases	LM	+		Detection of protease activity as indicator of BBB degradation	Early stage of research		+	[66]
PAFCin vivo	CTCs, Cell emboli	BM			Extremely high sensitivity, Theranostic capability	Detection of surface CTC receptors	+		[71]

CTCs—circulating tumor cells, LM—leptomeningeal metastasis; BM—brain metastasis; BT—brain tumor; FC—flow cytometry; ddRCP—droplet-digital polymerase chain reaction; ctDNA—cell free DNA; miRNA—microRNA; FISH—immunofluorescence in situ hybridization; ACGH—array comparative genomic hybridization; CGH—comparative genomic hybridization; NGS—next generation sequencing; PrAMA—proteolytic activity matrix assay; PAFC—photoacoustic flow cytometry.

3. In Vivo Diagnosis of CSF

Despite significant progress in neuroimaging in vivo (e.g., MRI, computed tomography [CT], radiography) [9,11,50,64,72–74], existing diagnosis, even advanced multi-modal imaging is not sufficient to make judgments about early LM and BM. The low spatial and temporal resolution of CT and MRI allows identification of only macroscopic changes in the CNS (e.g., metastases ≥10 mm by CT). Therefore, the diagnosis is typically based on the indirect signs of LM including pathological meningeal contrast enhancement at the MRI examination, which are often equivocal. In addition, a recent study has found that immunotherapy might be a source of MRI false positivity ('pseudomeningeosis') [73]. New generations of MRI, such as phase-contrast MRI, enable quantitative measurements of CSF flow but not suitable for detection of relatively fast moving single CTCs and particles due to slow time response [75]. The same limitation applies to intravital fluorescence microscopy which has been used for imaging CSF plasma (so-called, cisternography) but not single cells in CSF [76]. Furthermore, the translation of fluorescent neuroimaging to humans in vivo is problematic due to (1) cytotoxicity of fluorophores, (2) undesirable immune responses to tags and (3) assessing only superficial fluid flows due to strong influence of autofluorescent and scattering background.

Photoacoustic Flow Cytometry In Vivo

The most promising method for detecting CTCs in CSF is photoacoustic (PA) flow cytometry (PAFC), which compared to other in vivo diagnostic techniques, demonstrated ultra-sensitive molecular detection and counting of single cells in different body fluids (e.g., blood, lymph and CSF) [40,70,77–83]. The principle of multicolor PAFC is based on noninvasive (i.e., through intact skin) irradiation of the selected fluid with short laser pulses at different wavelengths followed by the detection of laser-induced acoustic waves (referred to as PA signals) using an ultrasound transducer placed on the skin (Figure 2a). PA methods provide higher sensitivity and resolution in deeper tissues (up to 2–3 cm, with potential up to 5–7 cm [70,84]) than other optical modalities. These benefits make possible detection of CTCs in CSF through the atlanto-occipital membrane. In PAFC, this allows distinguishing signals from single fast-moving particles (e.g., CTCs, exosomes, and emboli) at laser energies within the safety standards for humans [70,77,81,83,85,86]. In regards of CSF detection, PAFC has advantages compared to other in vivo methods. Specifically, the colorlessness and optical transparency of CSF, commonly accepted as a diagnostic limitation, provides low absorbance and, therefore, extremely low PA background signal, which significantly improves distinguishing stronly light absorbing objects [71]. It means that CTCs, exosomes or emboli with strong absorbing molecules (e.g., natural melanin or nanoparticles) are predominated over the absorption of CSF by a few orders of magnitude, especially in the near-infrared window of transparency for biotissues ("first window": 700–1100 nm). Based on this, some strong absorbing cells such as melanoma CTCs with natural intracellular high absorbing melanin as intrinsic non-toxic PA contrast agents, can be easily detected by PAFC in label-free mode. To detect low absorbing tumor-related CSF markers (e.g., breast cancer CTCs), they should be labeled by exogenous PA contrast agents conjugated with ligands (e.g., antibodies, peptides, or folic acid) against specific surface receptor(s). The key requirements for in vivo use of contrast agents include low toxicity and high PA contrast. Some of the best candidates are gold and magnetic nanoparticles [77,87,88].

The first successful demonstration of PAFC's capability to diagnose CSF tumor markers was reported using preclinical models of breast metastatic cancer (Figure 2b–d) [71]. It was shown that PAFC was able to detect CSF CTCs with 10–20 times higher sensitivity compared to in vitro methods. The most important finding is that some tumor-bearing mice without histologically detectable BM exhibited rare CSF CTCs (e.g., 1–3 signals every 40–60 min). The presence of blood CTCs in these mice suggests the possible origin of CSF CTCs to be from blood CTCs and indicates the potential of CSF CTCs as a predictive biomarker of BM. The obtained experimental evidence is in line with the aforementioned suggestion that blood and lymphatic CTCs might pass the compromised BBB and enter brain tissue, meninges and CSF to form BM and LM. This may serve as a scientific foundation for prognosis and prediction of LM and BM in patients.

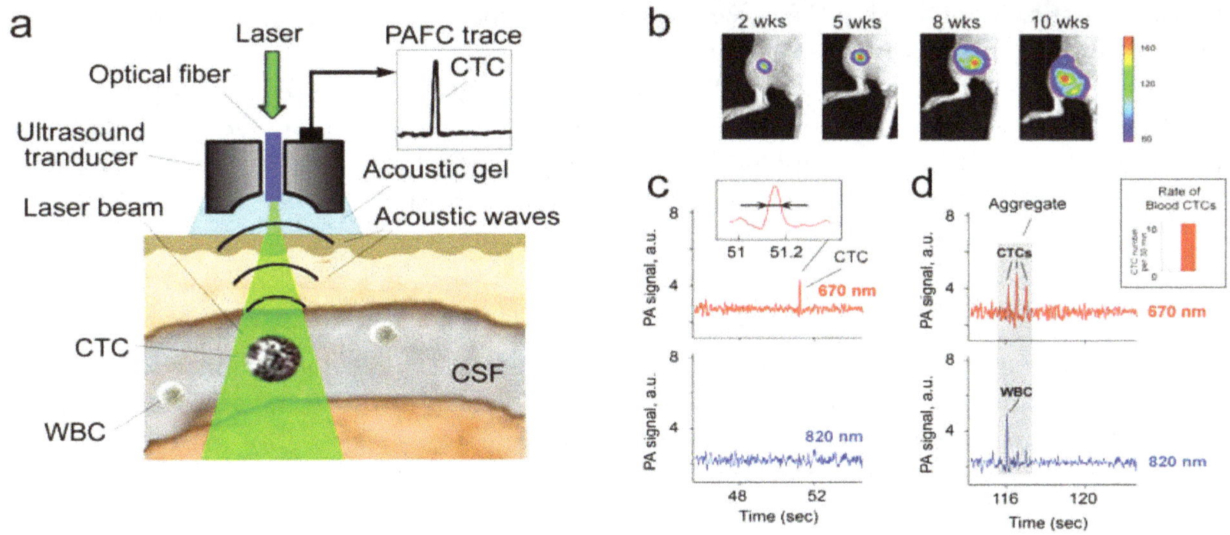

Figure 2. Assessment of circulating tumor markers in CSF in vivo with multicolor PAFC. (**a**) Principle of diagnosis with PAFC. (**b**) Intravital luminescence imaging of metastatic breast cancer progression in orthotopic xenograft mouse model after inoculation of human MDA-MB-231-luc2-GFP cells. (**c**) Two-color PAFC of the spontaneous CSF CTCs in vivo; inset: the photoacoustic signal width (indicated by arrows), which is associated with a single circulating tumor cell (CTC). (**d**) PAFC of circulating CTC-containing embolus in tumor-bearing mice; gray rectangle: aggregate of CSF-CTCs and leukocyte (WBC); insert: the blood CTC rate at the time of CSF monitoring.

Another interesting finding is the existence of CTC-containing emboli in CSF in vivo (Figure 2d). Identification of embolus is based on the width and shape of PA signal, assuming that embolus' multicellular structure produces a relatively wider PA signal containing a set of narrower peaks.

Overall, the success of preclinical studies together with the simplicity and safety of PAFC give confidence to rapidly translate this method into clinical practice. PAFC diagnosis of CSF in human subarachnoid space and spinal canal at a depth of 1–3 cm seems possible and was supported by the reports on high sensitivity and resolution of PA methods in deeper tissues. Recently, the clinical relevance of PAFC was successfully demonstrated in clinical trials with melanoma patients by detecting blood CTCs in 1–2 mm hand vessels at depth of 1-3 mm with a detection limit of 1 CTC/1000 mL (i.e.,10^3 –fold increased sensitivity compared to existing CTC assays) [40].

4. Future Directions

To date, crucial steps in increasing the survival of patients with LM and BM are (1) early diagnosis; (2) initiating preventive therapy such as targeted therapy of single CSF CTCs and their emboli and (3) assessing therapeutic efficacy in order to optimize an individual course of therapy.

4.1. Advance Diagnosis

Although many promising technologies to detect various CSF tumor markers during liquid biopsy have been reported, there is no standardized and validated assay that is currently ready to introduce for daily clinical practice as an advanced alternative or supplement of conventional cytology.

Novel approaches integrating unprecedented high sensitivity of in vivo flow cytometry and comprehensive molecular and genetic characterization of tumor markers in CSF in vitro are highly desired for clinical needs.

In addition, one of the possible future alternatives is CSF diagnosis in vivo using updated GILUPI CellCollector. This method was introduced in 2016 for EpCam-based detection of CTCs in blood by introduction of EpCAM-coated wire into a vein of the patient [89]. However, the invasive nature of the method and possibility of missing CTCs, which transit outside the wire, somewhat reduce enthusiasm of using GILUPI device for CSF assessment.

The new looks are also suggesting continuous cell exchange between CSF, blood, lymph and brain tissue [90–92] that should be considered at the diagnosis. The prognostic value of CTCs, if they are simultaneously tested in blood, lymph and CSF, would provide a new, highly sensitive and accurate prognostic biomarker of metastasis progression and therapy efficacy.

4.2. Therapeutic Perspectives

Minimal treatment options in current management of LM and BM lead to poor prognosis for patients due to low efficacy, late therapy initiation, use of common (i.e., not-personalized) therapeutic schematics and high toxicity. From this, one of the top future priorities is development of novel targeted and immune therapies. The molecular-targeted nanotechnology platform is highly promising for targeted drug delivery. For this purpose, nanoparticles should have high sensitivity, specificity and selectivity as well as safety, multifunctionality, multimodality, ability to penetrate BBB and high efficiency of drug delivery to tumor. Among existing nanoparticle-based drug cargoes, the most promising candidates include low toxic individual nanoparticles, high-contrast spasers, liposomes, polymer micelles, lipid micelles packaged with semiconducting polymer dots as simultaneous MRI and PA imaging and photodynamic and photothermal dual-modal therapeutic agents, layer by layer based composite structures (core-shells) and microcapsules (shells) and biocompatible natural magnetic nanoparticles [87,88,93–98]. The targeting could be achieved by surface modification using targeted molecules specific to CTCs, exosomes and emboli [88,99]. For example, a single injection of core-shells in CSF has shown the effectiveness of their use for the long-term delivery of painkillers in the treatment of persistent pain [100]. Potentially, these drug delivery systems may be effective for treating CTCs in CSF.

There is a high therapeutic potential of modern technologies for creating synthetic truncated antibodies [101] and scaffolds [102]. The revolutionary progress in genetic and protein engineering methods make it possible to directionally modify the molecular size, affinity, specificity and immunogenicity of an antibody, their derivatives and analogues, oriented to the use in the diagnosis and targeted therapy of cancer. Today, rational design and molecular engineering allow modelling of the compounds with preprogrammed properties and to create biotechnological producers of therapeutic medicines [102–106]. A promising direction is conjugation of these unique theranostic agents with nanoparticles. The advantages of using nanoparticles in these conjugates include developed surface of nanoparticles, which can be decorated with biocompatible functional moieties for targeted delivery; and diagnosis that guides and monitors effects of the nanoparticle-assisted therapy [107–110]. Recently, the design of a hybrid nanocomplex based on an upconversion nanoparticle (UCNP) was reported [111]. Owing to their unique photophysical properties, UCNPs are high-potential platform for theranostics complexes. Conversion of near-infrared light, which can deeply penetrate in biological tissue, to the higher photon energy visible, ultra-violet and near-infrared light is among UCNP's most useful properties. Two toxic agents—-beta-emitting radionuclide yttrium-90 and a highly efficient targeted toxin DARPin-exotoxin A from Pseudomonas aeruginosa—-were coupled to UCNP core to exert toxicity to cancer cells. As a result, on the one hand, the photophysical properties of hybrid nanocomplex enable background-free imaging of its distribution in cells and animals. On the other hand, specific delivery of UCNP complexes to cancer cells results in combined therapy by two toxic agents with markedly increased synergetic effect [111]. The design of the hybrid multifunctional nanoheterocomplex proves the principle "when the whole is greater than the sum of the parts."

The novel targeted CSF therapy may also use the advanced design of heterostructures based on the barnase:barstar pair [112]. The ribonuclease barnase and its inhibitor, barstar, are small stable proteins. They form extremely tight complex with a $K_d \sim 10^{-14}$ M. The strategy is applicable to any proteins or nanoparticles that can be functionally attached to the barstar and barnase, especially for production of heterooligomeric constructs because the extremely specific barnase barstar interaction eliminates reliably the mispairing problems. This universal platform is a promising alternative to commonly used chemical conjugation techniques in nanobiotechnology, theranostics and clinical applications. It provides a straightforward technology to design wide range of multifunctional nanoheterostructures for the highly efficient delivery of active agents to tumor cells for theranostics [112–117].

A very exciting future direction is the possibility of integration of in vivo molecular diagnosis, targeted therapy and estimation of therapeutic efficacy in one technological platform of PAFC [40,79,118]. PAFC's capability to identify a single high-absorbing CTC and immediately "kill" it through photothermal-indiced nanobubbles with photomechanical action on CTC membranes and vital intracellular structures was demonstrated for blood CTCs in experiments and, recently, in clinical research in blood circulation [40]. Furthermore, the following disappearance of the CTC-associated PA signals might serve as the criterion of effective therapy. These data bring hope that earliest rare CTCs might be identified and "killed" directly in CSF before colonization of brain tissue and formation of BM and LM.

It is expected that technological innovations and ongoing clinical trials would contribute to the finding of novel approaches to provide advances in BM and LM theranostics at the earliest possible stages before development of overt deadly lesions, to select patients with high risk of BM and LM for personalized therapy, to identify early disease progression and thereby improve survival rates of cancer patients.

Acknowledgments: We thank V.T., M.J., D.N. and many other colleagues who participated in this work and co-authored the cited joint papers.

References

1. Siegel, R.L.; Miller, K.D.; Jemal, A. Cancer statistics, 2019. *CA Cancer J. Clin.* **2019**, *69*, 7–34. [CrossRef] [PubMed]
2. Taillibert, S.; Chamberlain, M.C. Leptomeningeal metastasis. *Handb. Clin. Neurol.* **2018**, *149*, 169–204. [PubMed]
3. Steeg, P.S.; Camphausen, K.A.; Smith, Q.R. Brain metastases as preventive and therapeutic targets. *Nat. Rev. Cancer* **2011**, *11*, 352–363. [CrossRef] [PubMed]
4. Langley, R.R.; Fidler, I.J. The biology of brain metastasis. *Clin. Chem.* **2013**, *59*, 180–189. [CrossRef] [PubMed]
5. Figura, N.B.; Rizk, V.T.; Armaghani, A.J.; Arrington, J.A.; Etame, A.B.; Han, H.S.; Czerniecki, B.J.; Forsyth, P.A.; Ahmed, K.A. Breast leptomeningeal disease: A review of current practices and updates on management. *Breast Cancer Res. Treat.* **2019**, *177*, 277–294. [CrossRef] [PubMed]
6. Taylor, G.; Karlin, N.; Halfdanarson, T.R.; Coppola, K.; Grothey, A. Leptomeningeal Carcinomatosis in Colorectal Cancer: The Mayo Clinic Experience. *Clin. Colorectal. Cancer* **2018**, *17*, e183–e187. [CrossRef] [PubMed]
7. Blakeley, J.O.; Coons, S.J.; Corboy, J.R.; Kline Leidy, N.; Mendoza, T.R.; Wefel, J.S. Clinical outcome assessment in malignant glioma trials: Measuring signs, symptoms, and functional limitations. *Neuro Oncol.* **2016**, *18* (Suppl. 2), ii13–ii20. [CrossRef]
8. Sloan, A.E.; Nock, C.J.; Einstein, D.B. Diagnosis and treatment of melanoma brain metastasis: A literature review. *Cancer Control* **2009**, *16*, 248–255. [CrossRef] [PubMed]
9. Chamberlain, M.C.; Glantz, M.; Groves, M.D.; Wilson, W.H. Diagnostic tools for neoplastic meningitis: Detecting disease, identifying patient risk, and determining benefit of treatment. *Semin. Oncol.* **2009**, *36*, S35–S45. [CrossRef]
10. Nayar, G.; Ejikeme, T.; Chongsathidkiet, P.; Elsamadicy, A.A.; Blackwell, K.L.; Clarke, J.M.; Lad, S.P.; Fecci, P.E.

Leptomeningeal disease: Current diagnostic and therapeutic strategies. *Oncotarget* **2017**, *8*, 73312–73328. [CrossRef] [PubMed]

11. Pellerino, A.; Bertero, L.; Rudà, R.; Soffietti, R. Neoplastic meningitis in solid tumors: From diagnosis to personalized treatments. *Ther. Adv. Neurol. Disord.* **2018**, *11*, 1756286418759618. [CrossRef] [PubMed]
12. Boon, J.M.; Abrahams, P.H.; Meiring, J.H.; Welch, T. Lumbar puncture: Anatomical review of a clinical skill. *Clin. Anat.* **2004**, *17*, 544–553. [CrossRef] [PubMed]
13. Verheul, C.; Kleijn, A.; Lamfers, M.L.M. Cerebrospinal fluid biomarkers of malignancies located in the central nervous system. *Handb. Clin. Neurol.* **2017**, *146*, 139–169. [PubMed]
14. Hepnar, D.; Adam, P.; Žáková, H.; Krušina, M.; Kalvach, P.; Kasík, J.; Karpowicz, I.; Nasler, J.; Bechyně, K.; Fiala, T.; et al. Recommendations for cerebrospinal fluid analysis. *Folia Microbiol.* **2019**, *64*, 443–452. [CrossRef] [PubMed]
15. Nevel, K.S.; Wilcox, J.A.; Robell, L.J.; Umemura, Y. The Utility of Liquid Biopsy in Central Nervous System Malignancies. *Curr. Oncol. Rep.* **2018**, *20*, 60. [CrossRef]
16. Fontanilles, M.; Duran-Peña, A.; Idbaih, A. Liquid Biopsy in Primary Brain Tumors: Looking for Stardust! *Curr. Neurol. Neurosci. Rep.* **2018**, *18*, 13. [CrossRef]
17. Shankar, G.M.; Balaj, L.; Stott, S.L.; Nahed, B.; Carter, B.S. Liquid biopsy for brain tumors. *Expert Rev. Mol. Diagn.* **2017**, *17*, 943–947. [CrossRef]
18. Johanson, C.E.; Duncan, J.A.; Klinge, P.M.; Brinker, T.; Stopa, E.G.; Silverberg, G.D. Multiplicity of cerebrospinal fluid functions: New challenges in health and disease. *Cerebrospinal Fluid Res.* **2008**, *5*, 10. [CrossRef]
19. Sweetman, B.; Linninger, A.A. Cerebrospinal fluid flow dynamics in the central nervous system. *Ann. Biomed. Eng.* **2011**, *39*, 484–496. [CrossRef]
20. Huff, T.; Tadi, P.; Varacallo, M. Neuroanatomy, Cerebrospinal Fluid. In *StatPearls [Internet]*; StatPearls Publishing: Treasure Island, FL, USA, 2019.
21. Brinker, T.; Stopa, E.; Morrison, J.; Klinge, P. A new look at cerebrospinal fluid circulation. *Fluids Barriers CNS* **2014**, *11*, 10. [CrossRef]
22. Patel, A.S.; Allen, J.E.; Dicker, D.T.; Peters, K.L.; Sheehan, J.M.; Glantz, M.J.; El-Deiry, W.S. Identification and enumeration of circulating tumor cells in the cerebrospinal fluid of breast cancer patients with central nervous system metastases. *Oncotarget* **2011**, *2*, 752–760. [CrossRef] [PubMed]
23. Burns, T.F.; Wolff, A.C. Detection of circulating tumor cells in the cerebrospinal fluid: A new frontier. *Cell Cycle* **2012**, *11*, 203–204. [CrossRef]
24. Li, X.; Zhang, Y.; Ding, J.; Wang, M.; Li, N.; Yang, H.; Wang, K.; Wang, D.; Lin, P.P.; Li, M.; et al. Clinical significance of detecting CSF-derived tumor cells in breast cancer patients with leptomeningeal metastasis. *Oncotarget* **2017**, *9*, 2705–2714. [CrossRef] [PubMed]
25. Lin, X.; Fleisher, M.; Rosenblum, M.; Lin, O.; Boire, A.; Briggs, S.; Bensman, Y.; Hurtado, B.; Shagabayeva, L.; DeAngelis, L.M.; et al. Cerebrospinal fluid circulating tumor cells: A novel tool to diagnose leptomeningeal metastases from epithelial tumors. *Neuro Oncol.* **2017**, *19*, 1248–1254. [CrossRef]
26. Zorofchian, S.; Iqbal, F.; Rao, M.; Aung, P.P.; Esquenazi, Y.; Ballester, L.Y. Circulating tumour DNA, microRNA and metabolites in cerebrospinal fluid as biomarkers for central nervous system malignancies. *J. Clin. Pathol.* **2019**, *72*, 271–280. [CrossRef] [PubMed]
27. Touat, M.; Duran-Peña, A.; Alentorn, A.; Lacroix, L.; Massard, C.; Idbaih, A. Emerging circulating biomarkers in glioblastoma: Promises and challenges. *Expert Rev. Mol. Diagn.* **2015**, *15*, 1311–1323. [CrossRef]
28. Lee, J.S.; Melisko, M.E.; Magbanua, M.J.; Kablanian, A.T.; Scott, J.H.; Rugo, H.S.; Park, J.W. Detection of cerebrospinal fluid tumor cells and its clinical relevance in leptomeningeal metastasis of breast cancer. *Breast Cancer Res. Treat.* **2015**, *154*, 339–349. [CrossRef] [PubMed]
29. Ning, M.; Chunhua, M.; Rong, J.; Yuan, L.; Jinduo, L.; Bin, W.; Liwei, S. Diagnostic value of circulating tumor cells in cerebrospinal fluid. *Open Med.* **2016**, *11*, 21–24. [CrossRef] [PubMed]
30. Tu, Q.; Wu, X.; Le Rhun, E.; Blonski, M.; Wittwer, B.; Taillandier, L.; De Carvalho Bittencourt, M.; Faure, G.C. CellSearch technology applied to the detection and quantification of tumor cells in CSF of patients with lung cancer leptomeningeal metastasis. *Lung Cancer* **2015**, *90*, 352–357. [CrossRef]

31. Krol, I.; Castro-Giner, F.; Maurer, M.; Gkountela, S.; Szczerba, B.M.; Scherrer, R.; Coleman, N.; Carreira, S.; Bachmann, F.; Anderson, S.; et al. Detection of circulating tumour cell clusters in human glioblastoma. *Br. J. Cancer* **2018**, *119*, 487–491. [CrossRef]
32. Young, A.; Chapman, O.; Connor, C.; Poole, C.; Rose, P.; Kakar, A.K. Thrombosis and cancer. *Nat. Rev. Clin. Oncol.* **2012**, *9*, 437–449. [CrossRef] [PubMed]
33. Au, S.H.; Storey, B.D.; Moore, J.C.; Tang, Q.; Chen, Y.L.; Javaid, S.; Sarioglu, A.F.; Sullivan, R.; Madden, M.W.; O'Keefe, R.; et al. Clusters of circulating tumor cells traverse capillary-sized vessels. *Proc. Natl. Acad. Sci. USA* **2016**, *113*, 4947–4952. [CrossRef] [PubMed]
34. Mu, Z.; Wang, C.; Ye, Z.; Austin, L.; Civan, J.; Hyslop, T.; Palazzo, J.P.; Jaslow, R.; Li, B.; Myers, R.E.; et al. Prospective assessment of the prognostic value of circulating tumor cells and their clusters in patients with advanced-stage breast cancer. *Breast Cancer Res. Treat.* **2015**, *154*, 563–571. [CrossRef] [PubMed]
35. Wang, C.; Mu, Z.; Chervoneva, I.; Austin, L.; Ye, Z.; Rossi, G.; Palazzo, J.P.; Sun, C.; Abu-Khalaf, M.; Myers, R.E.; et al. Longitudinally collected CTCs and CTC-clusters and clinical outcomes of metastatic breast cancer. *Breast Cancer Res. Treat.* **2017**, *161*, 83–94. [CrossRef] [PubMed]
36. Au, S.H.; Edd, J.; Stoddard, A.E.; Wong, K.H.K.; Fachin, F.; Maheswaran, S.; Haber, D.A.; Stott, S.L.; Kapur, R.; Toner, M. Microfluidic Isolation of Circulating Tumor Cell Clusters by Size and Asymmetry. *Sci. Rep.* **2017**, *7*, 2433. [CrossRef]
37. Kulasinghe, A.; Schmidt, H.; Perry, C.; Whitfield, B.; Kenny, L.; Nelson, C.; Warkiani, M.E.; Punyadeera, C. A Collective Route to Head and Neck Cancer Metastasis. *Sci. Rep.* **2018**, *8*, 746. [CrossRef] [PubMed]
38. Gkountela, S.; Castro-Giner, F.; Szczerba, B.M.; Vetter, M.; Landin, J.; Scherrer, R.; Krol, I.; Scheidmann, M.C.; Beisel, C.; Stirnimann, C.U.; et al. Circulating Tumor Cell Clustering Shapes DNA Methylation to Enable Metastasis Seeding. *Cell* **2019**, *176*, 98–112. [CrossRef]
39. Kulasinghe, A.; Kapeleris, J.; Cooper, C.; Warkiani, M.E.; O'Byrne, K.; Punyadeera, C. Phenotypic Characterization of Circulating Lung Cancer Cells for Clinically Actionable Targets. *Cancers* **2019**, *11*, 380. [CrossRef]
40. Galanzha, E.I.; Menyaev, Y.A.; Yadem, A.C.; Sarimollaoglu, M.; Juratli, M.A.; Nedosekin, D.A.; Foster, S.R.; Jamshidi-Parsian, A.; Siegel, E.R.; Makhoul, I.; et al. In vivo liquid biopsy using Cytophone platform for photoacoustic detection of circulating tumor cells in patients with melanoma. *Sci. Transl. Med.* **2019**, *11*, eaat5857. [CrossRef]
41. Wrobel, J.K.; Toborek, M. Blood–brain Barrier Remodeling during Brain Metastasis Formation. *Mol. Med.* **2016**, *22*, 32–40. [CrossRef]
42. Deeken, J.F.; Loscher, W. The Blood-Brain Barrier a Overview of cerebrospinal fluid cytology and Cancer: Transporters, Treatment, and Trojan Horses. *Clin. Cancer Res.* **2007**, *13*, 1663–1674. [CrossRef] [PubMed]
43. Rahimi, J.; Woehrer, A. Overview of cerebrospinal fluid cytology. *Handb. Clin. Neurol.* **2017**, *145*, 563–571. [PubMed]
44. Pantel, K.; Alix-Panabières, C. The potential of circulating tumor cells as a liquid biopsy to guide therapy in prostate cancer. *Cancer Discov.* **2012**, *2*, 974–975. [CrossRef] [PubMed]
45. Pantel, K.; Alix-Panabières, C. Liquid biopsy: Potential and challenges. *Mol. Oncol.* **2016**, *10*, 371–373. [CrossRef] [PubMed]
46. Connolly, I.D.; Li, Y.; Gephart, M.H.; Nagpal, S. The "Liquid Biopsy": The Role of Circulating DNA and RNA in Central Nervous System Tumors. *Curr. Neurol. Neurosci. Rep.* **2016**, *16*, 25. [CrossRef]
47. Best, M.G.; Sol, N.; Zijl, S.; Reijneveld, J.C.; Wesseling, P.; Wurdinger, T. Liquid biopsies in patients with diffuse glioma. *Acta Neuropathol.* **2015**, *129*, 849–865. [CrossRef]
48. Pantel, K.; Alix-Panabières, C. Liquid biopsy and minimal residual disease—Latest advances and implications for cure. *Nat. Rev. Clin. Oncol.* **2019**, *16*, 409–424. [CrossRef]
49. Le Rhun, E.; Tu, Q.; De Carvalho Bittencourt, M.; Farre, I.; Mortier, L.; Cai, H.; Kohler, C.; Faure, G.C. Detection and quantification of CSF malignant cells by the CellSearch technology in patients with melanoma leptomeningeal metastasis. *Med. Oncol.* **2013**, *30*, 538. [CrossRef]
50. Jiang, B.Y.; Li, Y.S.; Guo, W.B.; Zhang, X.C.; Chen, Z.H.; Su, J.; Zhong, W.Z.; Yang, X.N.; Yang, J.J.; Shao, Y.; et al. Detection of Driver and Resistance Mutations in Leptomeningeal Metastases of NSCLC by Next-Generation Sequencing of Cerebrospinal Fluid Circulating Tumor Cells. *Clin. Cancer Res.* **2017**, *23*, 5480–5488. [CrossRef]

51. Hoon, D.S.; Kuo, C.T.; Wascher, R.A.; Fournier, P.; Wang, H.J.; O'Day, S.J. Molecular detection of metastatic melanoma cells in cerebrospinal fluid in melanoma patients. *J. Investig. Dermatol.* **2001**, *117*, 375–378. [CrossRef]
52. Olmedillas-López, S.; García-Arranz, M.; García-Olmo, D. Current and Emerging Applications of Droplet Digital PCR in Oncology. *Mol. Diagn. Ther.* **2017**, *21*, 493–510. [CrossRef] [PubMed]
53. Momtaz, P.; Pentsova, E.; Abdel-Wahab, O.; Diamond, E.; Hyman, D.; Merghoub, T.; You, D.; Gasmi, B.; Viale, A.; Chapman, P.B. Quantification of tumor-derived cell free DNA(cfDNA) by digital PCR (DigPCR) in cerebrospinal fluid of patients with BRAFV600 mutated malignancies. *Oncotarget* **2016**, *7*, 85430–85436. [CrossRef] [PubMed]
54. Hiemcke-Jiwa, L.S.; Minnema, M.C.; Radersma-van Loon, J.H.; Jiwa, N.M.; de Boer, M.; Leguit, R.J.; de Weger, R.A.; Huibers, M.M.H. The use of droplet digital PCR in liquid biopsies: A highly sensitive technique for MYD88 p.(L265P) detection in cerebrospinal fluid. *Hematol. Oncol.* **2018**, *36*, 429–435. [CrossRef] [PubMed]
55. Zhao, J.; Ye, X.; Xu, Y.; Chen, M.; Zhong, W.; Sun, Y.; Yang, Z.; Zhu, G.; Gu, Y.; Wang, M. EGFR mutation status of paired cerebrospinal fluid and plasma samples in EGFR mutant non-small cell lung cancer with leptomeningeal metastases. *Cancer Chemother. Pharmacol.* **2016**, *78*, 1305–1310. [CrossRef] [PubMed]
56. Ballester, L.Y.; Glitza Oliva, I.C.; Douse, D.Y.; Chen, M.M.; Lan, C.; Haydu, L.E.; Huse, J.T.; Roy-Chowdhuri, S.; Luthra, R.; Wistuba, I.I.; et al. Evaluating Circulating Tumor DNA From the Cerebrospinal Fluid of Patients with Melanoma and Leptomeningeal Disease. *J. Neuropathol. Exp. Neurol.* **2018**, *77*, 628–635. [CrossRef] [PubMed]
57. Ge, F.; Zhang, H.; Wang, D.D.; Li, L.; Lin, P.P. Enhanced detection and comprehensive in situ phenotypic characterization of circulating and disseminated heteroploid epithelial and glioma tumor cells. *Oncotarget* **2015**, *6*, 27049–27064. [CrossRef]
58. Jiao, X.D.; Ding, C.; Zang, Y.S.; Yu, G. Rapid symptomatic relief of HER2-positive gastric cancer leptomeningeal carcinomatosis with lapatinib, trastuzumab and capecitabine: A case report. *BMC Cancer* **2018**, *18*, 206. [CrossRef]
59. Lv, Y.; Mu, N.; Ma, C.; Jiang, R.; Wu, Q.; Li, J.; Wang, B.; Sun, L. Detection value of tumor cells in cerebrospinal fluid in the diagnosis of meningeal metastasis from lung cancer by immuno-FISH technology. *Oncol. Lett.* **2016**, *12*, 5080–5084. [CrossRef]
60. Ma, C.; Lv, Y.; Jiang, R.; Li, J.; Wang, B.; Sun, L. Novel method for the detection and quantification of malignant cells in the CSF of patients with leptomeningeal metastasis of lung cancer. *Oncol. Lett.* **2016**, *11*, 619–623. [CrossRef]
61. Magbanua, M.J.; Melisko, M.; Roy, R.; Sosa, E.V.; Hauranieh, L.; Kablanian, A.; Eisenbud, L.E.; Ryazantsev, A.; Au, A.; Scott, J.H.; et al. Molecular profiling of tumor cells in cerebrospinal fluid and matched primary tumors from metastatic breast cancer patients with leptomeningeal carcinomatosis. *Cancer Res.* **2013**, *73*, 7134–7143. [CrossRef]
62. Magbanua, M.J.; Roy, R.; Sosa, E.V.; Hauranieh, L.; Kablanian, A.; Eisenbud, L.E.; Ryazantsev, A.; Au, A.; Scott, J.H.; Melisko, M.; et al. Genome-wide copy number analysis of cerebrospinal fluid tumor cells and their corresponding archival primary tumors. *Genom. Data* **2014**, *2*, 60–62. [CrossRef] [PubMed]
63. Cordone, I.; Masi, S.; Summa, V.; Carosi, M.; Vidiri, A.; Fabi, A.; Pasquale, A.; Conti, L.; Rosito, I.; Carapella, C.M.; et al. Overexpression of syndecan-1, MUC-1, and putative stem cell markers in breast cancer leptomeningeal metastasis: A cerebrospinal fluid flow cytometry study. *Breast Cancer Res.* **2017**, *19*, 46. [CrossRef] [PubMed]
64. Milojkovic Kerklaan, B.; Pluim, D.; Bol, M.; Hofland, I.; Westerga, J.; van Tinteren, H.; Beijnen, J.H.; Boogerd, W.; Schellens, J.H.; Brandsma, D. EpCAM-based flow cytometry in cerebrospinal fluid greatly improves diagnostic accuracy of leptomeningeal metastases from epithelial tumors. *Neuro Oncol.* **2016**, *18*, 855–862. [CrossRef] [PubMed]
65. Gong, X.; Lin, D.; Wang, H.; Wang, Y.; Liu, B.; Wei, H.; Zhou, C.; Liu, K.; Wei, S.; Gong, B.; et al. Flow cytometric analysis of cerebrospinal fluid in adult patients with acute lymphoblastic leukemia during follow-up. *Eur. J. Haematol.* **2018**, *100*, 279–285. [CrossRef] [PubMed]
66. Conrad, C.; Dorzweiler, K.; Miller, M.A.; Lauffenburger, D.A.; Strik, H.; Bartsch, J.W. Profiling of metalloprotease activities in cerebrospinal fluids of patients with neoplastic meningitis. *Fluids Barriers CNS* **2017**, *14*, 22. [CrossRef] [PubMed]

67. Warkiani, M.E.; Khoo, B.; Wu, L.; Tay, A.K.; Bhagat, A.A.; Han, J.; Lim, C.T. Ultra-fast, label-free isolation of circulating tumor cells from blood using spiral microfluidics. *Nat. Protoc.* **2016**, *11*, 134–148. [CrossRef]
68. Zhou, J.; Kulasinghe, A.; Bogseth, A.; O'Byrne, K.; Punyadeera, C.; Papautsky, I. Isolation of circulating tumor cells in non-small-cell-lung-cancer patients using a multi-flow microfluidic channel. *Microsyst. Nanoeng.* **2019**, *5*, 8. [CrossRef]
69. Turetsky, A.; Lee, K.; Song, J.; Giedt, R.J.; Kim, E.; Kovach, A.E.; Hochberg, E.P.; Castro, C.M.; Lee, H.; Weissleder, R. On chip analysis of CNS lymphoma in cerebrospinal fluid. *Theranostics* **2015**, *5*, 796–804. [CrossRef]
70. Galanzha, E.I.; Shashkov, E.V.; Spring, P.; Suen, J.Y.; Zharov, V.P. In vivo noninvasive label-free detection and eradication of circulating metastatic melanoma cells using two-color photoacoustic flow cytometry with a diode laser. *Cancer Res.* **2009**, *69*, 7926–7934. [CrossRef]
71. Nedosekin, D.A.; Juratli, M.A.; Sarimollaoglu, M.; Moore, C.L.; Rusch, N.J.; Smeltzer, M.S.; Zharov, V.P.; Galanzha, E.I. Photoacoustic and photothermal detection of circulating tumor cells, bacteria and nanoparticles in cerebrospinal fluid in vivo and ex vivo. *J. Biophotonics* **2013**, *6*, 523–533. [CrossRef]
72. Barajas, R.F., Jr.; Cha, S. Imaging diagnosis of brain metastasis. *Prog. Neurol. Surg.* **2012**, *25*, 55–73. [PubMed]
73. Bier, G.; Klumpp, B.; Roder, C.; Garbe, C.; Preibsch, H.; Ernemann, U.; Hempel, J.M. Meningeal enhancement depicted by magnetic resonance imaging in tumor patients: Neoplastic meningitis or therapy-related enhancement? *Neuroradiology* **2019**, *61*, 775–782. [CrossRef] [PubMed]
74. Salehi Ravesh, M.; Huhndorf, M.; Moussavi, A. Non-contrast enhanced molecular characterization of C6 rat glioma tumor at 7 T. *Magn. Reason. Imaging* **2019**, *61*, 175–186. [CrossRef] [PubMed]
75. Sakhare, A.R.; Barisano, G.; Pa, J. Assessing test-retest reliability of phase contrast MRI for measuring cerebrospinal fluid and cerebral blood flow dynamics. *Magn. Reason. Med.* **2019**, *82*, 658–670. [CrossRef] [PubMed]
76. Shibata, Y.; Kruskal, J.B.; Palmer, M.R. Imaging of cerebrospinal fluid space and movement of hydrocephalus mice using near infrared fluorescence. *Neurol. Sci.* **2007**, *28*, 87–92. [CrossRef]
77. Galanzha, E.I.; Shashkov, E.V.; Kelly, T.; Kim, J.-W.; Yang, L.; Zharov, V.P. In vivo magnetic enrichment and multiplex photoacoustic detection of circulating tumour cells. *Nat. Nanotechnol.* **2009**, *4*, 855–860. [CrossRef]
78. Zharov, V.P. Ultrasharp nonlinear photothermal and photoacoustic resonances and holes beyond the spectral limit. *Nat. Photonics* **2011**, *5*, 110–116. [CrossRef]
79. Galanzha, E.I.; Zharov, V.P. Circulating tumor cell detection and capture by photoacoustic flow cytometry in vivo and ex vivo. *Cancers* **2013**, *5*, 1691–1738. [CrossRef]
80. Galanzha, E.I.; Viegas, M.G.; Malinsky, T.I.; Melerzanov, A.V.; Juratli, M.A.; Sarimollaoglu, M.; Nedosekin, D.A.; Zharov, V.P. In vivo acoustic and photoacoustic focusing of circulating cells. *Sci. Rep.* **2016**, *6*, 21531. [CrossRef]
81. Nolan, J.; Sarimollaoglu, M.; Nedosekin, D.A.; Jamshidi-Parsian, A.; Galanzha, E.I.; Kore, R.A.; Griffin, R.J.; Zharov, V.P. In Vivo Flow Cytometry of Circulating Tumor-Associated Exosomes. *Anal. Cell. Pathol.* **2016**, *2016*, 1628057. [CrossRef]
82. Tuchin, V.V.; Tarnok, A.; Zharov, V.P. In vivo flow cytometry: A horizon of opportunities. *Cytom. Part A* **2011**, *79A*, 737–745. [CrossRef] [PubMed]
83. Galanzha, E.I.; Zharov, V.P. Photoacoustic flow cytometry. *Methods* **2012**, *57*, 280–296. [CrossRef] [PubMed]
84. Wang, L.V.; Hu, S. Photoacoustic tomography: In vivo imaging from organelles to organs. *Science* **2012**, *335*, 1458–1462. [CrossRef] [PubMed]
85. Juratli, M.A.; Menyaev, Y.A.; Sarimollaoglu, M.; Siegel, E.R.; Nedosekin, D.A.; Suen, J.Y.; Melerzanov, A.V.; Juratli, T.A.; Galanzha, E.I.; Zharov, V.P. Real-Time Label-Free Embolus Detection Using In Vivo Photoacoustic Flow Cytometry. *PLoS ONE* **2016**, *11*, e0156269. [CrossRef] [PubMed]
86. Juratli, M.A.; Sarimollaoglu, M.; Nedosekin, D.A.; Melerzanov, A.V.; Zharov, V.P.; Galanzha, E.I. Dynamic Fluctuation of Circulating Tumor Cells during Cancer Progression. *Cancers* **2014**, *6*, 128–142. [CrossRef] [PubMed]
87. Kim, J.-W.; Galanzha, E.I.; Shashkov, E.V.; Moon, H.-M.; Zharov, V.P. Golden carbon nanotubes as multimodal photoacoustic and photothermal molecular agents. *Nat. Nanotechnol.* **2009**, *4*, 688–694. [CrossRef]
88. Kim, J.W.; Galanzha, E.I.; Zaharoff, D.A.; Griffin, R.J.; Zharov, V.P. Nanotheranostics of circulating tumor cells, infections and other pathological features in vivo. *Mol. Pharm.* **2013**, *10*, 813–830. [CrossRef]

89. Gorges, T.M.; Penkalla, N.; Schalk, T.; Joosse, S.A.; Riethdorf, S.; Tucholski, J.; Lücke, K.; Wikman, H.; Jackson, S.; Brychta, N.; et al. Enumeration and Molecular Characterization of Tumor Cells in Lung Cancer Patients Using a Novel In Vivo Device for Capturing Circulating Tumor Cells. *Clin. Cancer Res.* **2016**, *22*, 2197–2206. [CrossRef]
90. Garzia, L.; Kijima, N.; Morrissy, A.S.; De Antonellis, P.; Guerreiro-Stucklin, A.; Holgado, B.L.; Wu, X.; Wang, X.; Parsons, M.; Zayne, K.; et al. A Hematogenous Route for Medulloblastoma Leptomeningeal Metastases. *Cell* **2018**, *172*, 1050–1062. [CrossRef]
91. Tuchin, V.V.; Zharov, V.P.; Galanzha, E.I. Biophotonics for lymphatic theranostics in animals and humans. *J. Biophotonics* **2018**, *11*, e201811001. [CrossRef]
92. Louveau, A.; Smirnov, I.; Keyes, T.J.; Eccles, J.D.; Rouhani, S.J.; Peske, J.D.; Derecki, N.C.; Castle, D.; Mandell, J.W.; Lee, K.S.; et al. Structural and functional features of central nervous system lymphatic vessels. *Nature* **2015**, *523*, 337–341. [CrossRef] [PubMed]
93. Beziere, N.; Lozano, N.; Nunes, A.; Salichs, J.; Queiros, D.; Kostarelos, K.; Ntziachristos, V. Dynamic imaging of PEGylated indocyanine green (ICG) liposomes within the tumor microenvironment using multi-spectral optoacoustic tomography (MSOT). *Biomaterials* **2015**, *37*, 415–424. [CrossRef] [PubMed]
94. Galanzha, E.I.; Weingold, R.; Nedosekin, D.A.; Sarimollaoglu, M.; Nolan, J.; Harrington, W.; Kuchyanov, A.S.; Parkhomenko, R.G.; Watanabe, F.; Nima, Z.; et al. Spaser as a biological probe. *Nat. Commun.* **2017**, *8*, 15528. [CrossRef] [PubMed]
95. Zhang, D.; Wu, M.; Zeng, Y.; Liao, N.; Cai, Z.; Liu, G.; Liu, J. Lipid micelles packaged with semiconducting polymer dots as simultaneous MRI/photoacoustic imaging and photodynamic/photothermal dual-modal therapeutic agents for liver cancer. *J. Mater. Chem. B* **2016**, *4*, 589–599. [CrossRef]
96. Yashchenok, M.; Jose, J.; Trochet, P.; Sukhorukov, G.B.; Gorin, D.A. Multifunctional polyelectrolyte microcapsules as a contrast agent for photoacoustic imaging in blood. *J. Biophotonics* **2016**, *9*, 792–799. [CrossRef] [PubMed]
97. Novoselova, M.V.; Bratashov, D.N.; Sarimollaoglu, M.; Nedosekin, D.A.; Harrington, W.; Watts, A.; Han, M.; Khlebtsov, B.N.; Galanzha, E.I.; Gorin, D.A.; et al. Photoacoustic and fluorescent effects in multilayer plasmon-dye interfaces. *J. Biophotonics* **2019**, *12*, e201800265. [CrossRef]
98. Nima, Z.A.; Watanabe, F.; Jamshidi-Parsian, A.; Sarimollaoglu, M.; Nedosekin, D.A.; Han, M.; Watts, J.A.; Biris, A.S.; Zharov, V.P.; Galanzha, E.I. Bioinspired magnetic nanoparticles as multimodal photoacoustic, photothermal and photomechanical contrast agents. *Sci. Rep.* **2019**, *9*, 887. [CrossRef] [PubMed]
99. Zhang, H.K.; Chen, Y.; Kang, J.; Lisok, A.; Minn, I.; Pomper, M.G.; Boctor, E.M. Prostate-specific membrane antigen-targeted photoacoustic imaging of prostate cancer in vivo. *J. Biophotonics* **2018**, *11*, e201800021. [CrossRef]
100. Kopach, O.; Zheng, K.; Dong, L.; Sapelkin, A.; Voitenko, N.; Sukhorukov, G.B.; Rusakov, D.A. Nano-engineered microcapsules boost the treatment of persistent pain. *Drug Deliv.* **2018**, *25*, 435–447. [CrossRef]
101. Mironova, K.E.; Proshkina, G.M.; Ryabova, A.V.; Stremovskiy, O.A.; Lukyanov, S.A.; Petrov, R.V.; Deyev, S.M. Genetically encoded immunophotosensitizer 4D5scFV-miniSOG is a highly selective agent for targeted photokilling of tumor cells in vitro. *Theranostics* **2013**, *3*, 831–840. [CrossRef]
102. Deyev, S.M.; Lebedenko, E.N.; Petrovskaya, L.E.; Dolgikh, D.A.; Gabibov, A.G.; Kirpichnikov, M.P. Man-made antibodies and immunoconjugates with desired properties: Function optimization using structural engineering. *Rus. Chem. Rev.* **2015**, *84*, 1–26. [CrossRef]
103. Martsev, S.P.; Chumanevich, A.A.; Vlasov, A.P.; Dubnovitsky, A.P.; Tsybovsky, Y.I.; Deyev, S.M.; Cozzi, A.; Arosio, P.; Kravchuk, Z.I. Antiferritin single-chain Fv fragment is a functional protein with properties of a partially structured state: Comparison with the completely folded V(L) domain. *Biochemistry* **2000**, *39*, 8047–8057. [CrossRef] [PubMed]
104. Vorobyeva, A.; Bragina, O.; Altai, M.; Mitran, B.; Orlova, A.; Shulga, A.; Proshkina, G.; Chernov, V.; Tolmachev, V.; Deyev, S. Comparative Evaluation of Radioiodine and Technetium-Labeled DARPin 9_29 for Radionuclide Molecular Imaging of HER2 Expression in Malignant Tumors. *Contrast Media Mol. Imaging* **2018**, *2018*. [CrossRef] [PubMed]
105. Proshkina, G.M.; Shilova, O.N.; Ryabova, A.V.; Stremovskiy, O.A.; Deyev, S.M. A new anticancer toxin based on HER2/neu-specific DARPin and photoactive flavoprotein miniSOG. *Biochimie* **2015**, *118*, 116–122. [CrossRef] [PubMed]

106. Shipunova, V.O.; Nikitin, M.P.; Nikitin, P.I.; Deyev, S.M. MPQ-cytometry: A magnetism-based method for quantification of nanoparticle-cell interactions. *Nanoscale* **2016**, *8*, 12764–12772. [CrossRef] [PubMed]
107. Deyev, S.; Proshkina, G.; Ryabova, A.; Tavanti, F.; Menziani, M.C.; Eidelshtein, G.; Avishai, G.; Kotlyar, A. Synthesis, Characterization, and Selective Delivery of DARPin-Gold Nanoparticle Conjugates to Cancer Cells. *Bioconjug. Chem.* **2017**, *28*, 2569–2574. [CrossRef]
108. Balalaeva, I.V.; Zdobnova, T.A.; Krutova, I.V.; Brilkina, A.A.; Lebedenko, E.N.; Deyev, S.M. Passive and active targeting of quantum dots for whole-body fluorescence imaging of breast cancer xenografts. *J. Biophotonics* **2012**, *5*, 860–867. [CrossRef]
109. Zdobnova, T.A.; Stremovskiy, O.A.; Lebedenko, E.N.; Deyev, S.M. Self-Assembling Complexes of Quantum Dots and scFv Antibodies for Cancer Cell Targeting and Imaging. *PLoS ONE* **2012**, *7*, e48248. [CrossRef]
110. Deyev, S.; Proshkina, G.; Baryshnikova, O.; Ryabova, A.; Avishai, G.; Katrivas, L.; Giannini, C.; Levi-Kalisman, Y.; Kotlyar, A. Selective staining and eradication of cancer cells by protein-carrying DARPin-functionalized liposomes. *Eur. J. Pharm. Biopharm.* **2018**, *130*, 296–305. [CrossRef]
111. Guryev, E.L.; Volodina, N.O.; Shilyagina, N.Y.; Gudkov, S.V.; Balalaeva, N.Y.; Volovetskiy, A.B.; Lyubeshkin, A.V.; Sen', A.V.; Ermilov, S.A.; Vodeneev, V.A.; et al. Radioactive ($_{90}$Y) upconversion nanoparticles conjugated with recombinant targeted toxin for synergistic nanotheranostics of cancer. *Proc. Natl. Acad. Sci. USA* **2018**, *11*, 9690–9695.
112. Nikitin, M.P.; Zdobnova, T.A.; Lukash, S.V.; Stremovskiy, O.A.; Deyev, S.M. Protein-assisted self-assembly of multifunctional nanoparticles. *Proc. Natl. Acad. Sci. USA* **2010**, *107*, 5827–5832. [CrossRef] [PubMed]
113. Aghayeva, U.F.; Nikitin, M.P.; Lukash, S.V.; Deyev, S.M. Denaturation-Resistant Bifunctional Colloidal Superstructures Assembled via the Proteinaceous Barnase-Barstar Interface. *ACS Nano* **2013**, *7*, 950–961. [CrossRef] [PubMed]
114. Shipunova, V.O.; Zelepukin, I.V.; Stremovskiy, O.A.; Nikitin, M.P.; Care, A.; Sunna, A.; Zvyagin, A.V.; Deyev, S.M. Versatile Platform for Nanoparticle Surface Bioengineering Based on SiO_2-Binding Peptide and Proteinaceous Barnase*Barstar Interface. *ACS Appl. Mater. Interfaces* **2018**, *10*, 17437–17447. [CrossRef]
115. Shipunova, V.O.; Kotelnikova, P.A.; Aghayeva, U.F.; Stremovskiy, O.A.; Novikov, I.A.; Schulga, A.A.; Nikitin, M.P.; Deyev, S.M. Self-assembling nanoparticles biofunctionalized with magnetite-binding protein for the targeted delivery to HER2/neu overexpressing cancer cells. *J. Magn. Magn. Mater.* **2019**, *469*, 450–455. [CrossRef]
116. Sreenivasan, V.K.A.; Kelf, T.A.; Grebenik, E.A.; Stremovskiy, O.A.; Say, J.M.; Rabeau, J.R.; Zvyagin, A.V.; Deyev, S.M. A modular design of low background bioassays based on a high-affinity molecular pair barstar:barnase. *Proteomics* **2013**, *13*, 1437–1443. [CrossRef] [PubMed]
117. Grebenik, E.A.; Nadort, A.; Generalova, A.N.; Nechaev, A.V.; Sreenivasan, V.K.; Khaydukov, E.V.; Semchishen, V.A.; Popov, A.P.; Sokolov, V.I.; Akhmanov, A.S.; et al. Feasibility study of the optical imaging of a breast cancer lesion labeled with upconversion nanoparticle biocomplexes. *J. Biomed. Opt.* **2013**, *18*, 76004. [CrossRef] [PubMed]
118. Kim, J.-W.; Shashkov, E.V.; Galanzha, E.I.; Kotagiri, N.; Zharov, V.P. Photothermal antimicrobial nanotherapy and nanodiagnostics with self-assembling carbon nanotube clusters. *Lasers Surgery Med.* **2007**, *39*, 622–634. [CrossRef]

11

Detection of AR-V7 in Liquid Biopsies of Castrate Resistant Prostate Cancer Patients: A Comparison of AR-V7 Analysis in Circulating Tumor Cells, Circulating Tumor RNA and Exosomes

Mohammed Nimir [1,2], Yafeng Ma [1,2], Sarah A. Jeffreys [1,3], Thomas Opperman [1,2], Francis Young [1,2], Tanzila Khan [1,3], Pei Ding [1,3,4], Wei Chua [4], Bavanthi Balakrishnar [4], Adam Cooper [1,3,4], Paul De Souza [1,3,4,5] and Therese M. Becker [1,2,3,*]

1. Centre for Circulating Tumour Cell Diagnostics and Research, Ingham Institute for Applied Medical Research, 1 Campbell St, Liverpool, NSW 2170, Australia
2. South Western Clinical School, University of New South Wales, Goulburn St, Liverpool, NSW 2170, Australia
3. School of Medicine, Western Sydney University, Campbelltown, NSW 2560, Australia
4. Liverpool Hospital, Elizabeth St & Goulburn St, Liverpool, NSW 2170, Australia
5. School of Medicine, University of Wollongong, Wollongong, NSW 2522, Australia

* Correspondence: t.becker@unsw.edu.au

Abstract: Detection of androgen receptor (AR) variant 7 (AR-V7) is emerging as a clinically important biomarker in castrate resistant prostate cancer (CRPC). Detection is possible from tumor tissue, which is often inaccessible in the advanced disease setting. With recent progress in detecting AR-V7 in circulating tumor cells (CTCs), circulating tumor RNA (ctRNA) and exosomes from prostate cancer patients, liquid biopsies have emerged as an alternative to tumor biopsy. Therefore, it is important to clarify whether these approaches differ in sensitivity in order to achieve the best possible biomarker characterization for the patient. In this study, blood samples from 44 prostate cancer patients were processed for CTCs and ctRNA with subsequent AR-V7 testing, while exosomal RNA was isolated from 16 samples and tested. Detection of AR and AR-V7 was performed using a highly sensitive droplet digital PCR-based assay. AR and AR-V7 RNA were detectable in CTCs, ctRNA and exosome samples. AR-V7 detection from CTCs showed higher sensitivity and has proven specificity compared to detection from ctRNA and exosomes. Considering that CTCs are almost always present in the advanced prostate cancer setting, CTC samples should be considered the liquid biopsy of choice for the detection of this clinically important biomarker.

Keywords: prostate cancer; CTC; AR; AR-V7; ctRNA; exosome

---

1. Introduction

Advanced prostate cancer (PC) tends to be initially hormone sensitive and is treated with androgen deprivation therapy (ADT). However, resistance to first line therapy usually develops in approximately 20–40% of patients, referred to as castrate resistant prostate cancer (CRPC) [1]. Most commonly, PC cells become resistant through molecular changes of the androgen receptor (AR), such as mutations, gene amplification, and, more recently reported, the expression of transcript AR variants [2]. In particular, the expression of AR variant 7 (AR-V7), the most abundant and clinically relevant of all variants, has been implicated as a cause of CRPC [3,4]. The translated AR-V7 protein is truncated and lacks the ligand binding domain as well as sequences important for stability and maybe cellular localization [5,6]. Importantly, intracellular AR-V7 predominantly localizes to the nucleus and displays ligand independent transcriptional activity, which is thought to be fundamental in its ability to promote ADT resistance [3,6,7].

Due to its role in ADT resistance several therapies, such as Galeterone and EPI-506 have been developed to effectively target and reduce AR-V7 levels as shown in cell line studies, with clinical trials underway [3]. Consequently, expression of AR-V7 together with that of full-length AR (AR-FL) have emerged as clinically relevant molecular biomarkers for CRPC [8,9]. AR-V7 is detectable in tissue at the RNA and protein levels, and can be evaluated using RNA hybridization techniques and immunohistology [10,11]. AR-V7 expression is rare in hormone-sensitive PC but correlates with CRPC [12]. However, in the advanced PC setting, tissue biopsies are generally unavailable for biomarker testing and diagnostic decision making. Liquid biopsies have in recent years emerged as an alternative tumor source for biomarker testing [13]. To date, AR-V7 has been detectable in circulating tumor cells (CTCs) isolated from blood samples, whole blood mRNA, from ctRNA, tumor arisen cellular vesicles, so-called exosomes found in plasma or serum and even urine [14–19]. Significantly, AR-V7 detection in CTCs has been associated with non-response to novel anti-androgens such as abiraterone and enzalutamide [14,20]. In contrast, the response rates to taxane-based chemotherapy showed no significant difference between AR-V7-positive versus-negative CRPC patients [14]. As such, this has positioned AR-V7 as a potential predictive biomarker that can discriminate between the use of an anti-androgen therapy and taxane chemotherapy.

With AR-V7 emerging as a molecular biomarker of clinical importance, it is essential to determine the best strategy to detect this biomarker for predictive and prognostic purposes. Moreover, with blood biopsies emerging as a potential source to determine AR-V7 status, it is imperative to define which blood-based approach is the most sensitive and reliable. Herein we analyzed AR-V7 and AR-FL detectability in 44 PC patients using parallel blood samples for CTC isolation and ctRNA isolation. For 16 patients, parallel evaluation of AR-V7 and AR-FL status from exosomes was also possible (see Figure 1). Our data suggest that AR-V7 and AR-FL, while detectable from ctRNA and exosomal RNA, are most sensitively detected from CTC samples.

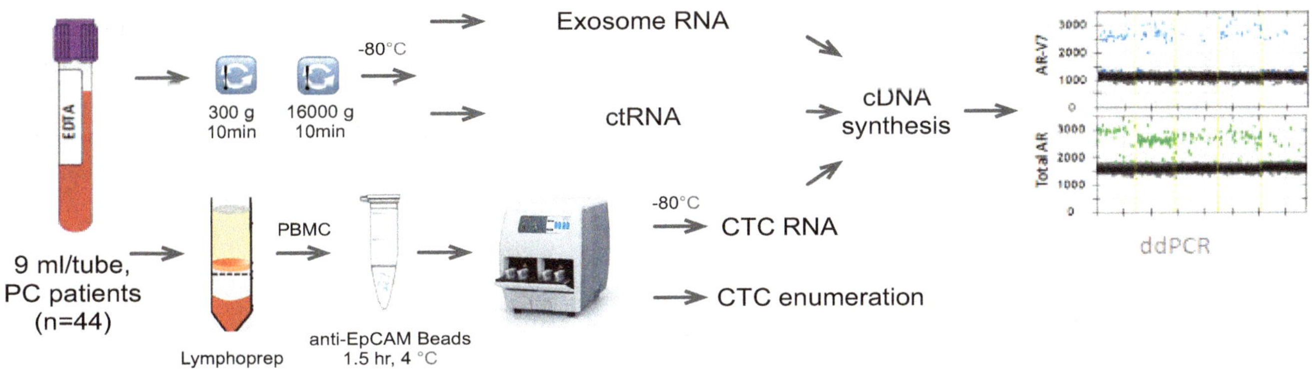

Figure 1. Work flow of the study. PC: prostate cancer, PBMC: peripheral blood mononuclear cells, ctRNA: circulating tumor ribonucleic acid, CTC: circulating tumor cell, cDNA: complementary deoxyribonucleic acid.

2. Materials and Methods

2.1. Patients

The study was undertaken in accordance with the Declaration of Helsinki with human ethics approval, HREC/13/LPOOL/158, from the South Western Sydney Local Health District Ethics Committee (approval Sep 2013, extension July 2018). A total of 44 patients with prostate cancer were recruited at Liverpool Public and St George Private Hospitals in Australia. Patient information is summarized in Table 1. Per patient, 3 × 9 mL blood draws into 9 mL EDTA vacutubes (BD, Franklin Lakes, NJ, USA) were collected, 2 for CTC isolation and 1 for plasma preparation.

Table 1. Patient Characteristics.

Patient	***n* = 44**
Mean age (years)	77 (55–94)
Gleason score (%)	
≥7	34 (77%)
<7	4 (9%)
N.A.	6 (14%)
Metastatic status (%)	41 (93%)
Bone metastases	35 (79%)
Lymph metastases	22 (50%)
Visceral metastases	13 (29%)
Definitive therapy (%)	
Radical Prostatectomy	12 (27%)
Radiotherapy	12 (27%)
Both	2 (4%)
None	22 (50%)
Past systemic therapy (%)	
None	2 (4%)
ADT	42 (95%)
Chemotherapy	17 (38%)
Novel antiandrogens	
Enzalutamide	14 (31%)
Abiraterone	9 (20%)
Both	4 (9%)

N.A.: data not available, ADT: androgen deprivation therapy.

2.2. *CTC Isolation*

Peripheral blood mononuclear cells (PBMCs) including CTCs were isolated in parallel from two blood samples using Lymphoprep™ with SepMate™ tube-based gradient centrifugation (STEMCELL Technologies, Vancouver, BC, Canada) as per the manufacturer's protocol. PBMCs were recovered and washed once with 5 mL phosphate-buffered saline (PBS) (Lonza, Basel, Switzerland), resuspended in 800 μL binding buffer with 40 μL FC buffer and 40 μL anti-EpCAM antibody coupled beads (IsoFlux CTC Enrichment kit, Fluxion Bioscience, San Francisco, CA, USA) and incubated for 90 min at 4 °C under gentle rotation. CTCs were then enriched using the standard IsoFlux CTC isolation protocol (Fluxion Bioscience, CA, USA). One CTC isolate underwent CTC enumeration and the other was immediately frozen at −80 °C until RNA extraction and gene expression analysis.

2.3. *CTC Enumeration*

IsoFlux enriched CTCs were enumerated with an IsoFlux CTC enumeration kit containing all the staining antibodies (Fluxion Bioscience, Alameda, CA, USA). Briefly, after 10 min fixation with fixation buffer, cells were blocked for 10 min with 10% normal donkey serum (NDS) and probed with rabbit anti-human CD45 antibody (1:100) followed by detection with tetramethylrhodamine (TRITC) conjugated donkey-anti rabbit IgG (1:200). Cells were then washed twice with binding buffer and permeabilized using 0.2% Trition 100 before probing with fluorescein (FITC) conjugated anti-human Pan-cytokeratin antibody (1:10). After two further washes, cells were transferred to SensoPlate™ 24-well glass bottom plates (Greiner Bio-one, Kremsmünster, Austria) with Hoechst dye included in the mounting media. Fluorescent microscopy (ALS CellCelector™, ALS, Jena, Germany) was used for CTC enumeration.

2.4. *Plasma Processing*

All plasma was processed within 4 h of blood draw. Whole blood was spun at 280× *g* for 10 min and the supernatant spun again using a microfuge at 16,000× *g* for 10 min at 4 °C. The supernatant was transferred and frozen at −80 °C until RNA extraction.

2.5. RNA Extraction and cDNA Synthesis

RNA from CTC samples was extracted with Total RNA Purification Micro Kit (Norgen Biotek, Thorold, ON, Canada) and eluted in 30 μL elution buffer. RNA from plasma was extracted with QIAamp circulating nucleic acid kit (QIAGEN, Hilden, Germany) and eluted in 65 μL elution buffer. Samples processed with this kit are referred to as ctRNA even for healthy controls that would only have normal cell free RNA. RNA from exosomes was extracted with Qiagen exoRNeasy serum/plasma Midi kit (QIAGEN, Hilden, Germany) and eluted in 30 μL. A total of 15 μL eluted RNA was reverse transcribed using SensiFAST cDNA synthesis kit (Bioline, Alexandria, Australia).

2.6. Droplet Digital PCR (ddPCR)

A total of 7 μL of cDNA of either CTC, ctRNA or exosomal samples was used for detection of AR-FL and AR-V7 transcripts by ddPCR as previously described [15]. Glyceraldehyde 3-phosphate dehydrogenase (GAPDH) transcript abundance was similarly evaluated (see Table 2 for primers, probes) and the optimized annealing temperature of 55 °C using the Bio-Rad QX200 ddPCR instrument (Bio-Rad, Hercules, CA, USA). Positive (cDNA from 22RV1 prostate cancer cells) and negative controls (DNase and RNase free H_2O) were included in all ddPCR experiments.

Table 2. GAPDH ddPCR primers and probe.

Name	Sequence
GAPDH-fwrd	CGGGAAGCTTGTCATCAATGG
GAPDH-rev	CTCCACGACGTACTCAGCG
GAPDH-probe	FAM-5′-TCTTCCAGGAGCGAGATCCCT-3-BQ1

2.7. Statistics

Data were analyzed with GraphPad Prism8 (GraphPad Software, San Diego, CA, USA). As a normal distribution was not assumed, non-parametric statistics was utilized. A Fisher exact test was used to test the significance between two categorical variables; $p < 0.05$ represents statistical significance. Unpaired comparison of CTC numbers between hormone sensitive prostate cancer (HSPC) and CRPC and AR copy numbers between healthy controls and patient samples was performed with the non-parametric Mann–Whitney test.

3. Results

3.1. Patients

Twelve HSPC and 32 CRPC patients at various stages of treatment were recruited, all patients had been originally diagnosed with subtype adenocarcinoma PC. Patient baseline characteristics are listed in Table 1. CTCs were enumerated and the expression levels of AR-FL and AR-V7 transcripts were detected in enriched CTC samples using our previously established ddPCR assay [15]. Extracted ctRNA and exosomal RNA were also tested by ddPCR for AR-FL and AR-V7 transcripts. To verify the quality of extracted RNA the reference gene GAPDH was tested for and detected in all samples (median GAPDH: 13212 copies/ml plasma, 2920-120606).

3.2. Prevalence of CTCs

CTCs were detected in 97.7% (43/44) of patients. CTC counts varied from 0 to 184 per 9 mL blood with no significant difference in CTC counts between HSCP and CRPC patients (Figure 2).

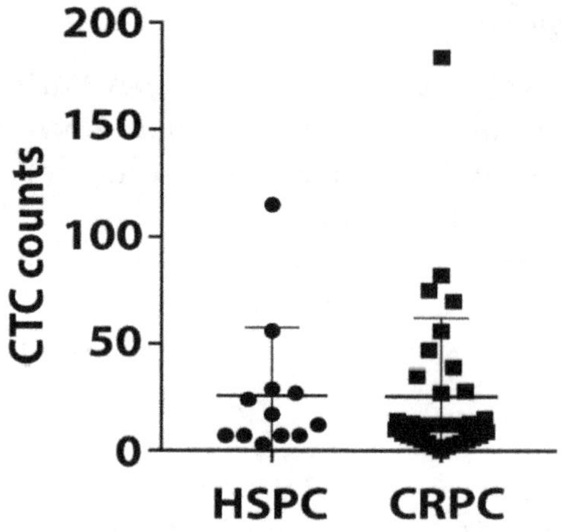

Figure 2. Circulating tumor cells (CTC) counts. CTCs were isolated using the IsoFlux CTC platform and enumerated. The range of CTC counts/9 mL blood for 12 HSPC and 32 CRPC patients is depicted. There is no significant difference of CTC numbers between HSPC and CRPC patients (p value = 0.59).

3.3. AR-V7 and AR-FL in CTC RNA and Plasma ctRNA

As expected, AR-FL and AR-V7 was only found in CTC processed blood samples with detectable CTCs. AR-FL was detected in 22 out of 44 patient CTC samples (50%). AR-FL copy number detected for CTC samples ranged from 0 to 13,714 copies/mL blood. A 62.5% share (20/32) of CRPC samples had AR-FL positive CTC samples, and only 2 HSPC patients, 16.7%, had low copy numbers of AR-FL (Figure 3, Supplementary Table S1). For AR-V7, 47.7% (21/44) of patient CTC samples tested positive with 0–146 copies/ml blood. A share of 53.1% (17/32) of CRPC patients had detectable AR-V7 (range 0–146 copies/mL blood), while AR-V7 was also identified in 4 HSPC patient CTC samples, at generally lower levels (range 0–4 copies/mL blood).

Interestingly, AR-FL detection from plasma-derived ctRNA was higher, with 70.5% (31/44) patients testing positive, although detection frequency was not significantly different from CTC samples (Fisher exact, $p = 0.13$). The detected AR-FL copy numbers found in ctRNA ranged from 0 to 180 per ml plasma and were overall less abundant compared to CTC samples. Detection of AR-FL in ctRNA samples was similarly common in HSPC patients at 66.7% (8/12) compared to CRPC patients at 71.9% (23/32), although copy numbers tended to be higher in CRPC samples ranging between 0 and 15.2 vs. 0–180, for HSPC and CRPC patients, respectively.

In contrast, AR-V7 detection in ctRNA was lower in comparison to CTC samples with only 15.9% (7/44) of patient ctRNA samples testing positive. The difference in AR-V7 detectability in CTCs vs. ctRNA was significant ($p = 0.003$). AR-V7 tends to be more frequently found in CRPC patient ctRNA 18.8% (6/32) vs. 8.3% (1/12) in HSPC patient ctRNA, but detectability and copy numbers are generally low for both CRPC and HSPC patients (range 0–8 per mL plasma). This data indicates that both AR-FL and AR-V7 are more readily detectable in CTC samples.

In the 44-patient cohort, AR-FL detection in CTCs correlated with CRPC ($p = 0.02$) while AR-V7 detection did not reach a significant correlation with CRPC ($p = 0.32$). In contrast, no correlation was found for AR-FL in ctRNA ($p = 0.73$) or AR-V7 ($p = 0.65$) (Supplementary Table S1).

The concordance of AR-FL positivity between CTC and ctRNA was 41% (18/44) while concordance of AR-V7 positivity was only 9.1% (4/44), indicating lack of sensitivity and possibly specificity for AR-V7 detection in ctRNA.

3.4. AR-V7 and AR-FL in Exosomal RNA, ctRNA and CTCs

From 16 of the analyzed patient samples, enough plasma was available to additionally extract exosomal RNA and analyze for AR-FL and AR-V7 transcripts.

Overall, 12.5% (2/16) of patients were positive for AR-V7 when analyzed using exosomal RNA, compared to 6.3% (1/16) being considered AR-V7 positive according to ctRNA analysis in that patient sub-cohort. AR-V7 was also detectable in all CTC samples from the 16 patient cohort that were found to have detectable AR-V7 either by ctRNA or exosomal. Importantly however, overall more (43.8%, 7/16) patients tested positive for AR-V7 by CTC RNA analysis. In contrast, AR-FL was detected in 68.8% (11/16) ctRNA samples, in 50% (8/16) CTC samples and 37.5% (6/16) of exosomal RNA samples (Table 3).

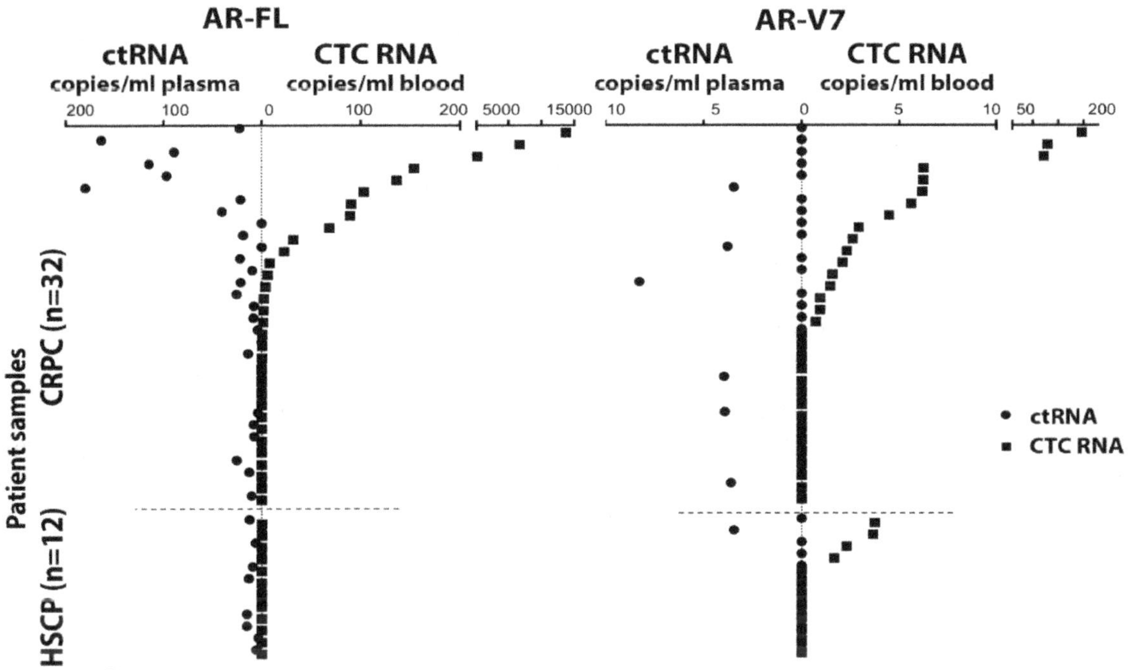

AR

	ctRNA	CTC RNA	Fishers exact
CRPC	23/32	20/32	0.60
HSPC	8/12	2/12	0.04*
Total	31/44	22/44	0.08

AR-V7

	ctRNA	CTC RNA	Fishers exact
CRPC	6/32	17/32	0.01**
HSPC	1/12	4/12	0.32
Total	7/44	21/44	0.003**

Figure 3. Detection of AR-FL (left) and AR-V7 (right) transcripts. Comparison of AR-FL and AR-V7 detection in CTC RNA and ctRNA from the same blood draw illustrated by mirrored scatter blot for the 44 patient samples (sorted in relation to detection in CTCs); summarized data are tabled below. *, ** indicate significance p-value ≤ 0.5 and ≤ 0.01, respectively.

Table 3. AR-FL/AR-V7 status in 16 patients tested from CTC, ctRNA and exosomal samples.

Patient	Status	AR-FL CTCs	AR-V7 CTCs	AR-FL ctRNA	AR-V7 ctRNA	AR-FL Exosomal	AR-V7 Exosomal
1	CRPC	−	−	−	−	+	−
2	CRPC	+	−	+	−	−	−
3	CRPC	+	−	+	−	−	−
4	CRPC	+	+	+	−	−	−
5	CRPC	−	+	+	−	+	+
6	CRPC	−	+	−	−	+	−
7	CRPC	+	+	+	−	+	+
8	CRPC	+	+	−	−	−	−
9	CRPC	+	−	+	−	−	−
10	HSPC	+	−	−	−	+	−
11	HSPC	+	−	+	−	−	−
12	HSPC	−	−	+	−	−	−
13	HSPC	−	+	+	−	−	−
14	HSPC	−	+	−	+	−	−
15	HSPC	−	−	+	−	+	−
16	HSPC	−	−	+	−	−	−

AR-V7 positive patients by CTC analysis are highlighted in grey.

3.5. Healthy Control Analysis and Implications for Assay Specificity

We previously have demonstrated that AR-FL and AR-V7 detection from CTC samples is not only highly sensitive but also highly specific. When establishing the ddPCR assay we found AR-V7 undetectable even in large numbers of healthy donor PBMCs and AR-FL detection was very low (the equivalent of 0.001 or 0.002 copies per cell were detected in only 2 of 6 healthy donor PBMC samples; considering the average residual lymphocyte number in a CTC sample that would be equivalent to 0.44 or 0.88 copies per ml blood) [15]. To evaluate sensitivity and specificity of our AR assays when screening ctRNA and exosomal RNA samples, we obtained blood from five healthy individuals, age and sex matched to our patient cohort, and extracted ctRNA and exosomal RNA for AR-FL and AR-V7 analysis. AR-FL was readily detectable in ctRNA from all healthy subjects tested with an average copy number of 14.1 per ml plasma (range: 9.6–21.5). Of note, AR-FL copy numbers in only 34.1% (15/44) of our patient ctRNA samples were above that healthy control copy number average. AR-FL was also detected in exosomal RNA from three of five healthy subjects, and crucially, the copy number range was comparable to that detected in the exosomal patient samples. AR-V7 was not as prevalent but still present with 11.1 copies/mL plasma for one healthy individual when analyzing exosomal RNA while we detected very similar copy numbers, 13.9 and 9.7 copies per ml plasma, in the two exosomal RNA AR-V7 positive patients. AR-V7 was not detected in healthy donor ctRNA samples in this small healthy control cohort. Overall, these data strongly suggest the presence of variable and sometimes high levels of AR mRNA in healthy male plasma and exosomes, which clearly impacts reliable detection of PC-derived AR-FL/AR-V7 from ctRNA and exosomes and indicates low specificity for tumor-derived AR when testing ctRNA and exosomal RNA (Figure 4).

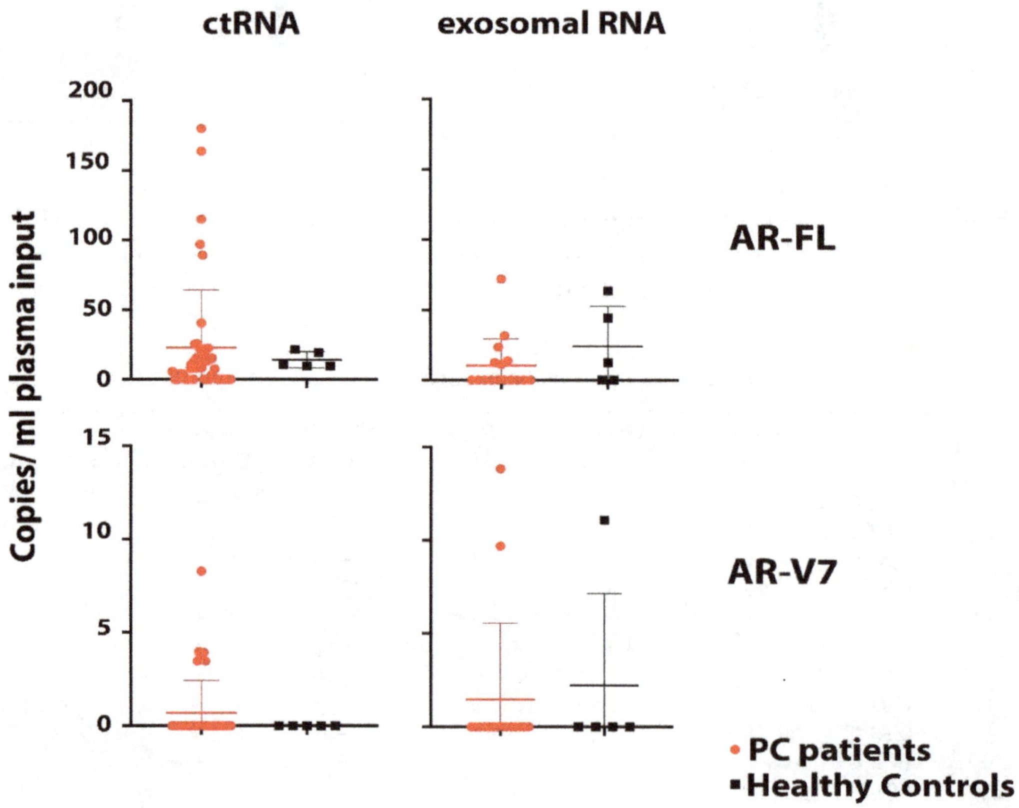

Figure 4. Comparison of AR-FL and AR-V7. AR-FL and AR-V7 detection in ctRNA (44 PC patients, 5 healthy controls) and exosomal RNA samples (16 PC patients and 5 healthy controls) indicates high background of AR transcript in plasma impacting specificity of tumor-derived AR-FL and AR-V7 detection. There is no significant difference of AR-FL and AR-V7 copy numbers detected in ctRNA and exosomal RNA between patients and healthy controls (p values: AR-FL ctRNA = 0.4; AR-FL exosomal = 0.32; AR-V7 ctRNA = 0.58; AR-V7 exosomal = 0.85).

4. Discussion

Advancements in molecular technology have aided the accurate detection and quantification of novel blood-based biomarkers paving the way for their use in clinical environments. AR-V7 is emerging as a biomarker with the potential as a predictive tool in treatment selection. Patients that have detectable AR-V7 are considered to be resistant to abiraterone and enzalutamide but respond to taxane chemotherapy, and potentially eligible for clinical trials of new generation ADT drugs. Advanced stage PC tissue samples are generally unavailable for testing and liquid biopsies are explored as a surrogate. Thus far, studies on AR-V7 have utilized various methods of detection, resulting in different degrees of correlation between data and disease parameters, highlighting that a clear "gold standard" still needs to be found [21]. While previously developing a sensitive and specific AR-FL/AR-V7 detection method for enriched CTC samples [15], here we compared this method for detection of AR-FL and AR-V7 from CTC samples and ctRNA samples of 44 PC patients and for a subset of patient samples we also were able to compare detection from exosomal RNA. Exosomes in particular have become an attractive source of tumor information. They are small double lipid membrane vesicles of endocytic origin, that contain proteins, nucleic acids and lipids released by cells. Since this includes cancer cells, exosomes can be extracted from liquid biopsies such as plasma, serum and urine to be tested for biomarker information. Importantly, both ctRNA and exosomal RNA require simpler processing protocols than CTCs with commercially available kits and thus AR-FL/AR-V7 testing using these tumor information sources would potentially be easier translated into a clinical setting.

However, our data indicate CTC-based detection is superior in sensitivity and specificity for AR-V7 with a detection rate of 48% as compared to 16% with ctRNA. Nevertheless, both assays demonstrated some association of AR-V7 detection with patients classified as castrate resistant, although our cohort was too small to find significant correlation. Previous studies which utilize CTC enrichment-based techniques reported broad detection rates ranging from 27% to 75% [15,20,22,23]. These studies demonstrated a link between baseline CTC derived AR-V7 status and disease burden, which was found to increase with subsequent lines of therapy. Our data follows this trend, with the majority of our AR-V7 patients having had at least two lines of previous treatment (Supplementary Table S1). Our study, with a relatively small study patient cohort, was mainly aimed at comparing the effectiveness of choosing simple blood-based tumor sources rather than CTC samples to screen for AR-V7. We achieved this aim and our data indicates that CTC sample testing for PC derived AR-V7 and AR-FL is more sensitive and specific and thus ultimately more reliable.

Interestingly, the few patients detected positive for AR-V7 by either ctRNA or exosomal RNA testing were also found to be AR-V7 positive by CTC testing. Some CTC AR-V7 positive patients were however not detected as positive by ctRNA or exosomal RNA testing, suggesting higher sensitivity of the AR-V7 testing in CTC samples. We also detected AR-V7 in one healthy donor exosomal RNA sample. The implications of that for the individual remain unclear and further follow up is not possible because the individual consented as a healthy volunteer. Since the AR-V7 levels were not negligible and ddPCR data showed convincing detection, we have to assume poorer specificity of AR-V7 detection from exosomes at this stage, in comparison to that shown in our previous study for CTC samples [15]. However, larger healthy control studies would be necessary to confirm this.

Not all CTC samples tested positive for AR-FL when either parallel ctRNA or exosomal RNA testing detected it. However, this finding has to be interpreted together with the fact that we found quite high levels of AR-FL transcripts in ctRNA and exosomal RNA samples from age- and sex-matched healthy control individuals. This may not be surprising as exosomes and cell free nucleic acids are thought to be released during normal tissue homeostasis, and it is quite conceivable that AR-FL transcript is released into the blood stream in that way from organs like the testis and prostate in male subjects. By analyzing AR-FL from CTC samples we seem to largely avoid such transcripts from non-tumor sources as AR expression in normal blood cells is known to be minimal or null [24].

There may be other issues underpinning how, in comparison to ctRNA-based detection, CTC-derived AR-V7 provides for a more reliable assay. Given that high CTC counts on their own have proven to be of prognostic value in PC and detection of CTCs indicate higher disease burden, CTC detection can be interpreted hand-in-hand with AR detection to exclude potential false positives, and in our hands our ddPCR assay never detected AR-FL or AR-V7 in the absence of CTCs in a parallel enumerated sample [25]. ctRNA, on the other hand, due to the high levels of nucleases in the blood, has to be considered highly unstable and is also far more susceptible to variations in pre-analytical handling. Meanwhile, CTCs in blood appear to demonstrate superior stability and we have demonstrated previously that AR-V7 copies can be detected from CTC samples from patient blood drawn in simple EDTA tubes and stored up to 48 h at room temperature, suggesting CTCs either protect the AR-V7 transcript and/or continue to express it in a drawn blood sample [26].

Since exosomes in blood are similarly believed to protect the intravesicular content including mRNAs it was important to also test AR-V7 and AR-FL detection from exosomes. We detected AR-V7 less frequently than in parallel CTC samples in only two patients. While both patients with detectable AR-V7 from exosomes are classified as castrate resistant, we also detected similar AR-V7 copy numbers in one healthy subject, challenging reliability of the finding in PC patients.

While our small patient cohort was enough to determine the best liquid biopsy entity for AR-V7 and the study was not intended to answer questions of AR-V7 biology, there are a few issues worth highlighting. Firstly, our sensitive assay detected AR-V7 in some HSPC patient CTC samples. This is not entirely unexpected and has been reported previously [27,28]. It will be interesting to see whether AR-V7 detection in HSPC patients may be an early predictor of developing CRPC, as it tends to be associated with longer time on treatment in our study (AR-V7 positive HSPC patients: between 24 month and 60 month on ADT versus AR-V7 negative HSPC patients: being maximally treated for 12 month with ADT (see Supplementary Table S1)). Secondly, AR-V7 was found in some patient CTC samples despite undetectability of AR-FL. This has been reported by others [29] and it would be interesting to investigate the impacts of AR-V7 potentially totally replacing AR-FL in these patients in future studies. Finally, there seems to be a trend which matches previous reports [30] that patients on second line ADT have more commonly detectable AR-V7 in CTCs. In our study, 10 of 14 (69%) CRPC patients receiving enzalutamide and 5 of 9 (55.5%) receiving abiraterone (some patients were treated consecutively with both) were AR-V7 positive by CTC testing.

Although the small patient cohort allowed us to answer the main question of which liquid biopsy entity is the better for detection of AR-V7 and AR-FL, a limitation of the study is that correlation with disease parameters was not as informative as a bigger cohort would have been. Larger cohorts of healthy control comparisons would be able to better define background AR-FL and AR-V7 in plasma.

5. Conclusions

This study compared AR-V7 and AR-FL detection in liquid biopsy (blood)-derived CTC samples, ctRNA samples and exosomes. Our data show that testing of these clinically highly relevant biomarkers for PC patients is most reliable performed from CTC samples in regards to sensitivity and specificity. It also should be noted that a recent report shows that at the protein level correlation with PC disease parameters is linked to the nuclear localization of AR-V7 protein in CTCs [30]. We usually analyze two parallel blood samples, one for CTC enumeration, and we have recently amended this protocol to incorporate AR-V7 immunocytostaining and cellular localization screening. The second blood sample is screened for AR-V7 and AR-FL transcripts by ddPCR. Both tests go hand-in-hand to confirm AR-V7 presence and highlight the importance of analyzing CTCs rather than other circulating tumor entities. Our future studies will clarify whether both CTC tests cooperatively show correlation with resistance to ADT.

Author Contributions: Conceptualization: T.M.B., P.D.S., Y.M.; methodology: Y.M., M.N., S.A.J., T.O., F.Y., T.K.; formal analysis: T.M.B., P.D.S., Y.M., M.N., P.D.; resources: T.M.B., P.D.S., P.D., W.C., B.B., A.C., writing—original draft preparation: T.M.B., Y.M., M.N., T.K.; writing—review and editing: T.M.B., P.D.S., Y.M., M.N., S.A.J., T.O., F.Y., T.K., W.C., B.B., P.D., A.C., supervision: T.M.B., P.D.S.; project administration: T.M.B., P.D.S.

Acknowledgments: Human ethics approval, HREC/13/LPOOL/158, was obtained and managed by the CONCERT Biobank.

References

1. Knudsen, K.E.; Scher, H.I. Starving the addiction: New opportunities for durable suppression of AR signaling in prostate cancer. *Clin. Cancer Res.* **2009**, *15*, 4792–4798. [CrossRef] [PubMed]
2. Wadosky, K.M.; Koochekpour, S. Molecular mechanisms underlying resistance to androgen deprivation therapy in prostate cancer. *Oncotarget* **2016**, *7*, 64447–64470. [CrossRef] [PubMed]
3. Antonarakis, E.S.; Armstrong, A.J.; Dehm, S.M.; Luo, J. Androgen receptor variant-driven prostate cancer: clinical implications and therapeutic targeting. *Prostate Cancer Prostatic Dis.* **2016**, *19*, 231–241. [CrossRef] [PubMed]
4. Li, Y.; Chan, S.C.; Brand, L.J.; Hwang, T.H.; Silverstein, K.A.; Dehm, S.M. Androgen receptor splice variants mediate enzalutamide resistance in castration-resistant prostate cancer cell lines. *Cancer Res.* **2013**, *73*, 483–489. [CrossRef] [PubMed]
5. Cao, B.; Qi, Y.; Zhang, G.; Xu, D.; Zhan, Y.; Álvarez, X.; Guo, Z.; Fu, X.; Plymate, S.R.; Sartor, O.; et al. Androgen receptor splice variants activating the full-length receptor in mediating resistance to androgen-directed therapy. *Oncotarget* **2014**, *5*, 1646–1656. [CrossRef]
6. Chan, S.C.; Li, Y.; Dehm, S.M. Androgen Receptor Splice Variants Activate Androgen Receptor Target Genes and Support Aberrant Prostate Cancer Cell Growth Independent of Canonical Androgen Receptor Nuclear Localization Signal*. *J. Boil. Chem.* **2012**, *287*, 19736–19749. [CrossRef] [PubMed]
7. Xu, D.; Zhan, Y.; Qi, Y.; Cao, B.; Bai, S.; Xu, W.; Gambhir, S.S.; Lee, P.; Sartor, O.; Flemington, E.K.; et al. Androgen receptor splice variants dimerize to transactivate target genes. *Cancer Res.* **2015**, *75*, 3663–3671. [CrossRef]
8. Henzler, C.; Li, Y.; Yang, R.; McBride, T.; Ho, Y.; Sprenger, C.; Liu, G.; Coleman, I.; Lakely, B.; Li, R.; et al. Truncation and constitutive activation of the androgen receptor by diverse genomic rearrangements in prostate cancer. *Nat. Commun.* **2016**, *7*, 13668. [CrossRef]
9. Conteduca, V.; Wetterskog, D.; Sharabiani, M.T.A.; Grande, E.; Fernandez-Perez, M.P.; Jayaram, A.; Salvi, S.; Castellano, D.; Romanel, A.; Lolli, C.; et al. Androgen receptor gene status in plasma DNA associates with worse outcome on enzalutamide or abiraterone for castration-resistant prostate cancer: a multi-institution correlative biomarker study. *Ann. Oncol.* **2017**, *28*, 1508–1516. [CrossRef]
10. Guedes, L.B.; Morais, C.L.; Almutairi, F.; Haffner, M.C.; Zheng, Q.; Isaacs, J.T.; Antonarakis, E.S.; Lu, C.; Tsai, H.; Luo, J.; et al. Analytic Validation of RNA In Situ Hybridization (RISH) for AR and AR-V7 Expression in Human Prostate Cancer. *Clin. Cancer Res.* **2016**, *22*, 4651–4663. [CrossRef]
11. Welti, J.; Rodrigues, D.N.; Sharp, A.; Sun, S.; Lorente, D.; Riisnaes, R.; Figueiredo, I.; Zafeiriou, Z.; Rescigno, P.; De Bono, J.S.; et al. Analytical Validation and Clinical Qualification of a New Immunohistochemical Assay for Androgen Receptor Splice Variant-7 Protein Expression in Metastatic Castration-resistant Prostate Cancer. *Eur. Urol.* **2016**, *70*, 599–608. [CrossRef] [PubMed]
12. Kallio, H.M.L.; Hieta, R.; Latonen, L.; Brofeldt, A.; Annala, M.; Kivinummi, K.; Tammela, T.L.; Nykter, M.; Isaacs, W.B.; Lilja, H.G.; et al. Constitutively active androgen receptor splice variants AR-V3, AR-V7 and AR-V9 are co-expressed in castration-resistant prostate cancer metastases. *Br. J. Cancer* **2018**, *119*, 347–356. [CrossRef] [PubMed]
13. Crowley, E.; Di Nicolantonio, F.; Loupakis, F.; Bardelli, A. Liquid biopsy: monitoring cancer-genetics in the blood. *Nat. Rev. Clin. Oncol.* **2013**, *10*, 472–484. [CrossRef] [PubMed]
14. Antonarakis, E.S.; Lu, C.; Luber, B.; Wang, H.; Chen, Y.; Nakazawa, M.; Nadal, R.; Paller, C.J.; Denmeade, S.R.; Carducci, M.A.; et al. Androgen Receptor Splice Variant 7 and Efficacy of Taxane Chemotherapy in Patients With Metastatic Castration-Resistant Prostate Cancer. *JAMA Oncol.* **2015**, *1*, 582–591. [CrossRef] [PubMed]
15. Ma, Y.; Luk, A.; Young, F.P.; Lynch, D.; Chua, W.; Balakrishnar, B.; De Souza, P.; Becker, T.M. Droplet Digital PCR Based Androgen Receptor Variant 7 (AR-V7) Detection from Prostate Cancer Patient Blood Biopsies. *Int. J. Mol. Sci.* **2016**, *17*, 1264. [CrossRef] [PubMed]
16. Hodara, E.; Morrison, G.; Cunha, A.T.; Zainfeld, D.; Xu, T.; Xu, Y.; Dempsey, P.W.; Pagano, P.C.; Bischoff, F.; Khurana, A.; et al. Multiparametric liquid biopsy analysis in metastatic prostate cancer. *JCI Insight* **2019**, *4*, 4. [CrossRef]

17. Todenhöfer, T.; Azad, A.; Stewart, C.; Gao, J.; Eigl, B.J.; Gleave, M.E.; Joshua, A.M.; Black, P.C.; Chi, K.N. AR-V7 Transcripts in Whole Blood RNA of Patients with Metastatic Castration Resistant Prostate Cancer Correlate with Response to Abiraterone Acetate. *J. Urol.* **2017**, *197*, 135–142. [CrossRef]
18. Woo, H.K.; Park, J.; Ku, J.Y.; Lee, C.H.; Sunkara, V.; Ha, H.K.; Cho, Y.K. Urine-based liquid biopsy: non-invasive and sensitive AR-V7 detection in urinary EVs from patients with prostate cancer. *Lab on a chip* **2018**, *19*, 87–97. [CrossRef]
19. Del Re, M.; Biasco, E.; Crucitta, S.; DeRosa, L.; Rofi, E.; Orlandini, C.; Miccoli, M.; Galli, L.; Falcone, A.; Jenster, G.W.; et al. The Detection of Androgen Receptor Splice Variant 7 in Plasma-derived Exosomal RNA Strongly Predicts Resistance to Hormonal Therapy in Metastatic Prostate Cancer Patients. *Eur. Urol.* **2017**, *71*, 680–687. [CrossRef]
20. Antonarakis, E.S.; Lu, C.; Wang, H.; Luber, B.; Nakazawa, M.; Roeser, J.C.; Chen, Y.; Mohammad, T.A.; Chen, Y.; Fedor, H.L.; et al. AR-V7 and Resistance to Enzalutamide and Abiraterone in Prostate Cancer. *N. Eng. J. Med.* **2014**, *371*, 1028–1038. [CrossRef]
21. Bernemann, C.; Schnoeller, T.J.; Luedeke, M.; Steinestel, K.; Boegemann, M.; Schrader, A.J.; Steinestel, J. Expression of AR-V7 in Circulating Tumour Cells Does Not Preclude Response to Next Generation Androgen Deprivation Therapy in Patients with Castration Resistant Prostate Cancer. *Eur. Urol.* **2017**, *71*, 1–3. [CrossRef] [PubMed]
22. Antonarakis, E.S.; Lu, C.; Luber, B.; Wang, H.; Chen, Y.; Zhu, Y.; Silberstein, J.L.; Taylor, M.N.; Maughan, B.L.; Denmeade, S.R.; et al. Clinical Significance of Androgen Receptor Splice Variant-7 mRNA Detection in Circulating Tumor Cells of Men With Metastatic Castration-Resistant Prostate Cancer Treated With First- and Second-Line Abiraterone and Enzalutamide. *J. Clin. Oncol.* **2017**, *35*, 2149–2156. [CrossRef] [PubMed]
23. Onstenk, W.; Sieuwerts, A.M.; Kraan, J.; Van, M.; Nieuweboer, A.J.; Mathijssen, R.H.; Hamberg, P.; Meulenbeld, H.J.; De Laere, B.; Dirix, L.Y.; et al. Efficacy of Cabazitaxel in Castration-resistant Prostate Cancer Is Independent of the Presence of AR-V7 in Circulating Tumor Cells. *Eur. Urol.* **2015**, *68*, 939–945. [CrossRef] [PubMed]
24. Mantalaris, A.; Panoskaltsis, N.; Sakai, Y.; Bourne, P.; Chang, C.; Messing, E.M.; Wu, J.H.D. Localization of androgen receptor expression in human bone marrow. *J. Pathol.* **2001**, *193*, 361–366. [CrossRef]
25. Onstenk, W.; De Klaver, W.; De Wit, R.; Lolkema, M.; Foekens, J.; Sleijfer, S. The use of circulating tumor cells in guiding treatment decisions for patients with metastatic castration-resistant prostate cancer. *Cancer Treat. Rev.* **2016**, *46*, 42–50. [CrossRef] [PubMed]
26. Luk, A.W.S.; Ma, Y.; Ding, P.N.; Young, F.P.; Chua, W.; Balakrishnar, B.; Dransfield, D.T.; De Souza, P.; Becker, T.M. CTC-mRNA (AR-V7) Analysis from Blood Samples—Impact of Blood Collection Tube and Storage Time. *Int. J. Mol. Sci.* **2017**, *18*, 1047. [CrossRef] [PubMed]
27. Hu, R.; Dunn, T.A.; Wei, S.; Isharwal, S.; Veltri, R.W.; Humphreys, E.; Han, M.; Partin, A.W.; Vessella, R.L.; Isaacs, W.B.; et al. Ligand-independent Androgen Receptor Variants Derived from Splicing of Cryptic Exons Signify Hormone Refractory Prostate Cancer. *Cancer Res.* **2009**, *69*, 16–22. [CrossRef]
28. Qu, Y.; Dai, B.; Ye, D.; Kong, Y.; Chang, K.; Jia, Z.; Yang, X.; Zhang, H.; Zhu, Y.; Shi, G. Constitutively Active AR-V7 Plays an Essential Role in the Development and Progression of Castration-Resistant Prostate Cancer. *Sci. Rep.* **2015**, *5*, 7654. [CrossRef]
29. El-Heliebi, A.; Hille, C.; Laxman, N.; Svedlund, J.; Haudum, C.; Ercan, E.; Kroneis, T.; Chen, S.; Smolle, M.; Rossmann, C.; et al. In Situ Detection and Quantification of AR-V7, AR-FL, PSA, and KRAS Point Mutations in Circulating Tumor Cells. *Clin. Chem.* **2018**, *64*, 536–546. [CrossRef]
30. Scher, H.I.; Lu, D.; Schreiber, N.A.; Louw, J.; Graf, R.P.; Vargas, H.A.; Johnson, A.; Jendrisak, A.; Bambury, R.; Danila, D.; et al. Association of AR-V7 on Circulating Tumor Cells as a Treatment-Specific Biomarker With Outcomes and Survival in Castration-Resistant Prostate Cancer. *JAMA Oncol.* **2016**, *2*, 1441–1449. [CrossRef]

12

Are Circulating Tumor Cells (CTCs) Ready for Clinical Use in Breast Cancer? An Overview of Completed and Ongoing Trials Using CTCs for Clinical Treatment Decisions

Fabienne Schochter *, Thomas W. P. Friedl *, Amelie deGregorio, Sabrina Krause, Jens Huober, Brigitte Rack and Wolfgang Janni

Department of Obstetrics and Gynecology, University Hospital Ulm, University of Ulm, 89075 Ulm, Germany; Amelie.deGregorio@uniklinik-ulm.de (A.d.G.); sabrina.krause@uniklinik-ulm.de (S.K.); jens.huober@uniklinik-ulm.de (J.H.); brigitte.rack@uniklinik-ulm.de (B.R.); wolfgang.janni@uniklinik-ulm.de (W.J.)

* Correspondence: fabienne.schochter@uni-ulm.de (F.S.); Thomas.Friedl@uniklinik-ulm.de (T.W.P.F.)

Abstract: In recent years, breast cancer treatment has become increasingly individualized. Circulating tumor cells (CTCs) have the potential to move personalized medicine another step forward. The prognostic relevance of CTCs has already been proven both in early and metastatic breast cancer. In addition, there is evidence that changes in CTC numbers during the course of therapy can predict treatment response. Thus, CTCs are a suitable tool for repeated treatment monitoring through noninvasive liquid biopsy. The next step is to evaluate how this information can be used for clinical decision making with regard to the extension, modification, or abandonment of a treatment regimen. This review will summarize the completed and ongoing clinical trials using CTC number or phenotype for treatment decisions. Based on current knowledge, CTCs can be regarded as a useful prognostic and predictive marker that is well suited for both risk stratification and treatment monitoring in breast cancer patients. However, there is still the need to provide sufficient and unequivocal evidence for whether CTCs may indeed be used to guide treatment decisions in everyday clinical practice. The results of the ongoing trials described in this review are eagerly awaited to answer these important questions.

Keywords: circulating tumor cells (CTCs); clinical trials; breast cancer; CTC-based treatment decisions

1. Introduction

In recent years, breast cancer treatment has become increasingly individualized. New targeted therapies have improved survival for many patients. With the aim of becoming even more patient-specific, circulating tumor cells (CTCs) are an interesting topic in translational oncology. CTCs represent rare cancer cells in the peripheral blood that have disseminated from the primary tumor (or metastatic sites) and play an important role in tumor progression and the formation of (new) metastases. Several recent reviews describe the biology of CTCs and discuss both their potential of being used as a prognostic and/or predictive marker and the challenges involved in incorporating CTCs in clinical practice [1,2].

CTCs have already proven their prognostic relevance in early breast cancer (EBC) and metastatic breast cancer (MBC) [3–5]. In addition, the SUCCESS-A and the ECOG-ACRIN study E5103 showed that the detection of persisting CTCs two years and even five years after (neo)adjuvant chemotherapy is related to an increased risk of recurrence [6–8]. In a multivariate analysis presented by Sparano

2018, CTC-positivity assessed after five years was the strongest predictor of late disease recurrence in patients with hormone receptor-positive breast cancer that had no signs of disease recurrence in the first five years after primary diagnosis [8].

If CTC-positivity predicts a worse clinical outcome [9] and CTC dynamics is a predictor of therapy response [10], the question remains whether this information can be used for forming treatment decisions. Trials are attempting to find a way to answer this challenging question using either CTC number or CTC phenotype as a criterion for therapy decisions. This article gives an overview of the current status (see Table 1) and—if available—results of all clinical breast cancer trials that involve interventions based on CTC number or phenotype.

Table 1. Clinical trials with CTC-based treatment decisions.

Trail	Status	Enrollment	Condition	Intervention	Primary Endpoints
Treat-CTC, NCT01548677 (phase II)	Closed after interim analysis	1317 enrolled 63 randomized (CTC-pos)	HER2-neg EBC with CTCs after CT	Trastuzumab iv 6 cycles vs. observation	CTC detection rate at week 18
SWOG S0500, NCT00382018 (phase III)	Completed	595 enrolled; 123 randomized (persisting high CTC count)	CT-resistant, CTC-pos MBC	Early switch in therapy vs. treatment until progression	OS, PFS
CirCe01, NCT01349842 (phase III)	Recruitment completed,	568 planed	CT-resistant, CTC-pos MBC	Early switch in therapy vs. treatment until progression	OS
STIC-CTC, NCT01710605 (phase III)	Completed	778 randomized	HR-pos and HER2-neg MBC	Decision CT or ET by clinical choice vs. CTC count	PFS, economic value
Circe TDM-1 NCT01975142 (phase II)	Closed after interim analysis	155 screened; 11 treated	HER2-neg MBC and HER2-pos CTCs	T-DM1	TRR
DETECT III, NCT01619111 (phase III)	Recruiting	120 planned, up to date: 105	HER2-neg MBC and HER2-pos CTCs	Standard treatment vs. Standard treatment + lapatinib	CTC clearance
DETECT IV, NCT02035813 (phase II)	Recruiting	Group A: 180 planned Up to date: 103 Group B: 120 planned Up to date: 107	HER2-neg MBC and HER2-neg CTCs	A: ET + ribociclib or everolimus B: eribulin	A: CTC clearance B: PFS
DETECT V, NCT02344472 (phase III)	Recruiting	270 planned Up to date 151	HR-pos, HER2-pos, MBC	trastuzumab/ pertuzumab + CT or ET with ribociclib	Tolerability, safety, and quality of life

CT = chemotherapy, CTC = circulating tumor cells, ET = endocrine treatment, HER2 = human epidermal growth factor receptor 2, HR = hormone receptor, EBC = early breast cancer, MBC = metastatic breast cancer, OS = overall survival, PFS = progression-free survival, TRR = tumor response rate, T-DM1 = trastuzumab emtansine, pos = positive, neg = negative

2. Material and Methods

On the database site ClinicalTrials.gov, 59 studies were retrieved by searching using the keywords "circulating tumor cells" and "breast cancer" at the time the search was performed (end of March

2019). Overall, 26 trials were initiated in Europe, 20 in the United States, 11 in Asian countries, and 2 in Brazil. In total, 31 interventional trials are registered, from which 10 are either recruiting or active but not recruiting. However, only 7 trials use either the presence, amount, or phenotype of CTCs in determining an intervention for the chosen treatment. One is an adjuvant trial, while the others describe trials in a metastatic setting. In the following, we will give an overview of these studies.

All clinical trials used the only FDA-cleared and current gold standard method for the enrichment and detection of CTCs, the CELLSEARCH® system (Menarini Silicon Biosystems, Bologna, Italy), which has been described in detail by Park Y et al. [11]. Briefly, the first step during analysis is the immunomagnetic enrichment of EpCAM-positive cells using antibody-coated ferrofluid nanoparticles. The EpCAM-enriched cells are then stained with antibodies specific for cytokeratins 8, 18, and 19 (epithelial cell markers) and for CD45 (leukocyte marker), as well as with the fluorescent nucleic acid dye 4,6-diamidino-2-phenylindole dihydrochloride (DAPI) for labeling of cell nuclei. CTCs are identified and counted using a semiautomated fluorescence-based microscope system that generates images of the stained cells, with CTCs being defined as cytokeratin-positive nucleated cells that lack *CD45* expression.

3. CTC-Based Clinical Trials in EBC

Treat-CTC

The TREAT-CTC study (NCT01548677) is the only study in the (neo)adjuvant setting in which treatment decisions are based on the presence of CTCs. This trial tried to address the question of an additional treatment possibility to eliminate CTCs persistent after (neo)adjuvant chemotherapy. A total of 1317 patients with HER2-negative EBC were screened for CTCs after completion of (neo)adjuvant chemotherapy and patients with at least 1 CTC/15 mL blood were randomized to either an additional treatment with trastuzumab (6 cycles of trastuzumab i.v.) or observation. In 95 (7.2%) of the patients, CTCs could be detected; 31 patients were randomized to trastuzumab treatment, while 32 patients were randomized to the observational control arm. The CTC-positivity rate was similar, and not significantly different in both arms after 18 weeks of treatment (17.2% vs. 13.8%); furthermore, no difference in disease-free survival could be seen [12]. Following the recommendation of the independent Data Monitoring Committee to stop the trial for futility, study recruitment was not continued after the first interim analysis. A possible explanation for this negative result might be that, while the HER2 status of CTCs was determined, HER2-positivity of CTCs was not required for study inclusion. In the majority of the patients (76%), the detected CTCs were HER2-negative. This is in accordance with the results of the NSABP-B47 trial that failed to show improved disease-free survival if trastuzumab is added to chemotherapy in patients with HER2-low (IHC 1+ or 2+ staining intensity) breast cancer [13]. Thus, both the NSABP B47 and the Treat CTC trial failed to confirm the hypothesis that women with early breast cancer showing low *HER2* expression might benefit from treatment with trastuzumab following adjuvant chemotherapy. Taken together, these results suggest that the failure of the Treat CTC trial was due to choosing an inappropriate treatment intervention for the targeted patient population rather than indicating a general failure of the concept of CTC-based intervention decisions.

4. CTC-Based Clinical Trials in MBC

In the MBC setting, trials that are based on CTC number usually use the cutoff of ≥5 CTCs which, initially, was not meant to be used for treatment decisions but was presented as a tool to separate the patients into two groups with different survival prospects. A recent retrospective pooled analysis including 2436 MBC patients confirmed the utility of the cutoff of ≥5 CTCs for risk stratification, as MBC patients could be separated into categories of either stage IV indolent (<5 CTCs) or stage IV aggressive (≥5 CTCs) with significantly longer overall survival in the group with <5 CTCs independently of clinical and molecular variables [14].

4.1. SWOG S0500

Since the first knowledge that a high count of CTCs predicts a worse clinical outcome and that changes in CTCs reflect therapy response, the question has been raised of whether MBC patients can be monitored and treated based on CTC dynamics. The first clinical phase III trial to investigate this hypothesis was initiated by the Southwest Oncology Group (SWOG). The SO500 study (NCT00382018) included 595 patients with MBC scheduled for first-line chemotherapy in the advanced setting. Before the start of first-line chemotherapy, patients were tested for CTCs. If patients did not have an increased CTC count (defined as less than 5 CTCs per 7.5 mL blood) at baseline (n = 276), they were treated according to physician's choice and no additional blood draws or interventions were performed. A total of 319 patients had an elevated CTC count (defined as five or more CTCs per 7.5 mL blood) before the start of first-line chemotherapy. A total of 288 of these high-risk patients were re-tested after the first cycle of first-line therapy, which was approximately 22 days after the first chemotherapy administration. Patients with a follow-up CTC count of less than 5 CTCs per 7.5 mL blood (n = 165) continued on the initially chosen chemotherapy regimen until progression. If the follow-up CTC level was persistently high (i.e., the CTC count remained at a level of five or more CTCs per 7.5 mL blood), the patients (n = 123) were included in the interventional part and randomly assigned to either continue the treatment until clinical and/or radiographic evidence of progression (n = 64) or switch early to another chemotherapy of the physician's choice (n = 59). The results showed no significant improvement in survival for an early change in treatment regime (median progression-free survival (PFS) 4.6 vs. 3.5 months, HR = 0.92; 95%CI = 0.64–1.32; median overall survival (OS) 1.5 vs. 10.7 months, HR = 1.00, 95%CI = 0.69–1.47). Nevertheless, the results confirmed the prognostic impact of CTCs: the median OS reached 35 months in patients with a low CTC count before treatment, 23 months in patients with a high CTC count before treatment but a low follow-up CTC count, and 13 months in patients with a persistently high CTC count. These differences were highly significant after adjusting for hormone receptor and HER2 status in a multivariable Cox model ($p < 0.001$). Overall, the study showed that for patients with persistently increased CTCs after one cycle of first-line chemotherapy, and that early switching to another standard chemotherapy regimen was not effective in terms of prolonged survival. Thus, persisting CTCs might represent a chemoresistant population of tumor cells which requires an alternative approach [15].

4.2. CirCe01

Another similar trial evaluating the response to chemotherapy by the CTC decrease after the first cycle is the CirCe01 study (NCT01349842). This French, multicenter, randomized phase III study included MBC patients that were CTC-positive (≥5 CTCs/7.5 mL) after two lines of chemotherapy. Patients were either followed by the conventional clinical and radiological assessments or by the determination of CTCs. Patients in the interventional (CTC-driven) arm were switched to another chemotherapy if CTC count did not decrease after one cycle of chemotherapy [16]. The aim was to switch non-responding patients early to avoid ineffective but toxic treatments. The recruitment has been completed, but the results are still pending.

4.3. STIC CTC

The French study group also initiated the STIC CTC Trial (NCT01710605). This phase III trial tries to answer the question of whether the choice between chemotherapy and endocrine therapy in first-line, HER2-negative, hormone receptor-positive MBC patients can be driven by the amount of CTCs. The patients were randomized to clinician's choice in the standard arm (i.e., the decision for chemotherapy or endocrine therapy was based on clinical assessment) or CTC-based decision in the interventional arm. The trial used the common cutoff of ≥5 CTCs in 7.5 mL peripheral blood; patients in the CTC-driven interventional arm with less than 5 CTCs received endocrine treatment, while patients with 5 or more CTCs were treated with chemotherapy (by the physician's choice). The results were presented at the San Antonio Breast Cancer Meeting 2018 [17]. Overall, 778 patients were randomized. In the clinician's choice arm, 72.6% and 27.4% of patients respectively received hormone therapy and chemotherapy, while in the CTC-driven arm, 62.6% and 37.4% of patients were respectively treated with endocrine therapy and chemotherapy. The primary endpoint of the study was met, as PFS of patients in the CTC decision-based arm (median 15.6 months) was not inferior to PFS of patients in the clinician's choice arm (median 14.0 months). The analysis focused further on the planned subgroup analysis of the discordant groups (CTCs high/clinical low risk; CTCs low/clinical high risk). Patients in the CTCs high/clinical low-risk subgroup who received a CTC-driven chemotherapy discordant to the physicians choice had a significantly longer PFS than the patients in the clinical-decision arm receiving endocrine therapy (HR = 0.62, 95%CI = 0.45–0.84; $p = 0.002$), with a non-significant trend toward longer OS (HR = 0.69, 95%CI = 0.43–1.11; $p = 0.12$). On the other side, patients in the CTCs low/clinical high-risk subgroup who were deescalated to a hormonal therapy from the clinically chosen chemotherapy due to having a low CTC count showed no significantly worse PFS compared to the patients treated with chemotherapy according to clinician's choice. In addition, when the patients from the two discordant subgroups were pooled, patients receiving chemotherapy showed significantly better PFS (HR = 0.66, 95%CI = 0.51–0.85; $p = 0.001$) and OS (HR = 0.65, 95%CI = 0.43–0.98; $p = 0.04$) than patients receiving endocrine therapy, challenging current treatment standards. The results of this trial are promising and indicate that including a CTC count in the decision algorithm for HER2-negative, hormone receptor-positive MBC patients might improve patient outcome in some cases. However, the results need to be confirmed for the new era of endocrine treatment regimen with CDK4/6 inhibitors.

4.4. *CirCe T-DM1*

The CirCe T-DM1 study (NCT01975142) is the first clinical trial not using the number but the phenotype of CTCs as a decision criterion. HER2-negative MBC patients needing a third or fourth line of therapy and with detected HER2-positive CTCs in baseline-screening were treated with the antibody–drug conjugate trastuzumab emtansine (TDM-1). The study was closed after the first interim analysis and presented at the ESMO 2017 [18]. A total of 155 patients were screened, 14 (9.0%) had HER2-positive CTCs and 11 patients were treated with TDM-1. Partial response was observed in only one patient; median PFS was 4.9 months (range: 1.8–10.1). Due to these results and the very low prevalence of HER2-positive CTCs (1.6% of the detected CTCs), the authors conclude that the tested therapeutic approach was not promising.

4.5. *DETECT Study Program*

The DETECT study concept is a comprehensive trial observing CTCs for liquid biopsy in MBC patients with different biologic tumor features. The study concept includes DETECT III, DETECT IV, and DETECT V. It investigates the efficacy of treatment decisions based on the presence and phenotype of CTCs.

Patients with HER2-negative MBC and HER2-positive CTCs are eligible for the DETECT III study (NCT01619111). In this phase III trial, 120 patients were randomized (1:1) to a standard endocrine (letrozole, anastrozole, or exemestane) or standard chemotherapeutic (docetaxel, paclitaxel, capecitabine, vinorelbine) treatment according to the physician's choice, with or without an additional HER2-targeted treatment with lapatinib. The primary endpoint of the DETECT III study is the efficacy as assessed by the CTC clearance rate, i.e., the proportion of patients with no evidence of CTCs in the blood after treatment.

Patients with HER2-negative MBC and only HER2-negative CTCs are being treated in the DETECT IV trial (NCT02035813), which is divided into two separate treatment and observation cohorts. In cohort A, patients with hormone receptor-positive breast cancer are treated with endocrine therapy (anastrozole, letrozole, exemestane, fulvestrant or tamoxifen; tamoxifen only for patients in combination with ribociclib) combined with the mTOR inhibitor everolimus or (after an amendment) the CDK4/6 inhibitor ribociclib. Cohort B includes patients with either triple-negative breast cancer or patients with hormone receptor-positive breast cancer that need a more aggressive treatment. Here, patients receive the mitotic inhibitor eribulin (halichondrin B analogue) as treatment in single-arm observation. The primary endpoints for cohort A and B are efficacy as assessed by the CTC clearance rate and PFS, respectively.

With DETECT V/CHEVENDO (NCT02344472), the study concept also offers a therapeutic option for patients with HER2-positive and hormone receptor-positive breast cancer, even though interventions are not based on CTCs. In this two-arm randomized phase III trial, all patients are treated with the dual HER2-targeted therapy with pertuzumab and trastuzumab, and are randomized to either in combination with chemotherapy (docetaxel, paclitaxel, capecitabine, vinorelbine) or endocrine therapy (fulvestrant, tamoxifen, letrozole, anastrozole, or exemestane) combined with the CDK4/6 inhibitor ribociclib. A modified adverse event score and the "quality-adjusted time without symptoms and toxicity" (Q-TWiST) method are used for assessing safety, tolerability and quality of life. In fact, the DETECT V trail is the only part in which the therapeutic decision is not driven by the number or phenotype of CTCs. However, one aim of the DETECT V trial is the development of an "endocrine responsiveness score" (ERS), based on the estrogen receptor and *HER2* expression of CTCs, to derive a predictive tool for the hormone receptor-positive, HER2-positive disease. The first results coming from the COMETI-2 study suggest that high rates of estrogen receptor are associated with a better response to endocrine treatment, while high rates of *HER2* expression are associated with a worse response [19]. The general goal of the endocrine responsiveness score is to identify patients with a predicted good response to endocrine therapy to avoid unnecessary chemotherapeutic treatments, which are generally associated with more adverse events and decreased quality of life compared to endocrine therapies.

In all trials of the DETECT study concept, CTCs are measured repetitively during treatment to obtain data on CTC dynamics and their possible role as a tool for treatment monitoring and early response assessment. Furthermore, the DETECT study concept is accompanied by a comprehensive translational research project ("DETECT-CTC") which is funded by the Deutsche Krebshilfe (German Cancer Aid). The main aim of DETECT-CTC is to apply innovative biomarkers and assays focusing on molecular characteristics of CTCs and circulating nucleic acids to analyze their potential function as a liquid biopsy tool for assessing the biological status of the advanced disease, and to determine their relevance for predicting treatment response and therapy monitoring in order to optimize treatment for patients with metastatic breast cancer. Various subprojects of DETECT-CTC focus on the molecular characterization of CTCs, circulating free DNA and microRNA in blood, the evaluation of DNA damage and repair markers on CTCs, the evaluation of the origins and molecular causes of resistance to endocrine therapy at the level of individual CTCs, the comparison of phenotypic expression of biomarkers between CTCs, disseminated tumor cells from bone marrow, primary tumor and metastases from the same patients, and the study of microevolution of resistant subclones in metastatic breast cancer through single-cell analysis of CTCs [20].

In a comprehensive analysis of a large cohort of DETECT screening patients with HER2-negative primary tumors, the rate of discordance in the HER2 phenotype between primary tumor and CTCs was assessed and, as such, a discordance might have far-reaching implications in terms of follow-up treatment and the addition of targeted therapies [21]. The analysis included data from 1123 patients with HER2-negative MBC, and CTC screening was performed on average 56 months after primary diagnosis. In blood samples from 711 (63.3%) of 1123 patients, at least one CTC was detected (median 7 CTCs). To assess the HER2 status of CTCs, cells were labeled with HER2 antibodies, and only CTCs with a strong IHC staining intensity for HER2 (IHC score 3+) were considered as HER2-positive. In 134 of the 711 CTC-positive patients, at least one HER2-positive CTC was detected, yielding a HER2 discordance between primary tumor and CTCs in 18.8% of patients. A multivariable logistic regression with presence of at least one HER2-positive CTC (yes vs. no) as binary response variable revealed that lobular tumor histology, positive hormone receptor status, younger age, and the presence of 5 or more CTCs significantly predicted discordance in HER2 phenotype between primary tumor and CTCs. The authors concluded that in view of the well-known tumor heterogeneity in MBC, CTC-based liquid biopsy might better reflect total tumor burden and heterogeneity of tumor biology than biopsy of a single metastatic site.

Interestingly, the rates of patients with HER2-positive CTCs seem to be lower in the French CirCe T-DM1 trial (9.0% of screened patients) as compared to the DETECT screening (18.8% of screened patients). Both trials used the CELLSEARCH® System for the detection of CTCs, and HER2-positivity as a criterion for patient inclusion was defined as the presence of at least one HER2-positive CTC both for the CirCe T-DM1 and the DETECT III trial. However, HER2-positivity was assessed based on single-cell FISH-analysis in CirCe T-DM1 and by an immunohistochemical antibody staining procedure in DETECT. Another difference between CirCe T-DM1 and DETECT is the fact that MBC patients could be included in the DETECT III trial regardless of treatment line, while for the CirCe T-DM1 trial only patients starting a third- or fourth-line systemic therapy were eligible. In CirCe T-DM1, the prevalence rate of HER2-positive CTCs out of all detected CTCs in included patients was very low with a median of 1.6%, and the French group concluded that the low prevalence rate of HER2-positive CTCs was an important reason for the failure of the study [18]. Results of the DETECT III trial are still pending, and it will be interesting to see whether the rate of HER2-positive CTCs out of all detected CTCs will be as low as the rate observed in the CirCe T-DM1 study.

5. Conclusions

A general finding of all trials evaluating the clinical utility for CTCs described here is that CTCs are rare cells which are not present in every patient, particularly so in patients with EBC. Nevertheless, there is ample evidence that CTCs—once detected—are a strong prognostic factor for reduced survival. However, at present, it is not clear how this knowledge can be transferred to a prediction of therapy response and an improved clinical outcome. The STIC CTC trial showed the first data for a positive effect of a CTC-based decision in a subgroup with a high count of CTCs before starting treatment. Thus, CTCs might represent a helpful early treatment monitoring tool and can be used in situations with uncertain therapy response. The awaited results of the ongoing trials described here will hopefully provide much needed data that help to answer the question regarding the clinical utility of CTCs. Furthermore, the identification of potential targets for more individualized treatment options might improve the use of CTCs in clinical practice.

References

1. Joosse, S.A.; Gorges, T.M.; Pantel, K. Biology, detection, and clinical implications of circulating tumor cells. *EMBO Mol. Med.* **2015**, *7*, 1–11. [CrossRef] [PubMed]
2. Pantel, K.; Alix-Panabieres, C. Liquid biopsy and minimal residual disease—Latest advances and implications for cure. *Nat. Rev. Clin. Oncol.* **2019**, *16*, 409–424. [CrossRef] [PubMed]
3. Bidard, F.-C.; Peeters, D.J.; Fehm, T.; Nolè, F.; Gisbert-Criado, R.; Mavroudis, D.; Grisanti, S.; Generali, D.; Garcia-Saenz, J.A.; Stebbing, J.; et al. Clinical validity of circulating tumour cells in patients with metastatic breast cancer: A pooled analysis of individual patient data. *Lancet Oncol.* **2014**, *15*, 406–414. [CrossRef]
4. Janni, W.J.; Rack, B.; Terstappen, L.W.M.M.; Pierga, J.-Y.; Taran, F.-A.; Fehm, T.; Hall, C.; De Groot, M.R.; Bidard, F.-C.; Friedl, T.W.; et al. Pooled Analysis of the Prognostic Relevance of Circulating Tumor Cells in Primary Breast Cancer. *Clin. Cancer Res.* **2016**, *22*, 2583–2593. [CrossRef] [PubMed]
5. Bidard, F.-C.; Michiels, S.; Riethdorf, S.; Mueller, V.; Esserman, L.J.; Lucci, A.; Naume, B.; Horiguchi, J.; Gisbert-Criado, R.; Sleijfer, S.; et al. Circulating Tumor Cells in Breast Cancer Patients Treated by Neoadjuvant Chemotherapy: A Meta-analysis. *J. Natl. Cancer Inst.* **2018**, *110*, 560–567. [CrossRef] [PubMed]
6. Janni, W.; Rack, B.K.; Fasching, P.A.; Haeberle, L.; Tesch, H.; Lorenz, R.; Schochter, F.; Tzschaschel, M.; De Gregorio, A.; Fehm, T.N.; et al. Persistence of circulating tumor cells in high risk early breast cancer patients five years after adjuvant chemotherapy and late recurrence: Results from the adjuvant SUCCESS A trial. *J. Clin. Oncol.* **2018**, *36*, 515. [CrossRef]
7. Trapp, E.; Janni, W.; Schindlbeck, C.; Jückstock, J.; Andergassen, U.; de Gregorio, A.; Alunni-Fabbroni, M.;

Tzschaschel, M.; Polasik, A.; Koch, J.G.; et al. Presence of Circulating Tumor Cells in High-Risk Early Breast Cancer During Follow-Up and Prognosis. *J. Natl. Cancer Inst.* **2019**, *111*, 380–387. [CrossRef] [PubMed]

8. Sparano, J.; O'Neill, A.; Alpaugh, K.; Wolff, A.C.; Northfelt, D.W.; Dang, C.T.; Sledge, G.W.; Miller, K.D. Association of Circulating Tumor Cells with Late Recurrence of Estrogen Receptor-Positive Breast Cancer: A Secondary Analysis of a Randomized Clinical Trial. *JAMA Oncol.* **2018**, *4*, 1700–1706. [CrossRef] [PubMed]
9. Rack, B.; Schindlbeck, C.; Jückstock, J.; Andergassen, U.; Hepp, P.; Zwingers, T.; Friedl, T.W.P.; Lorenz, R.; Tesch, H.; Fasching, P.A.; et al. Circulating tumor cells predict survival in early average-to-high risk breast cancer patients. *J. Natl. Cancer Inst.* **2014**, *106*. [CrossRef] [PubMed]
10. Pierga, J.Y.; Hajage, D.; Bachelot, T.; Delaloge, S.; Brain, E.; Campone, M.; Dieras, V.; Rolland, E.; Mignot, L.; Mathiot, C.; et al. High independent prognostic and predictive value of circulating tumor cells compared with serum tumor markers in a large prospective trial in first-line chemotherapy for metastatic breast cancer patients. *Ann. Oncol.* **2012**, *23*, 618–624. [CrossRef] [PubMed]
11. Park, Y.; Kitahara, T.; Urita, T.; Yoshida, Y.; Kato, R. Expected clinical applications of circulating tumor cells in breast cancer. *World J. Clin. Oncol.* **2011**, *2*, 303–310. [CrossRef] [PubMed]
12. Ignatiadis, M.; Litière, S.; Rothé, F.; Riethdorf, S.; Proudhon, C.; Fehm, T.; Aalders, K.; Forstbauer, H.; Fasching, P.A.; Brain, E.; et al. Trastuzumab versus observation for HER2 nonamplified early breast cancer with circulating tumor cells (EORTC 90091-10093, BIG 1-12, Treat CTC): A randomized phase II trial. *Ann. Oncol.* **2018**, *29*, 1777–1783. [CrossRef] [PubMed]
13. Fehrenbacher, L.; Cecchini, R.S.; Geyer, C.E.; Rastogi, P.; Costantino, J.P.; Atkins, J.N.; Polikoff, J.; Boileau, J.F.; Provencher, L.; Stokoe, C.; et al. Abstract GS1-02: NSABP B-47 (NRG oncology): Phase III randomized trial comparing adjuvant chemotherapy with adriamycin (A) and cyclophosphamide (C) â†' weekly paclitaxel (WP), or docetaxel (T) and C with or without a year of trastuzumab (H) in women with node-positive or high-risk node-negative invasive breast cancer (IBC) expressing HER2 staining intensity of IHC 1+ or 2+ with negative FISH (HER2-Low IBC). *Cancer Res.* **2018**, *78*, GS1–GS02.
14. Cristofanilli, M.; Pierga, J.-Y.; Reuben, J.; Rademaker, A.; Davis, A.A.; Peeters, D.J.; Fehm, T.; Nolé, F.; Gisbert-Criado, R.; Mavroudis, D.; et al. The clinical use of circulating tumor cells (CTCs) enumeration for staging of metastatic breast cancer (MBC): International expert consensus paper. *Crit. Rev. Oncol.* **2019**, *134*, 39–45. [CrossRef] [PubMed]
15. Smerage, J.B.; Barlow, W.E.; Hortobagyi, G.N.; Winer, E.P.; Leyland-Jones, B.; Srkalovic, G.; Tejwani, S.; Schott, A.F.; O'Rourke, M.A.; Lew, D.L.; et al. Circulating tumor cells and response to chemotherapy in metastatic breast cancer: SWOG S*J. Clin. Oncol.* **2014**, *32*, 3483–3489. [CrossRef] [PubMed]
16. Bidard, F.-C.; Fehm, T.; Ignatiadis, M.; Smerage, J.B.; Alix-Panabières, C.; Janni, W.; Messina, C.; Paoletti, C.; Müller, V.; Hayes, D.F.; et al. Clinical application of circulating tumor cells in breast cancer: Overview of the current interventional trials. *Cancer Metastasis Rev.* **2013**, *32*, 179–188. [CrossRef] [PubMed]
17. Bidard, F.; Jacot, W.; Dureau, S.; Brain, E.; Bachelot, T.; Bourgeois, H.; Goncalves, A.; Ladoire, S.; Naman, H.; Dalenc, F.; et al. Abstract GS3-07: Clinical utility of circulating tumor cell count as a tool to chose between first line hormone therapy and chemotherapy for ER+ HER2- metastatic breast cancer: Results of the phase III STIC CTC trial. *Cancer Res.* **2019**, *79*, GS3–GS07.
18. Bidard, F.-C.; Cottu, P.; Dubot, C.; Venat-Bouvet, L.; Lortholary, A.; Bourgeois, H.; Bollet, M.; Servent Hanon, V.; Luporsi-Gely, E.; Espie, M.; et al. Anti-HER2 therapy efficacy in HER2-negative metastatic breast cancer with HER2-amplified circulating tumor cells: Results of the CirCe T-DM1 trial. *Ann. Oncol.* **2017**, *28*. [CrossRef]
19. Paoletti, C.; Regan, M.M.; Liu, M.C.; Marcom, P.K.; Hart, L.L.; Smith, J.W.; Tedesco, K.L.; Amir, E.; Krop, I.E.; DeMichele, A.M.; et al. Abstract P1-01-01: Circulating tumor cell number and CTC-endocrine therapy index predict clinical outcomes in ER positive metastatic breast cancer patients: Results of the COMETI Phase 2 trial. *Cancer Res.* **2017**, *77*. [CrossRef]
20. Polasik, A.; Tzschaschel, M.; Schochter, F.; de Gregorio, A.; Friedl, T.W.; Rack, B.; Hartkopf, A.; Fasching, P.A.; Schneeweiss, A.; Mueller, V.; et al. Circulating Tumour Cells, Circulating Tumour DNA and Circulating MicroRNA in Metastatic Breast Carcinoma—What is the Role of Liquid Biopsy in Breast Cancer? *Geburtshilfe Frauenheilkd.* **2017**, *77*, 1291–1298. [CrossRef] [PubMed]
21. De Gregorio, A.; Friedl, T.W.; Huober, J.; Scholz, C.; De Gregorio, N.; Rack, B.; Trapp, E.; Alunni-Fabbroni, M.; Riethdorf, S.; Mueller, V.; et al. Discordance in Human Epidermal Growth Factor Receptor 2 (HER2) Phenotype Between Primary Tumor and Circulating Tumor Cells in Women With HER2-Negative Metastatic Breast Cancer. *JCO Precis. Oncol.* **2017**, 1–12. [CrossRef]

13

Detection of Androgen Receptor Variant 7 (*ARV7*) mRNA Levels in EpCAM-Enriched CTC Fractions for Monitoring Response to Androgen Targeting Therapies in Prostate Cancer

Claudia Hille [1], Tobias M. Gorges [1], Sabine Riethdorf [1], Martine Mazel [2], Thomas Steuber [3], Gunhild Von Amsberg [4], Frank König [5], Sven Peine [6], Catherine Alix-Panabières [2,†] and Klaus Pantel [1,*,†]

1 Department of Tumor Biology, University Medical Center Hamburg-Eppendorf, 20246 Hamburg, Germany; c.hille@uke.de (C.H.); t.gorges@uke.de (T.M.G.); s.riethdorf@uke.de (S.R.)
2 Laboratory of Rare Human Circulating Cells (LCCRH), University Medical Centre of Montpellier–UM EA2415, 34295 Montpellier, France; martine-mazel@chu-montpellier.fr (M.M.); c-panabieres@chu-montpellier.fr (C.A.-P.)
3 Martini Clinic, University Medical Center Hamburg-Eppendorf, 20246 Hamburg, Germany; steuber@uke.de
4 Department of Hematology and Oncology, University Medical Center Hamburg-Eppendorf, 20246 Hamburg, Germany; g.von-amsberg@uke.de
5 ATURO, Urology Practice, 14197 Berlin, Germany; frank.koenig@aturo.berlin
6 Department of Transfusion Medicine, University Medical Center Hamburg-Eppendorf, 20246 Hamburg, Germany; s.peine@uke.de
* Correspondence: pantel@uke.de.
† There authors contributed equally to this work.

Abstract: Expression of the androgen receptor splice variant 7 (*ARV7*) in circulating tumor cells (CTCs) has been associated with resistance towards novel androgen receptor (AR)-targeting therapies. While a multitude of ARV7 detection approaches have been developed, the simultaneous enumeration of CTCs and assessment of *ARV7* status and the integration of validated technologies for CTC enrichment/detection into their workflow render interpretation of the results more difficult and/or require shipment to centralized labs. Here, we describe the establishment and technical validation of a novel *ARV7* detection method integrating the CellSearch® technology, the only FDA-cleared CTC-enrichment method for metastatic prostate cancer available so far. A highly sensitive and specific qPCR-based assay was developed, allowing detection of *ARV7* and *keratin 19* transcripts from as low as a single *ARV7*$^+$/*K19*$^+$ cell, even after 24 h of sample storage. Clinical feasibility was demonstrated on blood samples from 26 prostate cancer patients and assay sensitivity and specificity was corroborated. Our novel approach can now be included into prospective clinical trials aimed to assess the predictive values of CTC/ARV7 measurements in prostate cancer.

Keywords: prostate cancer; biomarkers; circulating tumor cells; androgen receptor; ARV7; abiraterone; enzalutamide

---

1. Introduction

Prostate cancer (PCa) remains the second most commonly diagnosed cancer among men worldwide with an estimated 1.3 million new cases each year [1]. In contrast to other cancer types such as pancreatic cancer, routine preventive medical screens for PCa are accessible to a broad spectrum of the public and have been widely accepted, leading to a drastic increase of newly diagnosed PCa cases. Tissue biopsies are invasive and can be associated with adverse effects for the patient [2]. Furthermore, routine tissue biopsy is challenging in metastatic PCa (mPCa). In recent years, minimally invasive liquid biopsies, focusing on the identification of circulating tumor cells (CTCs) and circulating nucleic acids (ctDNA, miRNA) from whole blood samples, have gained tremendous attention [3–7]. While the prognostic relevance of CTCs in PCa, especially in the metastatic setting, has been thoroughly shown in large clinical trials [8–11], predictive value of CTC analysis and their clinical utility are still being debated [12–17]. While a multitude of therapeutic approaches exist, aimed at treating PCa in various disease stages, a subset of patients develop aggressive PCa subtypes that defy current therapeutic options. Therefore, simple detection of PCa is not sufficient and robust biomarkers are urgently needed to discern aggressive subtypes from clinically well treatable cancers, preferably without exposing patients to unnecessary tissue biopsies.

With the advent of novel hormone therapies such as enzalutamide and abiraterone and the emergence of innate and acquired resistance towards these therapies, the androgen receptor splice variant 7 (ARV7) has become a leading target of CTC research in PCa [17–19]. Multiple studies indicate that *ARV7* mRNA and ARV7 protein expression in CTCs is associated with resistance towards novel hormone therapies [20–25] and that *ARV7* expressing patients benefit more from taxane-based therapy [25–27]. This implicates ARV7 as a possible treatment selection biomarker for PCa patients prior to receiving novel hormone therapy (e.g., enzalutamide, abiraterone). Additionally, the ARV7 status is subject to change during therapy regimens [25,28,29], underlining the benefit of sequential sampling which becomes possible through liquid biopsy. ARV7 could therefore also represent a biomarker to monitor treatment response and predict upcoming therapy resistance.

While many approaches have been developed to assess ARV7 either on protein or mRNA level [20,24,30], only very few of these approaches allow for parallel CTC enumeration and morphological characterization while giving information on ARV7 status for individual CTCs [24,31], a limitation recently highlighted [32]. Additionally, even fewer were designed to use the only FDA-cleared CTC enrichment and detection technology shown to have clinical prognostic relevance in prostate cancer, the CellSearch® system [33]. Here, we aimed to develop a protocol for *ARV7* detection using the CellSearch® technology. With our novel workflow we were able to detect *ARV7* mRNA in as low as one CTC in 7.5 mL of whole blood.

2. Materials and Methods

2.1. Cancer Cell Lines

The human prostate cancer cell lines 22Rv1 (ATCC® CRL-2505), VCaP (ATCC® CRL-2876), LNCaP (ATCC® CRL-1740) and PC3 (ATCC® CRL-1345) were obtained from the American Type Culture Collection (ATCC, Manassas, VA, USA) and cultured according to ATCC recommendations.

LNCaP and 22Rv1 cells were cultured in RPMI 1640 medium, while the VCaP and PC3 cells were maintained in Dulbecco's Modified Eagle Medium (DMEM). Media were additionally fortified with 10% fetal calf serum (FCS) (Gibco—Life Technologies, Darmstadt, Germany), 1% L-glutamine (Gibco—Life Technologies, Darmstadt, Germany) and 1% penicillin/streptomycin (Gibco—Life Technologies, Darmstadt, Germany), as recommended by ATCC. Cells were cultured in 25 cm^2 flasks at 37 °C in a humidified atmosphere containing 5% CO_2.

2.2. Blood Collection and Processing

Male healthy donor (HD) and patient blood samples were acquired in accordance to the World Medical Association Declaration of Helsinki and the guidelines for experimentation with humans by the Chambers of Physicians of the State of Hamburg ("Hamburger Ärztekammer"). All patients gave informed, written consent prior to blood collection (Ethics Approval: PV3779). Samples were drawn from 26 metastatic prostate cancer (mPCa) patients into standard 7.5 mL ethylenediaminetetraacetic acid (EDTA) vacutainers or CellSave® (Menarini-Silicon Biosystems, Florence, Italy) preservation tubes respectively. Each patient therefore provided a matched sample of EDTA-KE (Sarstedt, Rheinbach, Germany) and CellSave® blood for further analysis. CTCs from EDTA blood samples were enriched via the CellSearch® Profile Kit (Menarini-Silicon Biosystems, Florence, Italy) and further analyzed for *ARV7* expression as described below. Samples collected into CellSave® blood preservation tubes were processed via the CellSearch® CXC-Kit (FITC labelled pan-keratin) [34]. Phycoerythrin labelled androgen receptor CellTracks Anti-Androgen Receptor (Janssen Diagnostics) antibody (10 μg/mL) was used for full-length AR (AR-FL) detection in the fourth channel of the CellSearch® for 12/26 mPCa patients. All analyses were performed by trained CellSearch® analysist. CTCs were defined as keratin positive and CD45 negative cells with a nuclear DAPI staining.

2.3. Spiking of Healthy Donor Blood

For spiking experiments, cell line cells were washed once with 1 x PBS (Gibco-Life Technologies, Darmstadt, Germany) and treated with 0.25% trypsin-EDTA (Gibco-Life Technologies, Darmstadt, Germany) for 5 min at 37 °C prior to being resuspended in culture medium. The cell suspension was centrifuged at 190× *g* for 5 min after which the supernatant was discarded and the cells were again resuspended in fresh culture medium. The cells were spread to a petri dish filled with corresponding medium, manually counted and picked under a light microscope. Defined cell counts were added directly to healthy donor blood samples.

2.4. Immunocytochemical Stainings on Cell Culture Plates

Cells were seeded into 24-well plates at the rate of 50,000 cells/well, and maintained at 37 °C in a humidified atmosphere containing 5% CO_2 until reaching 80% confluence. Cells were then fixed and permeabilized using IntraPrep Permeabilization Reagent (A07803, Beckman Coulter, Brea, CA, USA), and blocked with 10% Goat serum for 1 h at room temperature. Cells were subsequently incubated with (i) primary antibodies Anti-AR (AR-V7 specific) antibody [EPR15656] (Abcam, Cambridge, United Kingdom) at a final concentration of 10 μg/mL, or (ii) Rabbit IgG, monoclonal [EPR25A]-Isotype Control (Abcam, Cambridge, United Kingdom) (our negative control) at a final concentration of 10 μg/mL. All wells were also incubated with the anti-PanCK^{PE} (Menarini-Silicon Biosystems, Florence, Italy). Following this first incubation, cells were washed with 1% goat serum in PBS, incubated with the FITC-conjugated secondary antibody (1:20 in PBS containing 10% Goat serum), and washed twice with 1% goat serum in PBS.

In parallel, the presence of the androgen receptor (AR) was tested using the anti-AR^{AF488} [D6F11] XP Rabbit antibody (0.5 μg/mL, Ozyme, Saint Cyr L'Ecole, France); in the control wells, the Rabbit [DA1E] IgG^{AF488} XP isotype (0.5 μg/mL, Ozyme, Saint Cyr L'Ecole, France) was used. Cell imaging was obtained under 20x magnification using a Fluorescent Axio Observer microscope (Carl Zeiss, Oberkochen, Germany).

2.5. *Immunocytochemical Stainings on Cytospins*

Cell suspensions of selected prostate cancer cell lines (22Rv1, LNCap, PC3) were spun down on glass slides (190× *g*, 5 min) and dried at room temperature (RT) over night. Cells were subsequently fixed and permeabilized using the respective CellSearch CXC Kit reagents (Menarini-Silicon Biosystems, Florence, Italy) and blocked with 10% AB-Serum (BioRad, Rüdigheim, Germany). Primary antibodies targeting ARV7, 4 μg/mL of clone AG10008 (unlabeled, Precision, Columbia, Maryland, USA) and 6 μg/mL EPR15656 (unlabeled, Abcam, Cambridge, United Kingdom) were tested. Secondary antibodies were applied and contained a DAPI nuclear counterstain. Secondary rabbit-anti mouse (Alexa 546, polyclonal, Thermo Fisher Scientific, Dreieich, Germany) and mouse-anti-rabbit (Alexa 546, polyclonal, Thermo Fisher Scientific, Dreieich, Germany) antibodies were used. Cytospins were covered in Prolong Gold Antifade Reagent (Thermo Fisher Scientific, Dreieich, Germany) and cover slipped for analysis. Slides were manually assessed using a fluorescence microscope (Axioplan 2, Carl Zeiss, Oberkochen, Germany).

2.6. *Western Blots*

Cell lines (22Rv1, VCaP, LNCaP, and PC3) were cultured to 70% confluency, harvested in urea lysis buffer (9.8 M Urea, 15 mM EDTA, 30 mM Tris) and homogenized by ultrasonic treatment. Protein concentration was measured with the Pierce BCA Protein Assay Kit (Pierce, Rockford, Illinois, USA). 40 μg of total protein was applied for Western Blot analysis for each respective cell line alongside pre-stained peqGold protein marker-V (VWR, Erlangen, Germany). Proteins were separated according to size using a Laemmli buffer system and 8% polyacrylamide separation gel. Two ARV7 antibodies, mouse-AG10008 (Precision, Columbia, MD, USA; 2 μg/mL) and rabbit-EPR15656 (Abcam, Cambridge, United Kingdom; 1.5 μg/mL) were applied in dilutions according to the supplier's instruction manual in 5% milk powder. Alpha-tubulin was used as a loading control (Cell Signaling Technology, Danvers, MA, USA). Species specific secondary antibodies (horseradish peroxidase conjugated, DAKO, Glostrup, Germany) were applied at 1:10.000 dilution in 5% milk powder. Protein bands were visualized using SignalFire™Plus ECL reagent (Cell Signaling Technology, Danvers, MA, USA) and X-ray films (CEA, Hamburg, Germany) according to the instruction manual.

2.7. *RNA Extraction and cDNA Synthesis*

For cell line characterization and PCR establishment RNA was isolated from prostate cancer cell lines grown in a T25 culture flask at 70% confluency using the NucleoSpin® RNA isolation kit (Macherey-Nagel, Düren, Germany) according to manufacturer's instructions. RNA concentration and purity were controlled via NanoDrop 1000 spectrophotometer (Thermo Fisher Scientific, Dreieich, Germany) following isolation. 0.5 μg of RNA per cell line were used for DNA synthesis with the First Strand cDNA Synthesis Kit (Thermo Fisher Scientific, Dreieich, Germany) according to manufacturer's instructions. cDNA Synthesis was carried out in a PeqSTAR 96 Universal Gradient thermocycler (VWR International, Darmstadt, Germany).

Following CTC enrichment via the CellSearch® Profile Kit (Menarini-Silicon Biosystems, Florence, Italy) samples were transferred to a fresh 1.5 mL tube (Sarstedt, Rheinbach, Germany). To do so, a 1000 μL pipette tip was first coated with a solution of 0.1 mg/mL of BSA/PBS to circumvent binding and sticking of CTCs to the pipette surface. All RNA work was performed using sterile, DNA/RNA-free, filtered Biosphere® plus pipette tipps (Sarstedt, Rheinbach, Germany). The Profile® sample tube was washed with 500 μL of 1x DPBS (cell culture use) (Thermo Fisher Scientific, Dreieich, Germany), which was also added to the sample. Subsequently the sample was placed in a magnetic rack (Magnetcellect; R&D systems, Minneapolis, MN, USA) for 10 min. The supernatant was discarded, and the sample was washed with 1000 μL of 1x DPBS, followed by another 10 min attached to the magnetic rack. This step was repeated with 500 μL of 1x DPBS prior to resuspension of the Profile® beads in 150μL of lysis buffer (Dynabeads mRNA DIRECT Kit; Thermo Fisher Scientific, Dreieich, Germany). Samples were immediately frozen at −80 °C. Sample lysates were stored for a maximum of 14 days prior to RNA isolation and cDNA synthesis.

For RNA extraction, the Dynabeads mRNA DIRECT Kit (Thermo Fisher Scientific, Dreieich, Germany) was applied according to manufacturer's instructions. Following the last wash step with Wash buffer B, supernatant was removed, and beads were resuspended in 14.75 µL of Nuclease-free H_2O (Qiagen, Hilden, Germany) and placed in a PCR cycler at 75 °C for 5 min to ensure elution of mRNA from the beads. Subsequently cDNA was synthesized using the Sensiscript Reverse Transcription Kit (Qiagen, Hilden, Germany) with Recombinant Rnasin® (Promega, Mannheim, Germany) as an added RNase inhibitor. Primer addition was not necessary as the contained dynabeads function as oligo-dT primers. RNase inhibitor was limited to 0.25 µL, leading to a total mastermix of 5.25 µL added to each RNA sample (total reaction volume of 20 µL). Following cDNA synthesis, beads were removed via magnet and supernatant was transferred to a fresh PCR tube for subsequent qPCR analysis.

2.8. *Polymerase-Chain Reaction (PCR) Analysis*

For *AR-FL* and *ARV7* primer evaluation, 10 ng of cDNA of each prostate cancer cell line was applied per PCR. The PCR reaction conditions for initial primer testing were adapted from the original Antonarakis et al. publication by the Johns Hopkins Group [20]. Reactions were run in a PeqSTAR 96 Universal Gradient thermocycler (VWR International, Darmstadt, Germany).

PCR primer pairs (Sigma Aldrich, Steinheim, Germany) chosen for PCR targeted *AR-FL* (fw-CAGCCTATTGCGAGAGAGCTG, rev-GAAAGGATCTTGGGCACTTGC, fragment size of 125 bp) [20] and *ARV7* (Antonarakis et al. [20]: fw-CCATCTTGTCGTCTTCGGAAATGTTA, rev-TTTGAATGAGGCAAGTCAGCCTTTCT, fragment size of 125 bp; Guo et al. [35]: fw-CTACTCCGGACCTTACGGGGACATGCG, rev-TGCCAACCCGGAATTTTTCTCCC, fragment size of 314 bp; Liu et al. [36]: fw- CAGGGATGACTCTGGGAGAA, rev- GCCCTCTAGAGCCCTCATTT, fragment size of 112 bp; UKE: fw-AGAAAGGCTGACTTGCCTCA, rev- CGCCAGGTTTCTCCAGACTA, fragment size of 73 bp) gene sequences. Novel UKE primers were designed using the Primer 3 software [37]. Primers were aliquoted at stock concentrations of 100 µM with Nuclease-free H_2O (Qiagen, Hilden, Germany) and frozen at −20 °C. Final concentrations of 10 µM were applied to PCRs.

To visualize PCR products, they were mixed with DNA Gel loading dye (6x) (Thermo Fisher Scientific, Dreieich, Germany) and applied to 2% agarose gels containing GelRed® Nulceic Acid Gel Stain (Biotum, Fremont, CA, USA) at 1/µL per ml of agarose gel. The Quick-Load® 100 bp DNA Ladder (New England Biolabs, Frankfurt am Main, Germany) was used as a size standard. PCR fragments were visualized using the Gene Genius bioimaging system (Syngene, Bangalore, India).

2.9. *Quantitative Polymerase-Chain Reaction (qPCR) Analysis*

qPCRs were pipetted under a separate flow hood with sterile, DNA/RNA-free, filtered Biosphere® plus pipette tipps (Sarstedt, Rheinbach, Germany) and performed in a CFX96 Touch™ Real Time PCR Detection System (BioRad, Rüdigheim, Germany). Maxima SYBR-Green fluorescent dye (Thermo Fisher Scientific, Dreieich, Germany) was used for product detection. Amplification was performed under the following conditions: after an initial denaturation step (10 min at 95 °C), 40 amplification cycles were carried out, consisting of denaturation at 95 °C for 30 seconds, annealing at 60 °C for 30 s, and elongation for 30 s at 72 °C. A final elongation step at 72 °C (10 min) was followed by a melting curve analysis and storage of the samples at 4 °C. Data was summarized and converted into Excel files using the CFX Manager Software (BioRad, Rüdigheim, Germany). For qPCR analysis, two additional primer sets targeting *K19* (fw-CGAACCAAGTTTGAGACGGA; rev-GATCTGCATCTCCAGGTCGG; fragment size of 117 bp) and *Actin* (x) gene sequences were applied. Samples were applied in triplicates and average Cq values as well as standard deviations were calculated. Primers were aliquoted at stock concentrations of 100 µM with Nuclease-free H_2O (Qiagen, Hilden, Germany) and frozen at −20 °C. Final concentrations of 10 µM were applied to qPCRs.

Relative gene expression of *AR-FL* and *ARV7* in initial primer testing and cell line characterization was normalized from data sets by the comparative Cq method [38]. Briefly, the first amplification cycle showing significant increase of fluorescence signal over background level was defined as the cycle of quantification (Cq). Cq data of *AR-FL* and *ARV7* was normalized by subtracting the Cq value of *Actin* from the respective target gene for each cell line tested, generating a ΔCq value. Subsequently, the ΔΔCq values were calculated by subtracting the ΔCq of each specific gene calculated for the different cell lines (22Rv1, VCap, and LNCaP) from the ΔCq values calculated for gene expression in PC3 cells. Finally, ΔΔCq values were converted to log2 fold changes by applying $2^{-\Delta\Delta Cq}$. Ten nanograms of cDNA were applied per triplicate well.

Following RNA isolation and cDNA transcription from CellSearch® Profile Kit (Menarini-Silicon Biosystems, Florence, Italy) enriched samples, the 20 μL of cDNA mix was applied in triplicates for each gene (2 μL/well). No absolute quantification or normalization of genes was performed as levels of *Actin* gene expression is variable depending on background leucocyte cDNA co-amplified following CTC enrichment. Gene expression was confirmed when at least 2/3 triplicates showed detectable transcript levels under a Cq threshold of 35. Quality of the results was furthermore corroborated by melting curve analysis and subsequent visualization of amplified products on 2% agarose gels (see above).

3. Results

3.1. *Test of Commercially Available ARV7 Antibodies for Fourth CellSearch® Channel*

To allow assessment of ARV7 protein levels in parallel to CTC enumeration on a cell-specific level, we initially tested available ARV7 antibodies with the aim of adding them to the fourth channel of the CellSearch® system. Currently only few commercial antibodies are available, aimed at detecting ARV7 protein either via immunohistochemistry (IHC), immunocytochemistry (ICC) and/or western blot.

Three established prostate cancer cell lines were chosen for method establishment, each cell line representing a specific status of AR-full length (AR-FL) and ARV7 protein expression: 22Rv1 ($AR\text{-}FL^{+}/AR\text{-}V7^{+}$), LNCaP ($AR\text{-}FL^{+}/AR\text{-}V7^{+/-}$), and PC3 ($AR\text{-}FL^{-}/AR\text{-}V7^{-}$). First, the cell lines were characterized for AR-FL (Figure 1a), resulting in cell line specific nuclear ICC staining (22Rv1 and LNCaP) or absence of staining (PC3) for the full-length protein, seen in green. Next, we tested the anti-ARV7 antibody [EPR15656] described in literature to specifically stain nuclear ARV7 [25]. This antibody did not result in cell line specific staining results, as all three tested cell lines including the $ARV7^{-}$ PC3 cells showed green nuclear ARV7 staining (Figure 1b). Similar results were obtained using the antibody [EPR15656] as well as an additional commercially available ARV7 antibody on cell line cytospins (Figure S1). In western blot analysis the anti-ARV7 antibody [AG10008] by Precision showed cell line specific results, correctly detecting 22Rv1 and VCaP lysate as $ARV7^{+}$, LNCaP protein levels as below detection limit and identifying PC3 cells as $ARV7^{-}$ (Figure S1a). In contrast, the anti-ARV7 antibody [EPR15656], showed an unspecific western blot signal for PC3 cells (Figure S1a). In ICC both antibodies failed to correctly characterize the chosen prostate cancer cell lines, giving unspecific staining results (Figure S1b,c).

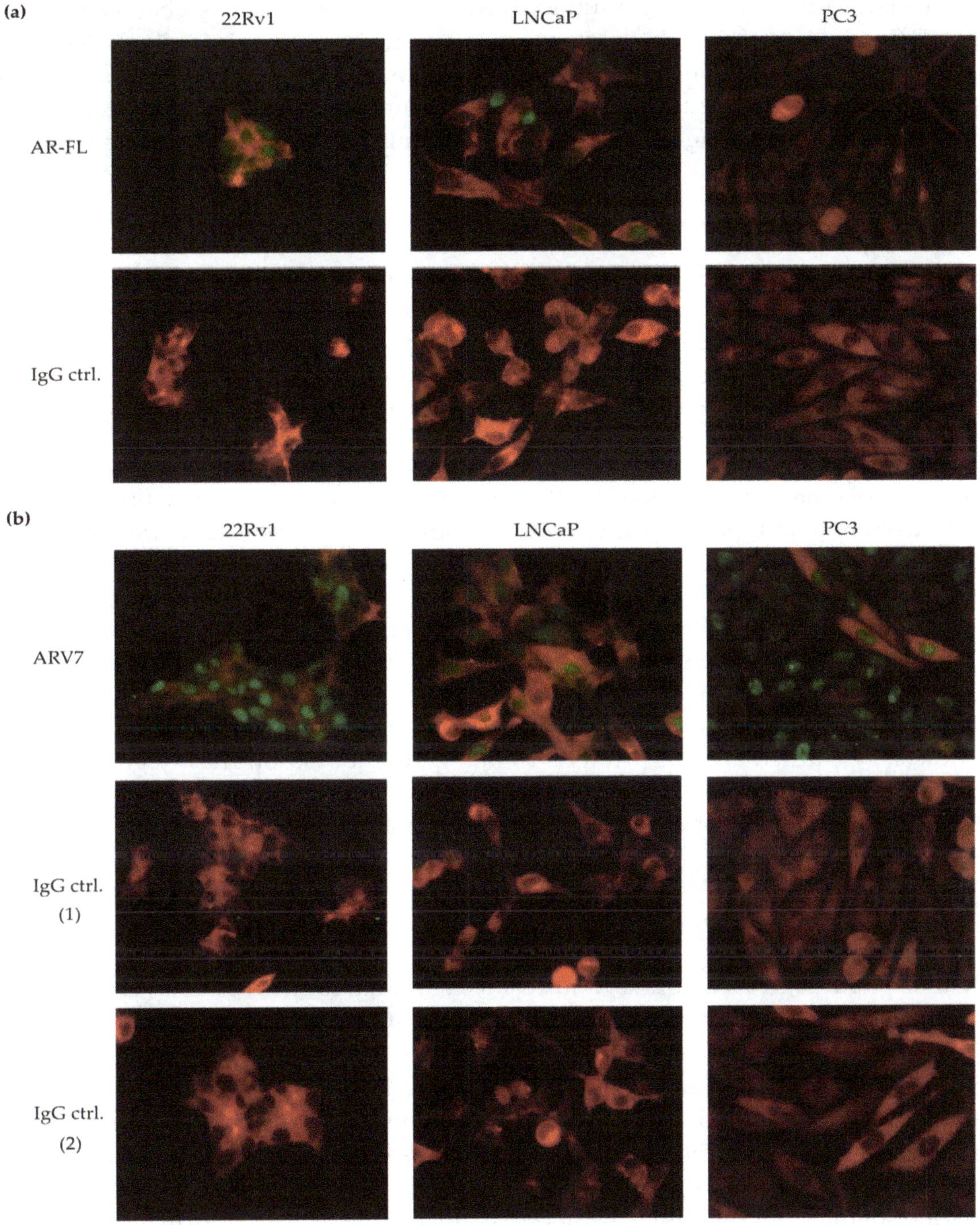

Figure 1. Immunocytochemical (ICC) staining of full-length androgen receptor (AR-FL) and ARV7 on three selected prostate cancer cell lines. Cells are stained for pan-keratin in red (anti-PanCK, CellSearch®, Menarini) in all images. (**a**) Upper panel: ICC staining performed using the anti-AR-FL antibody (7395S Ozyme) in green. AR-FL positive cells lines 22Rv1 and LNCaP show positive nuclear AR-FL staining, while PC3 cells remain unstained. Lower panel: ICC control staining using Rabbit [DA1E] IgG XP isotype (2975S Ozyme) in green showing the absence of unspecific staining on 22RV1, LNCaP and PC3 cells. (**b**) Upper panel: ICC staining performed using the anti-ARV7 antibody [EPR15656] (209491 Abcam) detected by a FITC-conjugated secondary antibody. A positive nuclear staining is observable on all three cell lines (in green), indicating unspecific signal of the antibody in PC3. Medium panel: ICC staining performed with the Rabbit IgG, monoclonal [EPR25A]-Isotype Control (172730 Abcam) detected with FITC-conjugated secondary antibody (97050, Abcam) showing negativity on 22RV1, LNCaP, and PC3 cells. Lower panel: ICC staining performed with FITC-conjugated secondary antibody (97050, Abcam) showing negativity on 22RV1, LNCAP, and PC3 cells.

In conclusion, none of the tested antibodies were deemed suitable for characterization of ARV7 protein on CTCs via the CellSearch® system. Additionally, the most intensively tested anti-ARV7 antibody [EPR15656] [25], described to give a specific nuclear and unspecific cytoplasmic staining did not show reliable results in our hands (Figure 1b), giving unspecific nuclear staining signals in ARV7$^-$ PC3 cells, even when neglecting the cytoplasmic staining and considering the described, relevant nuclear staining.

3.2. Development of a qPCR Based Assay to Detect ARV7 mRNA

As an alternative to protein-based detection we subsequently aimed at establishing a qPCR-based approach to detect *ARV7* on an mRNA level. We added an additional prostate cancer cell line to the analysis, to further confirm the robustness of our method. VCaP cells show similar *AR-FL* and *ARV7* expression profiles as 22RV1 cells (*AR-FL*$^+$/*AR-V7*$^+$) and were used as a second *ARV7*$^+$ cell line during method establishment. An overview of the *AR-FL* and *ARV7* status for all four cell lines is listed in Figure 2a.

Initially, we planned on using the *AR-FL* and *ARV7* primer sets already published [20] for our qPCR-based detection and then modifying the CTC pre-enrichment steps. However, when testing the primers using PCR according to the published protocol, it became clear that while the *AR-FL* primers showed specific bands at the correct expected size of 125 bp (Figure 2b), an additional, undescribed PCR fragment of around 250 bp was detected using the *ARV7* primers in 22Rv1 but not in LNCaP cells (Figure 2b). To ensure optimal primer quality for *ARV7* detection, additional *ARV7* primer sets described in literature [35,36] as well as an own design (UKE), were employed. To exclude that the unspecific PCR fragments detected were generated due to incorrect annealing temperature or incorrect cDNA synthesis, we tested all four primer sets in a gradient PCR on freshly generated 22Rv1 and LNCaP cDNA (Figure 2c). Again, an additional PCR product was detected for the Antonarakis [20] and Guo [35] primer sets across all annealing temperatures in 22Rv1 cells but not LNCaP cells (Figure 2c, lines 1,2).

This could represent an additional AR splice variant, similar to *ARV7* [30]. Using the original protocol of 40 amplification cycles [20,23] this additional transcript could come up in clinical samples, especially those with high CTC counts, and result in an unaccounted bias. In contrast, the Liu [36] and UKE primer sets, resulted in specific PCR fragments at 112 bp and 73 bp, respectively (Figure 2c, lines 3,4). The fragment signal intensity appeared slightly higher for the UKE primers (Figure 2c, line 4) in comparison to the Liu primers [36] (Figure 2c, line 3), which could indicate a higher amount of generated PCR product. However, this cannot be conclusively deduced from qualitative PCR. Decreasing the PCR cycles from 39 to 30 (Figure 2d), reduced the unspecific PCR signals down to hardly visible levels for the Antonarakis [20] and Guo [35] primer sets (Figure 2d). However, as quantitative PCR represents a much more sensitive method than qualitative PCR, both primer sets were discarded for further experiments. Both the Liu [36] and UKE primer sets displayed cell line specific PCR results and PCR fragments at correct sizes, resulting in further evaluation of these two primer sets via qPCR.

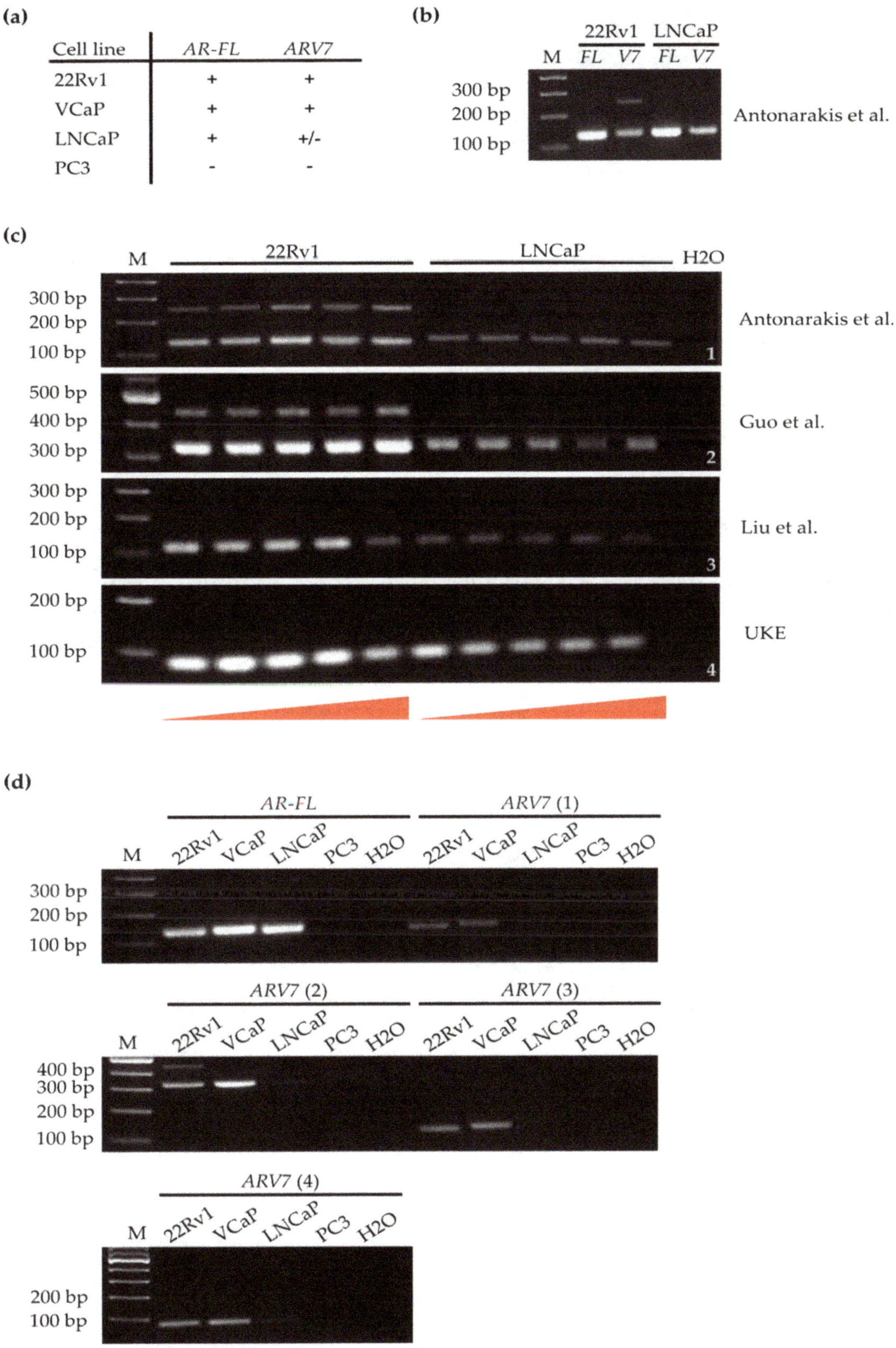

Figure 2. PCR-based detection of *AR-FL* and androgen receptor splice variant 7 (*ARV7*) in selected prostate cancer cell lines. Letter M indicating DNA ladder (marker) lanes. Ten nanograms of cDNA were analyzed for each PCR sample. (**a**) Schematic overview of *AR-FL* and *ARV7* positivity (+) and negativity (−) for established prostate cancer cell lines, as described in literature. (**b**) Agarose gels of a PCR detecting *AR-FL* and *ARV7* in cDNA isolated from 22RV1 and LNCaP cells. *ARV7* cDNA was detected using the primers described by Antonarakis et al. [20]. PCRs were performed for 39 cycles.

125 bp PCR products are expected for both *AR-FL* and *ARV7*. The *ARV7* PCR shows an additional, uncharacterized band at between 250–300 bp for 22RV1 cells, but not for LNCaP cells. (**c**) Agarose gel of a gradient PCR for *ARV7* on 22RV1 and LNCaP cDNA using different primer pairs. PCRs were performed for 39 cycles. Temperatures increasing from 58.5 °C to 65.5 °C, indicated by red triangles below gel images. Antonarakis (1) and Guo (2) primers both show secondary PCR bands on 22Rv1 cDNA (between 200–300bp and between 400–500 bp, respectively). Liu (3) and UKE (4) primers both give expected PCR bands for *ARV7* at 112 bp and 73 bp. Signal intensity appears higher, possibly indicating more generated PCR product, for UKE primers. (**d**) Agarose gels of PCRs detecting *AR-FL* and *ARV7* in cDNA of 22RV1, VCaP, LNCaP, and PC3 prostate cancer cell line cells. PCRs were performed for 30 cycles. AR-FL primer set, results in specific PCR signals in AR^+ and AR^- cell lines. *ARV7 (1)* corresponds to Antonarakis et al., *ARV7 (2)* corresponds to Guo et al., *ARV7 (3)* corresponds to Liu et al., and *ARV7 (4)* corresponds to our newly developed UKE primer sets.

Gene expression levels of *AR-FL* and *ARV7* (using the Liu and UKE primers) were assessed for 22Rv1, LNCaP, and VCaP cells in relation to their respective expression in PC3 cells (Figure 3a). As expected, both *AR-FL* and *ARV7* gene expression were dramatically increased in all three cell lines compared to PC3 cells. Additionally, the UKE primers showed most effective detection of *ARV7* (Figure 3a). All further experiments were therefore carried out using the newly designed UKE primers.

Apart from *AR-FL* and *ARV7*, *K19*, and *Actin* gene expression were also measured via qPCR. *Actin* functioning as a gene for normalization and a confirmation of successful cDNA synthesis, and *K19* as an established marker for CTC detection in blood [39,40] thus allowing confirmation of the presence of CTCs in future clinical samples. Figure 3b shows representative qPCR curves for all four cell lines (in different colors) for each gene. Due to the high sensitivity of qPCR analysis, *ARV7* expression can be detected at around 36 cycles for cDNA inputs generated from high PC3 cell counts (Figure 3b). This is an enormous difference to the approximately 22 cycles necessary for detection of $ARV7^+$ cell lines (Figure 3b). Despite the fact that such high CTC cell counts are extremely rarely to be expected in clinical samples, a cut-off of ≤35 cycles was established for gene expression to be counted as positive for the analyzed genes in all further analysis.

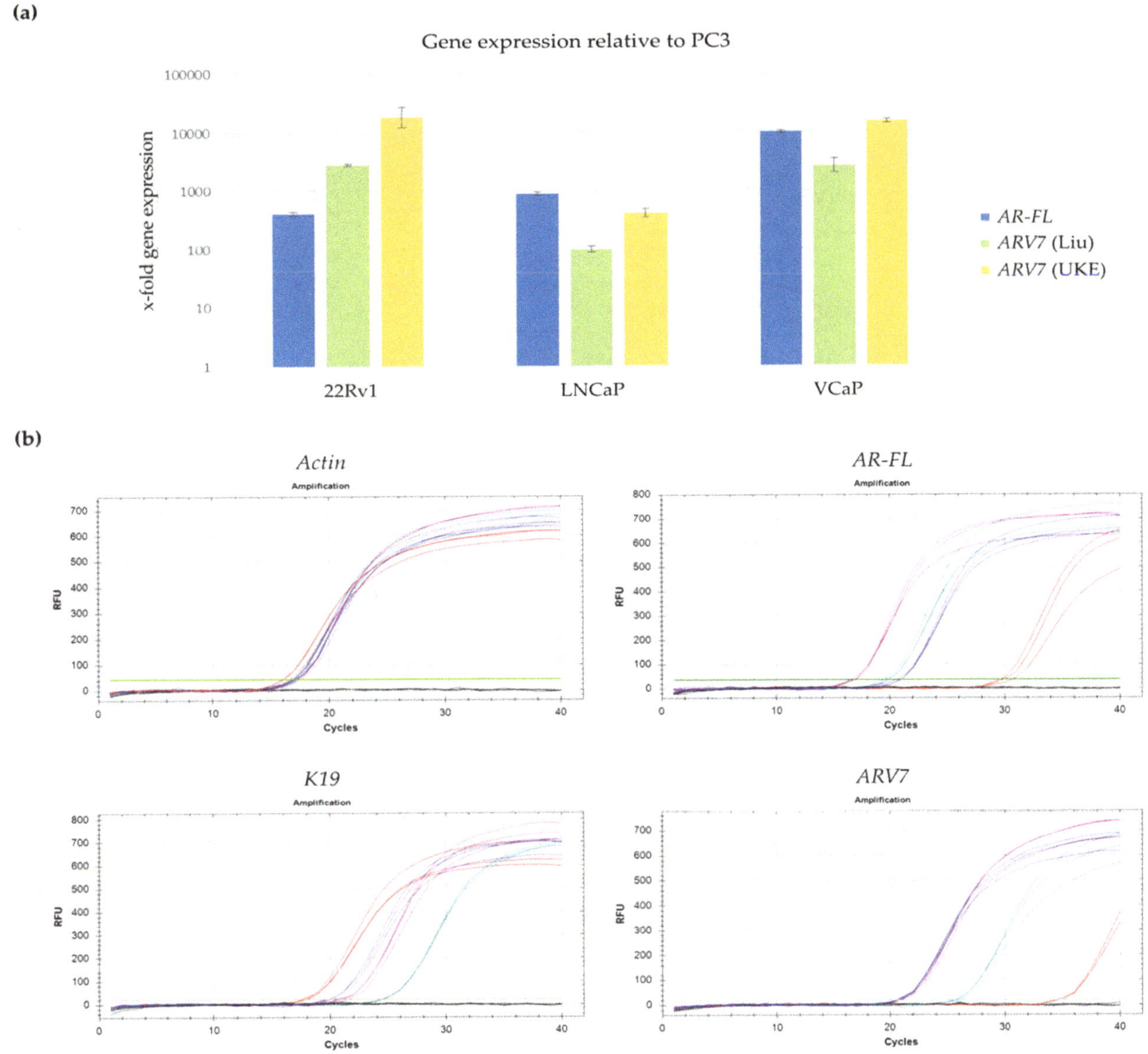

Figure 3. qPCR-based characterization of selected prostate cancer cell line cDNA. cDNA was generated from RNA isolated from 22RV1, VCaP, LNCaP, and PC3 cells and analyzed via qPCR. (**a**) Relative gene expression of *AR-FL* and *ARV7* using primers by Liu et al. and our newly developed primers (UKE). Gene expression was first normalized to actin and subsequently displayed relative to PC3 gene expression. Standard deviation is indicated as brackets. (**b**) Representative qPCR expression profiles for different target genes (*Actin, AR-FL, K19* and *ARV7*) across all four chosen cell lines: 22RV1 (blue), VCaP (purple), LNCaP (green), and PC3 (red). All samples were applied in triplicates.

3.3. Combining Profile-Kit-Based CTC Enrichment with ARV7 mRNA Detection

To allow for use of the CellSearch® system to isolate prostate cancer CTCs for *ARV7* detection on the one hand and enable parallel CTC quantification on the other, a two-armed approach was designed (Figure 4). 7.5 mL of whole blood was taken in parallel into standard EDTA tubes for RNA isolation and CellSave® blood preservation tubes for CTC enumeration, respectively. From EDTA blood, CTCs were enriched via the CellSearch® Profile Kit for subsequent RNA analysis. RNA was isolated and cDNA synthesized prior to analysis of *ARV7, K19,* and *Actin* via qPCR. In parallel CellSave® preserved blood was processed using the CellSearch® CXC Kit thus allowing for parallel AR-FL protein characterization in the fourth fluorescent channel of the device (Figure 4).

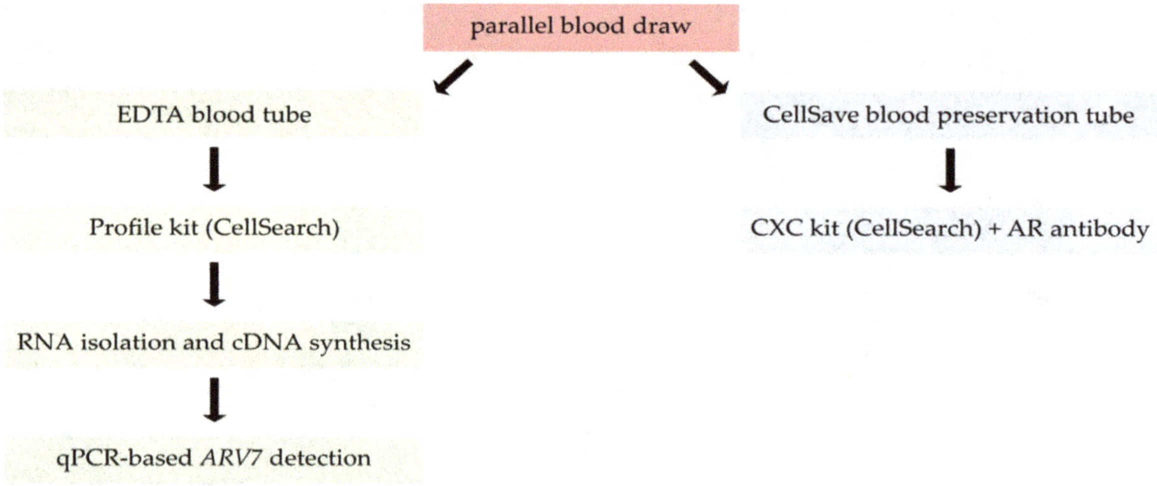

Figure 4. Schematic workflow of *ARV7* detection combined with parallel circulating tumor cell (CTC) enumeration.

"Mock" samples were generated to mimic clinical sample handling. Differing amounts of *ARV7*$^-$ and *ARV7*$^+$ cell line cells were manually spiked into healthy donor (HD) blood and directly processed by our workflow (Table 1). Following the qPCR run, generated products were applied to a gel electrophoresis allowing final confirmation of gene expression status (data not shown).

Table 1. Validation of protocol specificity and sensitivity. Titration experiments of spiked cell lines in blood from healthy donor (HD) samples. Indicated cell counts of *ARV7*$^+$ (22Rv1) and *ARV7*$^-$ (PC3) cells were manually spiked into HD blood and processed by our protocol. *ARV7* status is highlighted as "+" for positive and "−" for negative samples. Detection of gene expression was confirmed when at least 2/3 triplicates were positive in qPCR analysis. Unamplified qPCR samples are marked as N/D (not detected). HD samples were processed in parallel as a negative control for *ARV7* and *K19*. The bar (-) in the respective table column of detection indicates no further experiments were conducted.

Cell Line	*ARV7* Status	Cell Amount	Target	Detection	Detection (n = 2)	Detection (n = 3)
HD	−	0	*ARV7*	N/D	N/D	N/D
			K19	N/D	N/D	N/D
			Actin	yes	yes	yes
PC3	−	50	*ARV7*	N/D	-	-
			K19	yes	-	-
			Actin	yes	-	-
		100	*ARV7*	N/D	-	-
			K19	yes	-	-
			Actin	yes	-	-
22RV1	+	50	*ARV7*	yes	-	-
			K19	yes	-	-
			Actin	yes	-	-
		20	*ARV7*	yes	-	-
			K19	yes	-	-
			Actin	yes	-	-
		10	*ARV7*	yes	yes	yes
			K19	yes	yes	yes
			Actin	yes	yes	yes
		5	*ARV7*	yes	yes	yes
			K19	yes	yes	yes
			Actin	yes	yes	yes

All HD samples measured (n = 3) were *ARV7* and *K19* negative (Table 1). PC3 samples were negative for *ARV7* and positive for *K19*, confirming the specificity of the established assay. *ARV7* and *keratin 19* were still detectable down to 5 $ARV7^{+}$ 22RV1 cells using our protocol (n = 3), demonstrating high sensitivity (Table 1).

3.4. Assessment of Sample Storage Parameters

mRNA instability represents a common issue for RNA analysis. Sample processing time frames and optimal blood collection tubes therefore need to be carefully assessed to allow for reliable mRNA detection. As cells are not fixed in EDTA blood tubes, which is essential for subsequent RNA isolation, potential CTCs could deteriorate over time. This is especially crucial when calculating time frames for shipment of clinical samples. EDTA blood spiked with cell lines was left at room temperature (RT) for 24 h (Table 2) and 48 h (Table 3), respectively, to test processing windows. Following the qPCR run, generated products were applied to a gel electrophoresis allowing final confirmation of gene expression status (data not shown).

Table 2. Validation of protocol specificity and sensitivity after 24 h. Influence of sample storage on *ARV7* detection limits and assay robustness. *ARV7* status is highlighted as "+" for positive and "−" for negative samples. Detection of gene expression was confirmed when at least 2/3 triplicates were positive in qPCR analysis. N represents the number of repetitions performed per experimental setting. The ratio is defined as the frequency at which any specific gene was detected out of the N repetitions. N/D signifies no gene expression or gene expression above the set threshold of 35 cycles.

Cell Line	ARV7 Status	Cell Amount	Target	Detection	N	Ratio [detection/N]
HD	−	0	*ARV7* *K19* *Actin*	N/D N/D yes	3	3/3
22RV1	+	10	*ARV7* *K19* *Actin*	yes yes yes	1	1/1
		5	*ARV7* *K19* *Actin*	yes yes yes	2	2/2
		3	*ARV7* *K19* *Actin*	yes yes yes	4	1/4 3/4 4/4
		1	*ARV7* *K19* *Actin*	yes yes yes	3	1/3 2/3 3/3
VCaP	+	10	*ARV7* *K19* *Actin*	yes yes yes	2	2/2
		5	*ARV7* *K19* *Actin*	yes N/D yes	1	1/1

After 24 h of sample storage at RT, 5 $ARV7^{+}$ cells were still reliably detected using the assay (Table 2). This was confirmed on two $ARV7^{+}$ cell lines (22Rv1 and VCaP). Additionally, as low as 3 and down to 1 $ARV7^{+}$ cells were detectable (Table 2). With these low cell counts, detection frequency is more variable as cell enrichment from whole blood and extremely careful sample handling play crucial roles. Still, correct detection down to a single $ARV7^{+}$ cell is possible. After 48 h, detection of *ARV7* and *K19* transcripts is subject to even higher fluctuation and increased cell counts would be needed to robustly detect transcripts of interest from these samples (Table 3). The specificity of our assay was demonstrated as no signals for *ARV7* or *K19* were seen in blood samples from healthy individuals in EDTA blood tested for 24h (3/3) as well as 48 h (3/3) of sample storage (Tables 2 and 3).

Blood tube types vary and some may be more suitable for our assay than others. Therefore, we additionally tested the performance of AdnaCollect blood collection tubes (Qiagen, Hilden, Germany), designed for mRNA characterization by the AdnaTest Prostate Cancer (Qiagen, Hilden, Germany) with our assay. This tube has been used for PCR-based detection of RNA transcripts from whole blood and could therefore provide an alternative to EDTA, potentially prolonging the sample processing window. Again, different cell counts were spiked into HD blood, this time in AdnaCollect blood collection tubes, and processed after 48 h of storage with our protocol. In our hands, these tubes were able to detect *ARV7* in spiked samples, down to 5 $ARV7^{+}$ cells (Table 3). However, as *ARV7* and *K19* signals were seen in all three tested HD samples (Table 3) indicating low specificity, the use of this blood tube type was not further continued.

Table 3. Influence of sample tubes and sample storage times on ARV7 detection limits and assay specificity. *ARV7* status is highlighted as "+" for positive and "−" for negative samples. Detection of gene expression was confirmed when at least 2/3 triplicates were positive in qPCR analysis. N represents the number of repetitions performed per experimental setting. The ratio is defined as the frequency at which any specific gene was detected out of the N repetitions. N/D signifies no gene expression or gene expression above the set threshold of 35 cycles.

Tube	Cell Line	ARV7 Status	Cell Amount	Target	Detection	N	Ratio [detection/N]
EDTA	HD	−	0	*ARV7*	N/D	3	3/3
				K19	N/D		
				Actin	yes		
	22RV1	+	10	*ARV7*	N/D	1	1/1
				K19	yes		
				Actin	yes		
			5	*ARV7*	yes	3	2/3
				K19	yes		2/3
				Actin	yes		3/3
Adnagen	HD	−	0	*ARV7*	yes	3	1/3
				K19	yes		2/3
				Actin	yes		3/3
	22RV1	+	10	*ARV7*	N/D	1	1/1
				K19	N/D		
				Actin	yes		
			5	*ARV7*	yes	3	3/3
				K19	yes		1/3
				Actin	yes		3/3

Our protocol ensures specific detection of tumor cell transcripts in 7.5 mL of blood down to a single cell level even after 24 h of sample storage (Table 2). Conclusively, a sample preparation window of 24 h was determined for the evaluation of clinical samples taken into EDTA blood to allow for sample shipment while ensuring robust detection of *ARV7* from CTCs.

3.5. Clinical Feasibility of the Complete ARV7 Detection Workflow

The clinical feasibility of our assay was demonstrated by analyzing blood samples of 26 metastatic prostate cancer (mPCa) patients. Detailed clinical patient data is listed in Table S1. qPCR based *ARV7* analysis was performed within 24 h of sample collection from 7.5 ml of EDTA blood for all 26 patients. Parallel blood draws to assess CTC counts via CellSearch® were collected and processed for 23/26 patients. AR-FL staining in the fourth fluorescent channel was available for 12/23 patient samples processed via CellSearch® (Table 4).

Table 4. Correlation of qPCR results, AR-FL detection and CTC enumeration via CellSearch for 26 mPCa patients analyzed. Detection of a gene was confirmed when at least 2/3 triplicates were positive in qPCR analysis. N/D signifies no gene expression or gene expression above the set threshold of 35 cycles. CTC enumeration via CellSearch® was not conducted for the first three patient samples, indicated by a bar in the respective table column (-). This also applies to 14 samples collected regarding AR occurrence. The number of CTCs with detectable AR-FL expression is indicated in brackets.

	CellSearch			qPCR		
Sample	**CTC Count**	**AR (nucl.)**	**AR (cytopl.)**	***ARV7***	***K19***	***Actin***
UKE-1	-	-	-	yes	yes	yes
UKE-2	-	-	-	N/D	N/D	yes
UKE-3	-	-	-	N/D	N/D	yes
UKE-4	0	-	-	N/D	N/D	yes
UKE-5	0	-	-	N/D	N/D	yes
UKE-6	0	-	-	N/D	N/D	yes
UKE-7	0	0	0	N/D	N/D	yes
UKE-8	1	0	yes (1)	N/D	N/D	yes
UKE-9	1	yes (1)	0	N/D	N/D	yes
UKE-10	1	0	0	N/D	yes	yes
UKE-11	1	0	yes (1)	N/D	yes	yes
UKE-12	1	-	-	yes	yes	yes
UKE-13	1	0	0	yes	yes	yes
UKE-14	2	yes (1)	yes (1)	N/D	yes	yes
UKE-15	6	0	0	N/D	N/D	yes
UKE-16	6	-	-	N/D	N/D	yes
UKE-17	8	yes (3)	yes (4)	yes	yes	yes
UKE-18	9	0	yes (9)	yes	yes	yes
UKE-19	11	0	yes (11)	yes	yes	yes
UKE-20	11	0	yes (11)	yes	yes	yes
UKE-21	14	-	-	N/D	yes	yes
UKE-22	16	-	-	yes	yes	yes
UKE-23	22	-	-	yes	N/D	yes
UKE-24	80	-	-	yes	yes	yes
UKE-25	156	-	-	yes	yes	yes
UKE-26	398	-	-	yes	yes	yes

Of the patient samples analyzed via CellSearch® 86.2% (19/23) were found to have ≥1 CTC in 7.5 mL of blood. In 52.2% (12/23) of patients ≥5 CTCs were detected in 7.5 ml of whole blood, reaching the clinically prognostic cut-off value for worse overall survival for metastatic mPCa patients [8]. The median of detected CTCs for our cohort is 6 (range: 0–398 CTCs) and the average is 32 CTCs/7.5 mL of blood. *ARV7* mRNA was detected in 46.2% (12/26) of mPCa patients, *K19* was detected is 57.7% (15/26) of samples and *Actin* was detected in all samples (26/26), indicating effective cDNA transcription. Four measured patients were negative for the androgen receptor splice variant and positive for *K19* (e.g., samples UKE-10 and UKE-11). Additionally, one patient was positive for *ARV7* expression and negative for *K19* (UKE-23). No *ARV7* or *K19* gene expression was found in samples classified as CTC negative by the CellSearch® system. Evaluation of the first 26 clinical samples resulted in 42.3% of $ARV7^{+}/K19^{+}$ of all cases (11/26) and 52.6% of $ARV7^{+}/K19^{+}$ cases (10/19) with ≥1 detectable CTC. Representative CellSearch® images of AR-FL staining are shown in Figure 5.

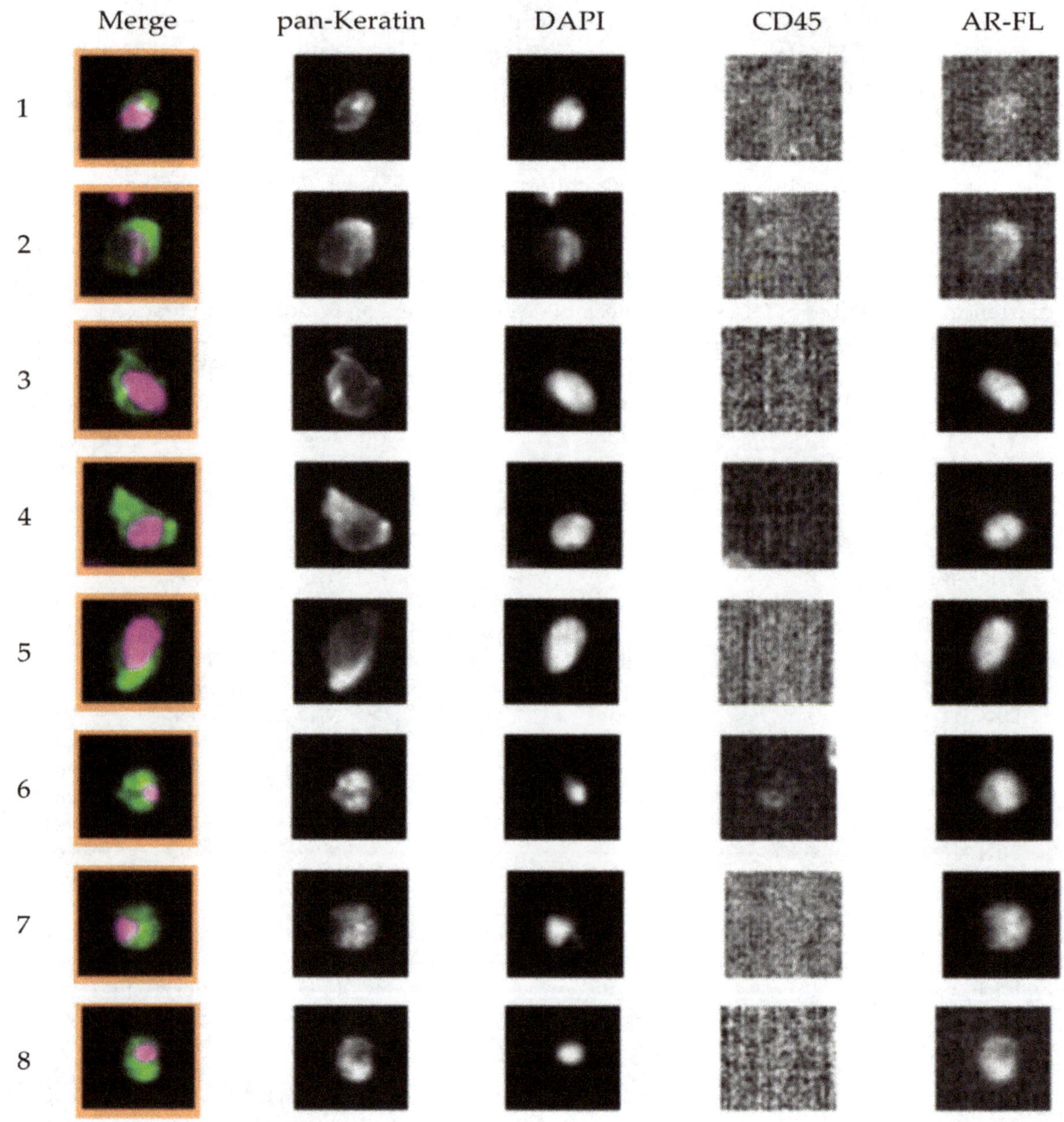

Figure 5. Representative CellSearch® images of CTCs and AR-FL staining from a single prostate cancer patient case. This patient had detectable AR-FL negative (**1**), weakly positive (**2**), nuclear AR-FL positive (**3–5**) and cytoplasmatically AR-FL positive (**6–8**) CTCs.

Only three of the 12 cases in which AR-FL protein staining was assessed in the CellSearch® (UKE-9, UKE-14, UKE-17) had detectable nuclear AR-FL protein levels (Table 4). Two of these three patients (UKE-14, UKE-17) had a mixed CTC population of nuclear and cytoplasmic AR-FL$^+$ CTCs. Of these three patients, two were *ARV7* negative with our assay (UKE-9, UKE-14). Additional five patients showed cytoplasmic AR-FL protein expression, more than half of these patients were ARV7 positive (3/5).

For the majority of patient samples tested (88.4% or 23/26), the CTC count as measured by the CellSearch® system was in accordance to *K19* detection in parallel samples (Table 4). Detection of *ARV7* was possible in 2/6 patients with only a single CTC detected in the patient's blood (UKE-12, UKE-13) confirming the assays sensitivity (Table 4). *K19* was detected in 4/6 patients with only a single CTC indicating careful and effective sample handling (Table 4).

4. Discussion

The CellSearch® Profile technology allows a reliable, standardized, and automated enrichment of EpCAM-positive cancer cells. ARV7 expression in CTCs of prostate cancer patients has been linked to resistance toward AR-targeted therapy, in particular enzalutamide and abiraterone [20,25]. Our novel approach ensures specific detection of *ARV7* transcripts in CTCs isolated by the CellSearch® system down to the single cell level. The specificity of our assay was indicated as no signals for *ARV7* or *K19* were seen in 9 blood samples from healthy, male individuals (Tables 1–3). Our protocol ensures specific detection of tumor cell transcripts in 7.5 mL of blood even after 24 h of sample storage (Table 2). Robust *ARV7* and *K19* detection is feasible in as low as 5 $ARV7^{+}/K19^{+}$ cells (Table 2). Transcript expression below 5 cells, even down to 1 $ARV7^{+}/K19^{+}$ cell, was possible (Table 2). The clinical feasibility of our assay and its high sensitivity (down to a single CTC) was demonstrated in a cohort of 26 mPCa patients (Table 4).

Antonarakis et al. linked *ARV7* mRNA expression on CTCs of mCRPC patients receiving enzalutamide and/or abiraterone therapy to lower PSA response rates, as well as shorter progression free and overall survival [20]. Following this initial study, conducted with a combination of bead-based CTC enrichment and subsequent qPCR multiplexing, the group confirmed their finding in a larger cohort of 202 CRPC patients [23]. In their study, CTC^{-} patients were found to have the best outcome (judged by best PSA-response, PSA progression-free survival, progression-free survival, and overall survival), followed by $CTC^{+}/ARV7^{-}$ and finally $CTC^{+}/ARV7^{+}$ patients [23]. Additionally, it was demonstrated that *ARV7* status can change in the course of hormone therapy [25,28,29] and that within one patient *ARV7* status on CTCs can be heterogeneous [41].

The CellSearch® system enables validated and automated enrichment of EpCAM-positive cancer cells [8,42–45]. Ideally, adding a specific and sensitive anti-ARV7 antibody to the fourth fluorescent channel of the CellSearch® device would therefore represent a valuable alternative to allow parallel CTC enumeration and the assessment of ARV7 status for each respective CTC. Unfortunately, detection of ARV7 protein using the CellSearch® technology was dramatically hampered by lacking specificity of most existing ARV7 antibodies (Figure 1, Figure S1). Recently, a novel commercially available antibody has been tested and validated for immunohistochemistry on primary tumor tissue, showing specific ARV7 staining results [32]. Whether this antibody might represent a promising novel candidate for immunocytochemical analysis and combination with CellSearch® needs to be investigated in future studies. However, so far most sources of CTC-related ARV7 information stems from RNA measurements.

The meaningful clinical impact of *ARV7* expression of CTCs [20,46] has led to the development of a multitude of different assays targeting ARV7 protein [25] or *ARV7* transcripts [30,31,47,48]. Primarily the developed methods are based on the analysis of pooled lysate of an enriched CTC fraction [26,30,48], only few perform whole blood gene expression analysis [47]. CTC are enriched by bead-based approaches [20,48], or the CellSearch® Profile kit and analyzed by subsequent qPCR or RNA-seq [26,30]. While these approaches effectively asses *ARV7* status, they give no additional information on the abundance of CTCs in a patient at the time point of blood draw. This could, however, prove to be valuable information allowing more precise interpretation of the qualitative $ARV7^{+}$ or $ARV7^{-}$ status of a patient. Without CTC count, an $ARV7^{-}$ status may refer to no available CTCs within the blood draw or to high amounts of $ARV7^{-}$ CTCs, respectively. The clinical information to be gained from both results is, however, very different, as no CTCs indicate good and ≥5 CTCs indicate poor outcome for the patient [8]. Multiplexing of additional genes such as prostate specific antigen (PSA) or prostate specific membrane antigen (PSMA), as well as epithelial genes is commonly used as a means of circumventing this issue and attempting to detect $ARV7^{-}$ CTCs [20,26,49]. While this is a feasible approach, it is limited by heterogeneous expression of these markers [31,41,49–51] and the required pre-amplification step can introduce bias.

Using our novel approach (Figure 4) information on both CTC count, AR-FL and *ARV7* status is collected. One could argue that the amount of CTCs present in the blood tube destined for *ARV7* assessment is also not directly assessed by our assay. However, studies have shown that CTC counts determined with the CellSearch® technology do not significantly fluctuate depending on circadian rhythm or serial blood draws [52,53], thereby indicating that stochastically, similar to equal CTC amounts would be expected in two sequential blood draws from the same patient at the same time (as is necessitated by our protocol). The importance of integrated CTC enumeration becomes apparent when looking at clinical cases such as UKE-23 (Table 4). While this patient had clearly detectable *ARV7* transcripts, he did not show *K19* positivity in our assay. Without the additional information of 22 CTCs being detected via CellSearch® analysis, interpretation of the qPCR results would have been impaired. This case also highlights the inert limitation of qPCR multiplexing, which lies in the before mentioned heterogeneity of gene and protein expression in CTCs [41,50,51]. In addition, CTC detection via the CellSearch® allows for morphological assessment of the CTCs in circulation and in our case, parallel characterization of AR-FL protein as well as its cellular location. Both represent important factors in resistance to androgen deprivation therapy [54]. The localization of the full length AR within the cell has been shown to be associated with disease progression on novel hormone therapies (e.g., enzalutamide and abiraterone) [55]. Therefore, it was critical for our assay to be able to distinguish both cytoplasmic and nuclear fractions of AR to support differentiation between "AR-on" and "AR-off" patients [54]. Apart from the AR-FL targeting antibody (by Janssen Diagnostics) used in this study, other well-established alternative antibodies have been published for AR-FL detection in the fourth channel of the CellSearch® [55].

To our knowledge, only two assays have been developed allowing parallel CTC enumeration and ARV7 protein [24,25] or transcript detection [31] on the same cell so far. El-Heliebi et al. isolated CTCs via the CellSearch® Profile kit or the size-dependent Parsortix™ platform (ANGLE plc, Guildford, UK) [56] and subsequently characterized them for *ARV7, AR-FL,* and *PSA* expression via in situ padlock probe technology [31]. This approach allows for absolute transcript quantification while keeping cell morphology intact and thereby enabling tumor cell enumeration [31]. In regards to CellSearch® Profile kit pre-enrichment, a single patient with high CTC load was included in this study to demonstrate general feasibility of the approach [31]. Additional technical validation will therefore be required to ensure sufficient sensitivity and specificity of this method for future clinical application.

The ARV7 assay most advanced in regards to clinical validation is the EPICs approach [24,25]. Here, the nuclear cell fraction is placed on slides, stained via ICC and automatically screened and evaluated. The assay focusses on nuclear ARV7 protein expression using the same antibody clone EPR15656 (Abcam) that we tested in our present study. While the EPICS approach allows for parallel CTC enumeration and ARV7 protein assessment, it requires sample shipment to a centralized lab in the US, a costs intensive approach when conducting larger clinical studies or when shipping patient samples for routine testing. A nuclear ARV7 staining has been postulated to be relevant to predict therapy outcome of AR-targeted therapies as well as taxanes in a cohort of 161 mCRPC patients, leading to a favorable coverage recommendation and certification of the approach in the state of California (USA) [57,58]. However, in our hands, the EPR15656 antibody did not result in specific nuclear staining signals for tested cancer cell line cells on chamber slides or cytospins (Figure 1, Figure S1).

While the *ARV7* detection assay established in this study is highly specific and sensitive, some limitations require mentioning. The main limitation is the fact that our assay does not allow for simultaneous morphological and molecular *ARV7* characterization of each single CTC. However, this is somewhat compensated by the use of a clinically validated CTC enrichment method, adding weight to the clinical relevance of the CTCs analyzed. Additionally, *ARV7* and *K19* transcript detection cannot be guaranteed down to a single CTC level in all patient samples. Nevertheless, we can secure determination of *ARV7* status from ≥5 CTCs which is the prognostic cut-off value for patients with metastatic prostate cancer. Due to the high sensitivity and specificity of our *ARV7* detection assay and the parallel nature of the CellSearch® CTC-enumeration, *K19* detection is not a mandatory prerequisite for robust *ARV7* assessment and result interpretation. However, we believe *K19* adds further valuable information in positive cases and represents an additional confirmation of successful CTC analysis.

Taken together, the use of a FDA-cleared enrichment technology, high assay sensitivity and specificity, a shipment window of 24 h and comparably low necessity of elaborate additional laboratory equipment (standard qPCR cycler) corroborate the value of our established method. Inclusion into prospective clinical trials will be now necessary to demonstrate clinical validity and utility. Furthermore, additional age-matched healthy donors and other control cohorts (e.g., prostatitis patients) should be included into future studies to further corroborate assay specificity. Head-to-head comparison with other ARV7/CTC technologies is desirable to assess to which extent different assays are redundant or complementary.

5. Conclusions

The novel workflow developed in this study allows for a semi-automated enrichment of CTCs followed by a qPCR assay measuring the *ARV7* status of CTCs. This approach can now be integrated into future clinical trials assessing treatment responses to antiandrogen therapies in prostate cancer patients.

Author Contributions: C.H. collected and primarily analyzed data, developed the methodology, visualized the results and wrote the manuscript draft. T.M.G. conceptualized and supervised the project and contributed to the method establishment. S.R. and M.M. collected and analyzed data. M.M. additionally participated in writing of the manuscript. T.S., G.v.A., F.K. and S.P. contributed clinical as well as healthy donor samples and clinical data. They furthermore aided in the clinical interpretation of the results. C.A.-P. and K.P. participated in writing of the manuscript and supported data analysis. K.P. furthermore supervised and conceptualized the project and acquired funding. All authors contributed to the editing and writing of the manuscript. Conceptualization, T.M.G., C.A.-P., and K.P.; Funding acquisition, K.P.; Investigation, C.H., S.R., M.M., T.S., G.v.A., F.K. and S.P.; Methodology, C.H. and T.M.G.; Supervision, T.M.G.; Visualization, C.H.; Writing—original draft, C.H. and K.P.; Writing—review and editing, C.H., T.M.G., S.R., M.M., T.S., G.v.A., F.K., S.P., C.A.-P., and K.P.

Acknowledgments: The authors acknowledge Cornelia Coith, Antje Andreas and Oliver Mauermann from UKE Hamburg for technical support.

References

1. World Cancer Research Fund Inthernational, American Institute for Cancer Research. Prostate Cancer. 2018. Available online: https://www.wcrf.org/dietandcancer/prostate-cancer (accessed on 30 July 2019).
2. Ladjevardi, S.; Auer, G.; Castro, J.; Ericsson, C.; Zetterberg, A.; Häggman, M.; Wiksell, H.; Jorulf, H. Prostate Biopsy Sampling Causes Hematogenous Dissemination of Epithelial Cellular Material. *Dis. Markers* **2014**, *2014*, 1–6. [CrossRef] [PubMed]
3. Alix-Panabières, C.; Pantel, K. Clinical Applications of Circulating Tumor Cells and Circulating Tumor DNA as Liquid Biopsy. *Cancer Discov.* **2016**, *6*, 479–491. [CrossRef] [PubMed]
4. Pantel, K.; Speicher, M.R. The biology of circulating tumor cells. *Oncogene* **2016**, *35*, 1216–1224. [CrossRef] [PubMed]
5. Bardelli, A.; Pantel, K. Liquid Biopsies, What We Do Not Know (Yet). *Cancer Cell* **2017**, *31*, 172–179. [CrossRef] [PubMed]
6. Alix-Panabieres, C.; Pantel, K. Circulating tumor cells: Liquid biopsy of cancer. *Clin. Chem.* **2013**, *59*, 110–118. [CrossRef] [PubMed]
7. Hille, C.; Pantel, K. Circulating tumour cells in prostate cancer. *Nat. Rev. Urol.* **2018**, *15*, 265–266. [CrossRef] [PubMed]
8. De Bono, J.S.; Scher, H.I.; Montgomery, R.B.; Parker, C.; Miller, M.C.; Tissing, H.; Doyle, G.V.; Terstappen, L.W.W.M.; Pienta, K.J.; Raghavan, D. Circulating tumor cells predict survival benefit from treatment in metastatic castration-resistant prostate cancer. *Clin. Cancer Res.* **2008**, *14*, 6302–6309. [CrossRef] [PubMed]

9. Scher, H.I.; Jia, X.Y.; de Bono, J.S.; Fleisher, M.; Pienta, K.J.; Raghavan, D.; Heller, G. Circulating tumour cells as prognostic markers in progressive, castration-resistant prostate cancer: A reanalysis of IMMC38 trial data. *Lancet. Oncol.* **2009**, *10*, 233–239. [CrossRef]
10. Scher, H.I.; Heller, G.; Molina, A.; Attard, G.; Danila, D.C.; Jia, X.; Peng, W.; Sandhu, S.K.; Olmos, D.; Riisnaes, R.; et al. Circulating Tumor Cell Biomarker Panel As an Individual-Level Surrogate for Survival in Metastatic Castration-Resistant Prostate Cancer. *J. Clin. Oncol.* **2015**, *33*, 1348–1355. [CrossRef]
11. Heller, G.; Fizazi, K.; McCormack, R.; Molina, A.; MacLean, D.; Webb, I.J.; Saad, F.; de Bono, J.S.; Scher, H.I. The Added Value of Circulating Tumor Cell Enumeration to Standard Markers in Assessing Prognosis in a Metastatic Castration-Resistant Prostate Cancer Population. *Clin. Cancer Res.* **2017**, *23*, 1967–1973. [CrossRef]
12. Miller, M.C.; Doyle, G.V.; Terstappen, L.W. Significance of Circulating Tumor Cells Detected by the CellSearch System in Patients with Metastatic Breast Colorectal and Prostate Cancer. *J. Oncol.* **2010**, *2010*, 617421. [CrossRef] [PubMed]
13. Goodman, O.B.; Symanowski, J.T.; Loudyi, A.; Fink, L.M.; Ward, D.C.; Vogelzang, N.J. Circulating Tumor Cells as a Predictive Biomarker in Patients With Hormone-sensitive Prostate Cancer. *Clin. Genitourin. Cancer* **2011**, *9*, 31–38. [CrossRef] [PubMed]
14. Danila, D.C.; Fleisher, M.; Scher, H.I. Circulating tumor cells as biomarkers in prostate cancer. *Clin. Cancer Res.* **2011**, *17*, 3903–3912. [CrossRef] [PubMed]
15. Gorges, T.M.; Pantel, K. Circulating tumor cells as therapy-related biomarkers in cancer patients. *Cancer Immunol. Immunother.* **2013**, *62*, 931–939. [CrossRef] [PubMed]
16. Singhal, U.; Wang, Y.; Henderson, J.; Niknafs, Y.S.; Qiao, Y.; Gursky, A.; Zaslavsky, A.; Chung, J.-S.; Smith, D.C.; Karnes, R.J.; et al. Multigene Profiling of CTCs in mCRPC Identifies a Clinically Relevant Prognostic Signature. *Mol. Cancer Res.* **2018**, *16*, 643–654. [CrossRef]
17. Pantel, K.; Hille, C.; Scher, H.I. Circulating Tumor Cells in Prostate Cancer: From Discovery to Clinical Utility. *Clin. Chem.* **2019**, *65*, 87–99. [CrossRef] [PubMed]
18. Luo, J.; Attard, G.; Balk, S.P.; Bevan, C.; Burnstein, K.; Cato, L.; Cherkasov, A.; De Bono, J.S.; Dong, Y.; Gao, A.C.; et al. Role of Androgen Receptor Variants in Prostate Cancer: Report from the 2017 Mission Androgen Receptor Variants Meeting. *Eur. Urol.* **2018**, *73*, 715–723. [CrossRef]
19. Antonarakis, E.S.; Armstrong, A.J.; Dehm, S.M.; Luo, J. Androgen receptor variant-driven prostate cancer: Clinical implications and therapeutic targeting. *Prostate Cancer Prostatic Dis.* **2016**, *19*, 231–241. [CrossRef]
20. Antonarakis, E.S.; Lu, C.; Wang, H.; Luber, B.; Nakazawa, M.; Roeser, J.C.; Chen, Y.; Mohammad, T.A.; Chen, Y.; Fedor, H.L.; et al. AR-V7 and Resistance to Enzalutamide and Abiraterone in Prostate Cancer. *New Engl. J. Med.* **2014**, *371*, 1028–1038. [CrossRef]
21. Antonarakis, E.S.; Luo, J. Blood Based Detection of Androgen Receptor Splice Variants in Patients with Advanced Prostate Cancer. *J. Urol.* **2016**, *196*, 1606–1607. [CrossRef]
22. Lokhandwala, P.M.; Riel, S.L.; Haley, L.; Lu, C.; Chen, Y.; Silberstein, J.; Zhu, Y.; Zheng, G.; Lin, M.-T.; Gocke, C.D.; et al. Analytical Validation of Androgen Receptor Splice Variant 7 Detection in a Clinical Laboratory Improvement Amendments (CLIA) Laboratory Setting. *J. Mol. Diagn.* **2017**, *19*, 115–125. [CrossRef] [PubMed]
23. Antonarakis, E.S.; Lu, C.; Luber, B.; Wang, H.; Chen, Y.; Zhu, Y.; Silberstein, J.L.; Taylor, M.N.; Maughan, B.L.; Denmeade, S.R.; et al. Clinical Significance of Androgen Receptor Splice Variant-7 mRNA Detection in Circulating Tumor Cells of Men With Metastatic Castration-Resistant Prostate Cancer Treated With First- and Second-Line Abiraterone and Enzalutamide. *J. Clin. Oncol.* **2017**, *35*, 2149–2156. [CrossRef]
24. Scher, H.I.; Graf, R.P.; Schreiber, N.A.; Jayaram, A.; Winquist, E.; McLaughlin, B.; Lu, D.; Fleisher, M.; Orr, S.; Lowes, L.; et al. Assessment of the Validity of Nuclear-Localized Androgen Receptor Splice Variant 7 in Circulating Tumor Cells as a Predictive Biomarker for Castration-Resistant Prostate Cancer. *JAMA Oncol.* **2018**, *4*, 1179–1186. [CrossRef]
25. Scher, H.I.; Lu, D.; Schreiber, N.A.; Louw, J.; Graf, R.P.; Vargas, H.A.; Johnson, A.; Jendrisak, A.; Bambury, R.; Danila, D.; et al. Association of AR-V7 on Circulating Tumor Cells as a Treatment-Specific Biomarker With Outcomes and Survival in Castration-Resistant Prostate Cancer. *JAMA Oncol.* **2016**, *2*, 1441–1449. [CrossRef] [PubMed]
26. Onstenk, W.; Sieuwerts, A.M.; Kraan, J.; Van, M.; Nieuweboer, A.J.; Mathijssen, R.H.; Hamberg, P.;

Meulenbeld, H.J.; De Laere, B.; Dirix, L.Y.; et al. Efficacy of Cabazitaxel in Castration-resistant Prostate Cancer Is Independent of the Presence of AR-V7 in Circulating Tumor Cells. *Eur. Urol.* **2015**, *68*, 939–945. [CrossRef] [PubMed]

27. Antonarakis, E.S.; Lu, C.; Luber, B.; Wang, H.; Chen, Y.; Nakazawa, M.; Nadal, R.; Paller, C.J.; Denmeade, S.R.; Carducci, M.A.; et al. Androgen Receptor Splice Variant 7 and Efficacy of Taxane Chemotherapy in Patients With Metastatic Castration-Resistant Prostate Cancer. *JAMA Oncol.* **2015**, *1*, 582–591. [CrossRef] [PubMed]
28. Bernemann, C.; Schnoeller, T.J.; Luedeke, M.; Steinestel, K.; Boegemann, M.; Schrader, A.J.; Steinestel, J. Expression of AR-V7 in Circulating Tumour Cells Does Not Preclude Response to Next Generation Androgen Deprivation Therapy in Patients with Castration Resistant Prostate Cancer. *Eur. Urol.* **2017**, *71*, 1–3. [CrossRef]
29. Nakazawa, M.; Lu, C.; Chen, Y.; Paller, C.J.; Carducci, M.A.; Eisenberger, M.A.; Luo, J.; Antonarakis, E.S. Serial blood-based analysis of AR-V7 in men with advanced prostate cancer. *Ann. Oncol.* **2015**, *26*, 1859–1865. [CrossRef]
30. De Laere, B.; Van Dam, P.-J.; Whitington, T.; Mayrhofer, M.; Diaz, E.H.; Eynden, G.V.D.; Vandebroek, J.; Del-Favero, J.; Van Laere, S.; Dirix, L.; et al. Comprehensive Profiling of the Androgen Receptor in Liquid Biopsies from Castration-resistant Prostate Cancer Reveals Novel Intra-AR Structural Variation and Splice Variant Expression Patterns. *Eur. Urol.* **2017**, *72*, 192–200. [CrossRef]
31. El-Heliebi, A.; Attard, G.; Balk, S.P.; Bevan, C.; Burnstein, K.; Cato, L.; Cherkasov, A.; De Bono, J.S.; Dong, Y.; Gao, A.C.; et al. In Situ Detection and Quantification of AR-V7, AR-FL, PSA, and KRAS Point Mutations in Circulating Tumor Cells. *Clin. Chem.* **2018**, 64. [CrossRef]
32. Sharp, A.; Welti, J.C.; Lambros, M.B.; Dolling, D.; Rodrigues, D.N.; Pope, L.; Aversa, C.; Figueiredo, I.; Fraser, J.; Ahmad, Z.; et al. Clinical Utility of Circulating Tumour Cell Androgen Receptor Splice Variant-7 Status in Metastatic Castration-resistant Prostate Cancer. *Eur. Urol.* **2019**. [CrossRef]
33. Riethdorf, S.; O'Flaherty, L.; Hille, C.; Pantel, K. Clinical applications of the CellSearch platform in cancer patients. *Adv. Drug Deliv. Rev.* **2018**, *125*, 102–121. [CrossRef] [PubMed]
34. Riethdorf, S.; Fritsche, H.; Müller, V.; Rau, T.; Schindlbeck, C.; Rack, B.; Janni, W.; Coith, C.; Beck, K.; Jänicke, F.; et al. Detection of Circulating Tumor Cells in Peripheral Blood of Patients with Metastatic Breast Cancer: A Validation Study of the CellSearch System. *Clin. Cancer Res.* **2007**, *13*, 920–928. [CrossRef] [PubMed]
35. Guo, Z.; Yang, X.; Sun, F.; Jiang, R.; Linn, D.E.; Chen, H.; Chen, H.; Kong, X.; Melamed, J.; Tepper, C.G.; et al. A novel androgen receptor splice variant is up-regulated during prostate cancer progression and promotes androgen depletion-resistant growth. *Cancer Res.* **2009**, *69*, 2305–2313. [CrossRef] [PubMed]
36. Liu, L.L.; Xie, N.; Sun, S.; Plymate, S.; Mostaghel, E.; Dong, X. Mechanisms of the androgen receptor splicing in prostate cancer cells. *Oncogene* **2014**, *33*, 3140–3150. [CrossRef] [PubMed]
37. Untergasser, A.; Cutcutache, I.; Koressaar, T.; Ye, J.; Faircloth, B.C.; Remm, M.; Rozen, S.G. Primer3—New capabilities and interfaces. *Nucleic Acids Res.* **2012**, *40*, e115. [CrossRef]
38. Livak, K.J.; Schmittgen, T.D. Analysis of relative gene expression data using real-time quantitative PCR and the 2(-Delta Delta C(T)) Method. *Methods* **2001**, *25*, 402–408. [CrossRef]
39. Stathopoulou, A.; Angelopoulou, K.; Perraki, M.; Georgoulias, V.; Malamos, N.; Lianidou, E. Quantitative RT-PCR luminometric hybridization assay with an RNA internal standard for cytokeratin-19 mRNA in peripheral blood of patients with breast cancer. *Clin. Biochem.* **2001**, *34*, 651–659. [CrossRef]
40. Stathopoulou, A.; Gizi, A.; Perraki, M.; Apostolaki, S.; Malamos, N.; Mavroudis, D.; Georgoulias, V.; Lianidou, E.S. Real-time quantification of CK-19 mRNA-positive cells in peripheral blood of breast cancer patients using the lightcycler system. *Clin. Cancer Res.* **2003**, *9*, 5145–5151.
41. Gorges, T.M.; Kuske, A.; Röck, K.; Mauermann, O.; Müller, V.; Peine, S.; Verpoort, K.; Novosadova, V.; Kubista, M.; Riethdorf, S.; et al. Accession of Tumor Heterogeneity by Multiplex Transcriptome Profiling of Single Circulating Tumor Cells. *Clin. Chem.* **2016**, *62*, 1504–1515. [CrossRef]
42. Cristofanilli, M.; Budd, G.T.; Ellis, M.J.; Stopeck, A.; Matera, J.; Miller, M.C.; Reuben, J.M.; Doyle, G.V.; Allard, W.J.; Terstappen, L.W.; et al. Circulating Tumor Cells, Disease Progression, and Survival in Metastatic Breast Cancer. *New Engl. J. Med.* **2004**, *351*, 781–791. [CrossRef] [PubMed]
43. Cohen, S.J.; Punt, C.J.; Iannotti, N.; Saidman, B.H.; Sabbath, K.D.; Gabrail, N.Y.; Picus, J.; Morse, M.; Mitchell, E.;

Miller, M.C.; et al. Relationship of Circulating Tumor Cells to Tumor Response, Progression-Free Survival, and Overall Survival in Patients With Metastatic Colorectal Cancer. *J. Clin. Oncol.* **2008**, *26*, 3213–3221. [CrossRef] [PubMed]

44. Ligthart, S.T.; Coumans, F.A.W.; Attard, G.; Cassidy, A.M.; De Bono, J.S.; Terstappen, L.W.M.M. Unbiased and Automated Identification of a Circulating Tumour Cell Definition That Associates with Overall Survival. *PLoS ONE* **2011**, *6*, e27419. [CrossRef] [PubMed]
45. Scholtens, T.M.; Schreuder, F.; Ligthart, S.T.; Swennenhuis, J.F.; Greve, J.; Terstappen, L.W. Automated identification of circulating tumor cells by image cytometry. *Cytometry A* **2012**, *81A*, 138–148. [CrossRef] [PubMed]
46. Markowski, M.C.; Silberstein, J.L.; Eshleman, J.R.; Eisenberger, M.A.; Luo, J.; Antonarakis, E.S. Clinical Utility of CLIA-Grade AR-V7 Testing in Patients With Metastatic Castration-Resistant Prostate Cancer. *JCO Precis. Oncol.* **2017**, 1–9.
47. Todenhöfer, T.; Azad, A.; Stewart, C.; Gao, J.; Eigl, B.J.; Gleave, M.E.; Joshua, A.M.; Black, P.C.; Chi, K.N. AR-V7 Transcripts in Whole Blood RNA of Patients with Metastatic Castration Resistant Prostate Cancer Correlate with Response to Abiraterone Acetate. *J. Urol.* **2017**, *197*, 135–142. [CrossRef] [PubMed]
48. Tommasi, S.; Pilato, B.; Carella, C.; Lasorella, A.; Danza, K.; Vallini, I.; De Summa, S.; Naglieri, E. Standardization of CTC AR-V7 PCR assay and evaluation of its role in castration resistant prostate cancer progression. *Prostate* **2018**, *79*, 54–61. [CrossRef] [PubMed]
49. Sieuwerts, A.M.; Mostert, B.; Van Der Vlugt-Daane, M.; Kraan, J.; Beaufort, C.M.; Van, M.; Prager, W.J.; De Laere, B.; Beije, N.; Hamberg, P.; et al. An In-Depth Evaluation of the Validity and Logistics Surrounding the Testing of AR-V7 mRNA Expression in Circulating Tumor Cells. *J. Mol. Diagn.* **2018**, *20*, 316–325. [CrossRef]
50. Gorges, T.M.; Riethdorf, S.; Von Ahsen, O.; Nastały, P.; Röck, K.; Boede, M.; Peine, S.; Kuske, A.; Schmid, E.; Kneip, C.; et al. Heterogeneous PSMA expression on circulating tumor cells—A potential basis for stratification and monitoring of PSMA-directed therapies in prostate cancer. *Oncotarget* **2016**, *7*, 34930–34941. [CrossRef] [PubMed]
51. Scher, H.I.; Graf, R.P.; Schreiber, N.A.; McLaughlin, B.; Jendrisak, A.; Wang, Y.; Lee, J.; Greene, S.; Krupa, R.; Lu, D.; et al. Phenotypic Heterogeneity of Circulating Tumor Cells Informs Clinical Decisions between AR Signaling Inhibitors and Taxanes in Metastatic Prostate Cancer. *Cancer Res.* **2017**, *77*, 5687–5698. [CrossRef]
52. Martín, M.; García-Sáenz, J.A.; Casas, M.L.M.D.L.; Vidaurreta, M.; Puente, J.; Veganzones, S.; Rodríguez-Lajusticia, L.; De La Orden, V.; Oliva, B.; De La Torre, J.-C.; et al. Circulating tumor cells in metastatic breast cancer: timing of blood extraction for analysis. *Anticancer. Res.* **2009**, *29*, 4185–4187. [PubMed]
53. Tibbe, A.G.J.; Miller, M.C.; Terstappen, L.W.M.M. Statistical considerations for enumeration of circulating tumor cells. *Cytom. Part. A* **2007**, *71*, 154–162. [CrossRef]
54. Miyamoto, D.T.; Lee, R.J.; Stott, S.L.; Ting, D.T.; Wittner, B.S.; Ulman, M.; Smas, M.E.; Lord, J.B.; Brannigan, B.W.; Trautwein, J.; et al. Androgen receptor signaling in circulating tumor cells as a marker of hormonally responsive prostate cancer. *Cancer Discov.* **2012**, *2*, 995–1003. [CrossRef] [PubMed]
55. Crespo, M.; Van Dalum, G.; Ferraldeschi, R.; Zafeiriou, Z.; Sideris, S.; Lorente, D.; Bianchini, D.; Rodrigues, D.N.; Riisnaes, R.; Miranda, S.; et al. Androgen receptor expression in circulating tumour cells from castration-resistant prostate cancer patients treated with novel endocrine agents. *Br. J. Cancer* **2015**, *112*, 1166–1174. [CrossRef] [PubMed]
56. Hvichia, G.; Parveen, Z.; Wagner, C.; Janning, M.; Quidde, J.; Stein, A.; Müller, V.; Loges, S.; Neves, R.; Stoecklein, N.; et al. A novel microfluidic platform for size and deformability based separation and the subsequent molecular characterization of viable circulating tumor cells. *Int. J. Cancer* **2016**, *138*, 2894–2904. [CrossRef]
57. Services, C.f.M.M. 2018. Available online: https://med.noridianmedicare.com/documents/10546/6990981/MolDX+Circulating+Tumor+Cell+Marker+Assays+LCD/8eaf89f0-9970-455c-b048-ebaeaf42bd7d (accessed on 30 July 2019).
58. Services C.f.M.M. 2018. Available online: https://www.cms.gov/medicare-coverage-database/details/lcd-details.aspx?LCDId=37914&ver=2&Cntrctr=All&UpdatePeriod=796&bc=AQAAEAAAAAAA& (accessed on 30 July 2019).

14

Clinical Impact of Circulating Tumor Cells in Patients with Localized Prostate Cancer

Lucile Broncy [1] and Patrizia Paterlini-Bréchot [1,2,*]

1 INSERM Unit 1151, Faculté de Médecine, Université Paris Descartes, 75014 Paris, France
2 Laboratoire de Biochimie A, Hôpital Necker-Enfants Malades, 75015 Paris, France
* Correspondence: patrizia.paterlini@inserm.fr.

Abstract: The main issue concerning localized prostate cancers is the lack of a suitable marker which could help patients' stratification at diagnosis and distinguish those with a benign disease from patients with a more aggressive cancer. Circulating Tumor Cells (CTC) are spread in the blood by invasive tumors and could be the ideal marker in this setting. Therefore, we have compiled data from the literature in order to obtain clues about the clinical impact of CTC in patients with localized prostate cancer. Forty-three publications have been found reporting analyses of CTC in patients with non-metastatic prostate cancer. Of these, we have made a further selection of 11 studies targeting patients with clinical or pathological stages T1 and T2 and reporting the clinical impact of CTC. The results of this search show encouraging data toward the use of CTC in patients with early-stage cancer. However, they also highlight the lack of standardized methods providing a highly sensitive and specific approach for the detection of prostate-derived CTC.

Keywords: prostate cancer (PCa); circulating tumor cells (CTC); liquid biopsy

1. Introduction

Globally, prostate cancer (PCa) is the most commonly diagnosed type of cancer in men, with 180,890 newly diagnosed cases and 26,120 deaths in the United States (US) in 2016 [1] and estimated 164,690 new cases and 29,430 deaths in 2018 [2]. As the fifth leading cause of cancer death worldwide, PCa accounted globally for 1.6 million newly diagnosed cases and 366,000 deaths in 2015 [1]. Increased risk factors for PCa include genetic predisposition, family history of prostate or breast cancers and older age, with a median age at diagnosis of 72 years [3]. Accordingly, the steady increase in PCa incidence in the US since 1950 appears related to an overall increase in life expectancy. In the US, the lifetime risk of being diagnosed with prostate cancer is of 11% and the lifetime risk of dying of prostate cancer is 2.5% [4]. Importantly, in autopsies of men who died of other causes, more than 20% of men aged 50 to 59 years and over 33% of men aged 70 to 79 years were found to have prostate cancer [5]. At diagnosis, 79% of prostate cancer cases were localized; in 12%, the cancer had spread to regional lymph nodes, and 5% of patients had distant metastasis. The 5-year relative survival rate for localized and regional prostate cancer is 100%, compared with 29.8% for metastatic cases.

PCa is recognized as a genetically heterogeneous disease [6] comprising a large scope of malignancies, from indolent localized cancers that may never progress to rapidly progressing castration-resistant PCa. Currently, the diagnosis of PCa is based on the pathological evaluation of tissue biopsy but the treatment options are determined by risk stratification based on both Gleason score and serum PSA level [7]. For example, high-risk PCa was defined by D'Amico as PSA ≥20 ng/mL and/or biopsy Gleason Score ≥8 and/or clinical stage ≥2c [8].

Elevated serum prostate-specific antigen (PSA) levels are often seen in the context of PCa but can also reflect other prostatic diseases such as benign prostatic hyperplasia, prostatic infection, and prostatic infarction [9]. Relatedly, the low specificity of the PSA screening test has raised concerns in the scientific community regarding the over-diagnosis of PCa [10]. Since the implementation of the PSA screening test into clinical practices in the 1990s, a significant shift towards localized PCa at diagnoses has been observed, with >95% of diagnoses being of clinically localized PCa [11]. Furthermore, only approximately 40 to 50% of patients with elevated PSA testing undergoing a biopsy have prostate cancer. However, a recent study evaluating the ERSPC and PLCO clinical trials has shown that PSA testing reduces the mortality due to PCa by approximately 30% [12]. In cases of localized PCa, radical prostatectomy remains the gold standard treatment option [7]. Importantly, up to 30% of patients treated with radical prostatectomy eventually develop recurrence [13]. More performant risk stratification and prognostic markers are urgently needed to improve the management of patients with localized prostate cancer and identify cases with a high risk of progression.

Circulating tumor cells (CTC) are cells that detach from the primary or secondary tumor sites and invade the bloodstream. As primary actors of the metastatic dissemination, CTC represent a very promising biomarker to aid cancer diagnosis, treatment decision and patient follow-up [14]. In fact, CTC could provide a valuable complement to PSA or other tests with the aim to identify patients with more aggressive cancers. The prognostic value of CTC collected by the epithelial marker-dependent method CellSearch has been established in the context of metastatic PCa [15]. However, the clinical utility of CTC in the context of localized PCa remains unclear. Here, we review studies on non-metastatic prostate cancer to evaluate the potential clinical utility of CTC in localized PCa.

2. Materials and Methods

The present review was prepared by selecting English-written research papers describing the detection and/or characterization of circulating tumor cells in the context of localized prostate cancer. To that end, we searched PubMed using the following keywords: "circulating tumor cells" or "circulating cancer cells" and "localized prostate cancer" or "non-metastatic prostate cancer" or "early-stage prostatic carcinoma". Reviews and studies on liquid biopsy that did not concern localized PCa, as well as studies that did not report on CTC were excluded from the systematic review. Preclinical models, as well as methods for detection of disseminated tumor cells (DTC) in the bone marrow of localized PCa patients, were also excluded. A total of 43 studies were included in the systematic review, as shown in Table 1. It should be noted that several selected studies also reported on locally advanced prostate cancer (T3 and T4) cases. The American Cancer Society defines localized PCa as clinical or pathological tumor stages T1 and T2. PCa with the T1 stage corresponds to a clinically unapparent tumor that is neither palpable nor visible by imaging while T2 corresponds to a tumor that is confined within the prostate and that is either palpable or visible by imaging or demonstrated in radical prostatectomy [16]. In contrast, a pathological tumor stage T3 defines a tumor that has extended through the prostatic capsule and T4 defines a tumor which is invading adjacent organs such as the bladder, sphincter or rectum [16].

Table 1. Studies on circulating tumor cells in patients with non-metastatic prostate cancer.

Study [ref]	N° of Patients (pT Stage)	CTC Method	Cutoff	N° of CTC+ Patients (%)	Median CTC/mL (Range/mL)	Comments	Clinical Impact of CTC
Moreno 1992 [17]	4 (ND [1])	PSA RT-PCR	NA [2]	0 (0%)	NA	17 negative controls (0% positive). Only patients with lymph nodes or metastases tested positive.	ND
Israeli 1995 [18]	13 (7 T2, 5 T3, 1 T4)	PSA & PSMA RT-PCR	NA	PSMA 7 (54%) PSA 0 (0%)	NA	6 years after radical prostatectomy, with undetectable PSA serum levels, patients are PSMA positive	ND
Olsson 1996 [19]	100 (59 T1-T2, 41 T3–T4	PSA RT-PCR	NA	74% (76% T1–T2, 71% T3–T4)	NA	Potential surgical failure defined as a tumor at the surgical margin or extending into the seminal vesicle	Yes
Ennis 1997 [20]	227 (72 T1c, 129 T2, 26 T3)	PSA RT-PCR	NA	61 (26.9%)	NA	Patients treated with prostatectomy had a higher rate of RT-PCR positivity than patients treated with radiation	Yes
Oefelein 1999 [21]	101 (T1–T3a)	PSA RT-PCR	NA	22 (22%)	NA	Median follow-up 22 months	No
Okegawa 1999 [22]	31 (T2–T3)	PSA & PSMA RT-PCR	NA	ND	NA	RT-PCR performed before radical prostatectomy	Yes
Sabile 1999 [23]	10 (ND)	PSA RT-PCR	NA	4 (40%)	NA	Density gradient has higher isolation efficiency than epithelial marker-dependent immunocapture	ND
Mejean 2000 [24]	99 (37 T1, 52 T2, 8 T3, 2 T4)	PSA RT-PCR	NA	33 (33%)	NA	RT-PCR performed preoperatively, 92 controls included (2% scored positive)	Yes
Llanes 2000 [25]	25 (T1–T2b)	PSA RT-PCR	NA	7 (28%)	NA	The best predictors of extraprostatic disease were the biopsy Gleason score and the PSA level.	No
Slawin 2000 [26]	228 (154 T1–T2, 47 T3a, 16 T3b, 11 T4)	hK2 RT-PCR	NA	57 (25%)	NA	14 healthy controls (14% positive). RT-PCR performed before prostatectomy. Association with the risk of metastasis to pelvic lymph nodes ($P = 0.028$).	Yes
Shariat 2002 [27]	224 (T1–T2)	PSA RT-PCR	NA	54 (24%)	NA	RT-PCR performed preoperatively.	No
Bianco 2002 [28]	96 AAM (35 T1, 61 T2) 150 CAM (62 T1, 88 T2)	PSA RT-PCR	NA	26 (27%) AAM 34 (23%) CAM	NA	RT-PCR performed preoperatively.	Yes in AAM
Hara 2002 [29]	44 (26 T1, 15 T2, 2 T3, 1 T4)	PSA, PSMA & PSCA RT-PCR	NA	1 (2.3%) PSA3 (6.8%) PSMA 1 (2.3%) PSCA	NA	RT-PCR performed preoperatively. Positive PSA result in 1 prostatitis case, positive PSMA result in 1 prostatitis and 1 benign prostatic hyperplasia case.	Yes
Thomas 2002 [30]	141 (118 T1c/T2a, 18 T2b/c, 5 T3a)	PSA & PSMA RT-PCR	NA	73 (51.8%)	NA	Only initial PSA and biopsy Gleason score were independent predictors of biochemical failure.	No

Table 1. *Cont.*

Study [ref]	N° of Patients (pT Stage)	CTC Method	Cutoff	N° of CTC+ Patients (%)	Median CTC/mL (Range/mL)	Comments	Clinical Impact of CTC
Gewanter 2003 [31]	161 (121 T1–T2, 39 T3, 1 Tx)	PSA RT-PCR	NA	22 (20%)	NA	29 months follow-up. Only post-treatment testing predicted for clinical relapse.	No
Fizazi 2007 [32]	83 (38 T1c, 38 T2, 7 T3)	EpCAM + telomerase PCR	NA	58 (70%)	NA	Preoperative CTC detection; 22 healthy controls (0% positive)	ND
Davis 2008 [33]	97 (78 T2, 19 T3)	CellSearch	≥1 CTC /22.5 mL	20 (21%)	0.18 (0.04–2.62)	Preoperative CTC detection; 4 of 20 healthy controls positive for CTC (20%)	ND
Helo 2009 [34]	129 (71 T2, 43 T3a, 13 T3b, 2 T4)	PSA & PSMA RT-PCR	≥80 mRNA/mL	3 (2.6%)	NA	19 healthy controls (0% positive). RT-PCR performed 6 months after surgery for 42 patients and before surgery for 85 patients	No
Maestro 2009 [35]	26 (ND)	CellSearch + CellSpotter Analyzer	≥2 CTC /7.5 mL	4 (15.4%)	ND	106 healthy controls (0% positive)	ND
Eschwe-ge 2009 [36]	155 (T2–T3)	PSA & PSMA RT-PCR	NA	57 (37%)	NA	Preoperative CTC detection; 100 healthy controls (0% positive).	Yes
Giesing 2010 [37]	129 (T1–T4)	Filtration + PSA & AOX RT-PCR	NA	42 (32.5%)	NA	The AOX test was tumour predicting With a positive predictive value of 69% and a negative predictive value of 92%,	Yes
Stott 2010 [38]	19 (T2–T3a)	Microfluidic (EpCAM)	≥14 CTC /mL	8 (42%)	95 (38–222)	6/8 patients with a decline of CTC 24 h after prostatectomy	ND
Joung 2010 [39]	103 (25 T1–T2b, 78 T2c–T3)	PSCA RT-PCR	NA	17 (16.5%)	NA	RT-PCR performed before surgery	Yes
Yates 2012 [40]	92 (61 T1, 31 T2)	PSA & PSMA RT-PCR	NA	63 (68.5%) PSA 68 (78.9%) PSMA	NA	Blood samples taken 1 day preoperatively and 7 days postoperatively	Yes
Lowes 2012 [41]	26 (11 T2, 15 T3)	CellSearch	≥1 CTC /7.5 mL	19 (73%)	ND	Blood drawn before radiation therapy; 7 healthy controls included (0% positive)	Yes
Khurana 2013 [42]	10 (5 T2c, 5 T3a)	CellSearch	≥1 CTC /7.5 mL	1 (10%)	0 (0–0.13)	Blood drawn preoperatively. Very low CTC numbers.	No
Thalgott 2013 [43]	20 (ND)	CellSearch	≥1 CTC /7.5 mL	1 (5%)	0 (0–0.13)	15 healthy controls (0% positive). Shorter overall survival observed only for metastatic patients with ≥ 3 CTC	No
Loh 2014 [44]	36 (9 T1, 14 T2, 13 T3)	CellSearch	≥1 CTC /7.5 mL	5 (14%)	0 (0–0.4)	Blood drawn before therapy. Median follow-up 42 months.	No
Kolostova 2014 [45]	55 (45 T2, 10 T3)	MetaCell®filtration	≥1 CTC /8 mL	28 (52%)	ND	CTC were cultured in vitro for downstream applications for 7–28 days. The captured cancer cells displayed plasticity.	ND
Shao 2014 [46]	40 (26 T2, 13 T3, 1 Tx)	Near-infrared dyes + FACS	≥1 CTC /7.5 mL	39 (97.5%)	10 (0–439)	Blood samples collected preoperatively. Live CTC evidenced by staining with heptamethine carbocyanine dyes	No
Pal 2015 [47]	35 (32 T1–T2, 3 T3)	CellSearch	≥1 CTC /22.5 mL	16 (45%)	0 (0–0.1)	Blood samples drawn before and after surgery. Median follow-up 510 days.	No
Thalgott 2015 [48]	15 (1 T2, 14 T3)	CellSearch	≥1 CTC /20 mL	3 (20%)	0 (0–0.2)	15 healthy controls (0% positive). Median follow-up 44.3 months.	No

Table 1. *Cont.*

Study [ref]	N° of Patients (pT Stage)	CTC Method	Cutoff	N° of CTC+ Patients (%)	Median CTC/mL (Range/mL)	Comments	Clinical Impact of CTC
Meyer 2016 [49]	152 (95 T2, 40 T3a, 17 T3b)	CellSearch	≥1 CTC /7.5 mL	17 (11%)	0.13 (0.13–13.3)	Blood samples collected preoperatively. Median follow-up 48 months.	No
Toden-höfer 2016 [50]	50 (37 T2, 13 T3)	Microfluidic (size, deforma-bility)	≥1 CTC /2 mL	25 (50%)	4.5 (0.5–208.5)	Pancytokeratin positive CTC showed expression of androgen receptor.	No
Kuske 2016 [51]	86 (37 T1, 45 T2, 4 T3)	CellSearch EPISPOT CellCollector	≥1 CTC /7.5 mL	37% CS 54.9% CC 58.7% EPI	0.24 (0.13–1.3) CS 0.32 (0.13–1.6) CC 0.4 (0.13–1.7) EPI	Blood drawn preoperatively. CTC detected by EPISPOT correlated to tumor stage, no correlation found with CellSearch (CS) or CellCollector (CC)	Yes
Tsumura 2017 [52]	59 (26 T1c–T2a, 15 T2b-c, 17 T3, 1 T4)	CellSearch	≥1 CTC /7.5 mL	0 (0%) preoperative 7 (11.8%) intraoperative	ND	Blood drawn both before and during surgery, with detection of CTC only during surgery.	No
Garcia 2017 [53]	16 (ND)	AR-V7 protein in serum samples	AR-V7 protein detection	3 (18.7%)	NA	CD133 expression in CTC was higher among AR-V7 positive cases vs. AR-V7 negative	ND
Puche-Sanz 2017 [54]	86 (T1–T2)	CK immune-magnetic	≥1 CTC /10 mL	16 (18.6%)	0 (0–0.4)	Blood samples collected before biopsy. Analysis of AR expression in tumor tissue.	Yes
Awe 2017 [55]	41 (T1–T4)	Filtration + immunostaining CK, CD45, AR	≥1 CTC /3 mL	41 (100%)	ND	Blood samples collected before prostatectomy	ND
Renier 2017 [56]	1 (ND)	Microfluidic vortex chip (size-based) + immuno-staining CK, CD45, PSA	>3.37 CTC/7.5 mL = >0.45 CTC/mL	1 (100%)	1.5	Some double positive cells (CK+, CD45+) found but counted as WBC. Some cells did not express epithelial markers (CK, PSA) but mesenchymal instead (Vim, N-cad)	ND
Russo 2018 [57]	47 (31 T2, 16 T3a)	AdnaTest Prostate-Cancer	0.15 ng /µL for AR, c-kit, c-met, ALDH1, TYMS. 0.25 ng /µL for Akt-2 & PI3Kα	12 (25.5%)	NA	Blood samples drawn before prostatectomy. No healthy controls tested.	Yes
Miyamoto 2018 [58]	34 (22 T1, 11 T2, 1 T3)	CTC-iChip + WTA + multiplex (8 genes) ddPCR	Mean CTC in healthy + 2SD of CTC score in healthy	ND	NA	Blood samples collected before surgery; 34 age-matched healthy donors included.	Yes
Murray 2018 [59]	241 (181 low risk + 60 intermed-iate risk)	Density gradient + PSA ICC	≥1 CTC/8 mL	37 low risk (20.4%) 26 intermedi-ate risk (43.3%)	ND	Blood samples collected 3 months after radiotherapy and stored 48 h at 4 °C.	Yes

[1] ND = Not described. [2] NA = Not applicable. pT stage = Pathological tumor stage. CTC = Circulating tumor cells. PSA = Prostate-specific antigen. PSMA = Prostate-specific membrane antigen. PSCA = Prostate stem cell antigen. AAM = African Americans. CAM = Caucasian Americans. RT-PCR = Reverse transcription-polymerase chain reaction. EpCAM = Epithelial cell adhesion molecule. AOX = Antioxydant genes. FACS = Fluorescence-activated cell sorting. AR = Androgen receptor. AR-V7 = Androgen receptor splice variant seven. CK = Cytokeratins. Vim = Vimentin. N-Cad = N-Cadherin. WTA = Whole transcriptome amplification. ddPCR = droplet digital polymerase chain reaction. ICC = Imunocytochemistry. 2SD = Two times the standard deviation.

3. CTC Detection in Non-Metastatic Prostate Cancer

In light of the published data, the extensive variability of CTC detection results in the context of localized PCa appears related to the diversity of distinct methods used for CTC collection/detection. Therefore, the present review will classify the published results depending on the CTC collection/detection methods used.

3.1. CTC Detection by RT-PCR in Non-Metastatic Prostate Cancer

Reverse transcription-polymerase chain reaction (RT-PCR) is commonly used to generate amplified cDNA from target mRNA [60]. Therefore, the RT-PCR results largely depend on the specific mRNA that is targeted. As a surrogate test for CTC detection in localized PCa, RT-PCR has mainly been used to target the mRNA of prostate-specific antigen (PSA), prostate-specific membrane antigen (PSMA) and prostate stem cell antigen (PSCA). The only study having compared those three targets as surrogate markers of CTC in localized PCa reported that the detection of PSCA mRNA in blood was the most accurate preoperative predictor of disease-free survival (DFS), probably because PSCA was the only mRNA not detected in 71 non-malignant disorders (PSA detected in 1 and PSMA in 2 of 71 non-malignant disorders) [29]. However, the very low detection rate of PSCA mRNA in that study (detected in only 1 of 43 localized PCa patients) begs caution when interpreting the results. Additionally, Joung et al. reported no association of PSCA mRNA detection with clinical variables on a larger cohort of localized PCa patients [39].

Most studies have focused on PSA and PSMA detection in localized PCa and have yielded somewhat contradictory results. Sabile et al. reported that the density gradient separation of mononuclear cells had a higher isolation efficiency than epithelial marker-dependent immunocapture for CTC detection based on PSA RT-PCR [23]. However, by using density gradient separation and PSA RT-PCR, Moreno et al. failed to detect CTC in 4 patients with localized PCa, possibly owing to distinct RT-PCR primer sequences [17]. By studying patients with a mean follow-up of 13.6 months, Olsson et al. determined that PSA mRNA detection was a significant predictor of disease recurrence after prostatectomy [19] while Mejean et al. found a statistical association of PSA-positive RT-PCR with metastasis and recurrence after a follow-up of 26 months [24]. Interestingly, the latter study also tested 11 patients with prostatitis and found positive PSA RT-PCR results in 2 of 11 cases (18%). Other studies including longer follow-up periods reported no correlation of PSA mRNA detection with clinical variables such as overall survival (OS), progression-free survival (PFS), Gleason score, tumor stage or preoperative serum PSA level [20,21,25,27,30,31,34]. Israeli et al. reported that PSMA RT-PCR was more sensitive than PSA RT-PCR in detecting hematogenous tumor cell dissemination but their results were not correlated to clinical variables [18]. Okegawa et al. compared the detection of PSA and PSMA mRNAs as prognostic indicators in a small cohort of 31 localized PCa and determined that PSMA mRNA detection in blood was a significant predictor of PFS after a mean of 16.7 months of follow-up [22]. In contrast, studies including larger casistics and longer follow-up times reported no significant correlation of PSMA mRNA detection with clinical variables [30,34]. Eschwege et al. argued that more specific PSMA RT-PCR primers should be used and that only dual PSA-PSMA-positive blood samples could be considered as reflecting the presence of CTC in blood [36]. Interestingly, the latter study included a rather large cohort of 155 localized PCa patients, more than 100 healthy controls, none of which tested positive for both PSA and PSMA, and 5-year follow-up data showing that the preoperative detection of both PSA and PSMA mRNAs in blood was an independent prognostic factor of disease recurrence [36]. Similarly, Yates et al. showed that both PSA and PSMA mRNA detection improved the prediction of biochemical recurrence over Kattan nomogram [40]. Slawin et al. took a slightly different approach by amplifying the human KLK2 gene, coding for an androgen-regulated protein (hK2) that has an 80% amino acid sequence identity with PSA [26]. Although the Authors determined that RT-PCR-hK2 results allowed for the prediction of lymph node-positive disease, the positivity of their test in 14% of 14 healthy controls indicates a lack of specificity of CTC detection via hK2 RT-PCR [26]. Multiplex RT-PCR approaches may be more efficient in identifying hematogenous prostatic cell dissemination. Yet, the major shortcoming of any RT-PCR approach as a surrogate marker for CTC detection is the related inability to count and further characterize CTC from blood. Furthermore, methodological variability related to different cell extraction methods, primers used for RT-PCR, controls of specificity and sensitivity and the timing of sample collection and storage is expected to account for the heterogeneity of the results obtained and their clinical relevance.

3.2. CTC Detection by CellSearch in Non-Metastatic Prostate Cancer

The CellSearch method uses the epithelial cell adhesion molecule (EpCAM) to capture circulating cells and defines CTC as nucleated cells (DAPI+) of epithelial (CK+) and non-hematopoietic (CD45−) origin, which, in fact, better corresponds to a definition of circulating epithelial cells (CEpC). It is important to note that CEpC have been found in the blood of patients with benign colon diseases [61] and benign pancreatic diseases [62]. The lack of specificity of the CellSearch method is exemplified by the finding of CTC in up to 20% of healthy donors tested [33].

The preoperative detection of CTC by CellSearch in localized PCa has been reported in 0% to 73% of patients, depending on the study (see Table 1). The fact that distinct cutoff values were used to define CTC positivity in those studies complicates the task to compare their results. For example, Davis et al. and Pal et al. chose to place the cutoff at 1 CTC per 22.5 mL of blood, corresponding to the finding of at least 1 CTC in 3 CellSearch samples of 7.5 mL each [33,47]. In contrast, the majority of studies using CellSearch to detect CTC in localized PCa have used a cutoff of 1 CTC in 7.5 mL [41–44,49,51]. None of the ten studies using CellSearch to detect CTC in localized PCa have reported a significant correlation of CTC numbers with clinical variables such as OS, PFS, Gleason score, tumor stage or preoperative serum PSA level.

3.3. CTC Detection by Other Marker-Dependent Methods in Non-Metastatic Prostate Cancer

A recent study comparing CellSearch with another EpCAM-dependent method (CellCollector) and the EPISPOT assay (based on the negative enrichment of CTC by leukocyte depletion) reported that only CTC detected by EPISPOT in 58.7% of patients were significantly correlated with clinical parameters such as PSA serum values ($p < 0.0001$) and the clinical tumor stage ($p = 0.04$) [51]. Using the EpCAM-dependent immune-magnetic enrichment of CTC followed by telomerase detection via an enzyme-linked immunosorbent assay, Fizazi et al. detected CTC in 70% of 83 localized PCa patients without false-positive results in 22 healthy controls tested [32]. Unfortunately, the latter study did not include any prognostic evaluation. The EpCAM-dependent microfluidic isolation of CTC has been reported by Stott et al., showing the detection of up to 222 CTC per mL of blood tested and a decline of CTC numbers in 6 of 8 patients 24 h after radical prostatectomy [38]. However, in the latter study, the finding of CTC in healthy controls implied the need for a cutoff value at 14 CTC per mL of blood. By using the AdnaTest, relying on EpCAM and MUC-1 antigens for immune-magnetic isolation and on RT-PCR of Androgen Receptor (AR), c-kit, c-met, ALDH1 and TYMPS for CTC detection, Russo et al. failed to demonstrate a significant association of CTC with clinical parameters [57]. In contrast, Puche-Sanz et al. used cytokeratin-mediated immune-magnetic enrichment of CTC and reported a significant correlation of CTC detection with AR expression in the tumor tissue [54]. The assessment of the AR-V7 splice variant protein in plasma, performed through a capillary nano-immunoassay platform was proposed by Garcia et al. as a surrogate marker for CTC in localized PCa patients [53]. Interestingly, the authors reported a significant correlation of AR-V7 detection with preoperative serum PSA levels and the expression of the stem cell marker CD133. However, the latter study did not perform a longitudinal follow-up of localized PCa patients for further prognostic evaluation [53]. Murray et al. took a different approach by using density gradient isolation and detection of CTC by PSA immunocytochemistry on a large cohort of localized PCa patients [59]. Importantly, the authors reported a significant correlation between CTC detection and clinical variables such as PFS after a long follow-up period of 15 years. However, Murray et al. stored the blood samples at 4 °C during 48 h before analyzing them, which could significantly impact the CTC detection results. Furthermore, the authors neither provided counting of the CTC nor exemplar CTC images.

3.4. CTC Detection Following Size-Based Isolation Methods in Non-Metastatic Prostate Cancer

To date, few studies have used size-based separation methods to study CTC from localized PCa patients. Giesing et al. were the first to use blood filtration followed by RT-PCR of PSA and a selection of antioxidant genes (AOX) to detect CTC in 42 localized PCa patients [37]. The authors determined that the detection of antioxidant gene expression in CTC could predict tumor diagnosis with 86% sensitivity and 82% specificity. A few years later, Kolostova et al. used MetaCell®filtration followed by a short-term in vitro culture to identify CTC in 28 of 55 localized PCa patients [45]. Unfortunately, no correlation was found between CTC detection and the clinical parameters. Similarly, Todenhöfer et al. failed to demonstrate a significant correlation with the clinical parameters of CTC detected by fluorescence imaging (EpCAM+ & CD45−) following microfluidic enrichment based on cell-size and deformability [50]. Interestingly, Renier et al. used a similar size-based microfluidic enrichment of CTC followed by the immunofluorescent detection of cytokeratins (CK), PSA and CD45 and reported that some cells did not express epithelial markers (CK) but mesenchymal markers instead (Vim, N-cad), thereby pointing to the process of epithelial to mesenchymal transition (EMT) in circulating prostate cells [56]. Awe et al. also reported on distinct subpopulations of CTC following size-based enrichment and immunostaining for cytokeratins, CD45 and the androgen receptor (AR) but they did not evaluate the clinical impact of those CTC [55]. Efficient risk stratification of localized PCa patients by means of a liquid biopsy was only recently achieved by Miyamoto et al. using the size-based microfluidic enrichment of CTC followed by whole transcriptome amplification and multiplex droplet digital PCR of a panel of 8 genes [58]. By using the differential weighting of 6 genes from the panel, the authors could predict early prostate cancer dissemination in localized disease [58].

4. Analysis of the Clinical Value of CTC Detection in Localized Prostate Cancer (Stages T1, T2)

Among the 43 studies targeting non-metastatic patients with PCa included in the present review, 31 investigated the potential clinical impact of CTC detection, looking for a statistical correlation between the detection of CTC and PCa clinical and/or pathological characteristics. Studies reporting on PCa with early (T1–T2) and advanced (T3–T4) stages but without a separate statistical analysis of T1–T2 cancers were further excluded. The details of the remaining 11 studies reporting on the analysis of the clinical value of CTC detection in localized (T1–T2) PCa are shown in Table 2. For clarity, the diagnostic value refers to a test's ability to identify a disease or a specific condition, with degrees of specificity and sensitivity to express its confidence and accuracy [63]. The predictive value refers to a test's ability to predict the patient´s response to a specific treatment while its prognostic value identifies risks of progression of the disease independently of a specific treatment [64].

Table 2. The clinical value of Circulating Tumor Cells' (CTC) detection in patients with localized prostate cancer.

Study [ref]	Method	N° of Patients	N° of CTC+ Patients (%)	Mean Follow-Up Period	Type of Clinical Value (*P* Value)	Comments
Olsson 1996 [19]	PSA RT-PCR	100 (cT1–cT2c)	74 (74%)	13.6 months	Predictive of surgical failure ($P < 0.0286$)	Correlation of RT-PCR results before prostatectomy with disease recurrence after prostatectomy.
Ennis 1997 [20]	PSA RT-PCR	156 (cT1–cT2)	ND	ND	Prognostic ($P < 0.0001$)	Correlation of RT-PCR results with pathological stage and prediction of extra-capsular disease.
Mejean 2000 [24]	PSA RT-PCR	79 (cT1–cT2)	ND	26 months	Predictive ($P < 0.04$)	CTC detection associated with the development of metastases and risk of relapse after prostatectomy.
Slawin 2000	hK2 RT-PCR	154 (pT1–pT2)	ND	ND	Prognostic ($P = 0.028$)	Association with the risk of metastasis to pelvic lymph nodes.
Bianco 2002 [28]	PSA RT-PCR	96 (35 pT1, 61 pT2)	25 (27%) African Americans	33 months	Prognostic ($P = 0.01$)	Association with tumor stage and recurrence in African-Americans.
Yates 2012 [40]	PSA & PSMA RT-PCR	92 (61 pT1, 31 pT2)	63 (68.5%) PSA 68 (78.9%) PSMA	72 months	Predictive ($P = 0.03$)	Improved prediction of biochemical recurrence.
Puche-Sanz 2017 [54]	CK immune-magnetic	86 (pT1–pT2)	16 (18.6%)	ND	Theranostic & Diagnostic ($P = 0.03$)	Expression of AR in tumor tissue correlated significantly with presence of CTC in blood. Diagnosis of PCa by CTC has a 14.2% sensitivity and a 78.4% specificity.
Llanes 2000 [25]	PSA RT-PCR	25 (pT1–pT2b)	7 (28%)	ND	Not significant	The best predictors of extraprostatic disease were the biopsy Gleason score and the PSA level.
Shariat 2002 [27]	PSA RT-PCR HK2L	224 (pT1–pT2)	54 (24%)	52.9 months	Not significant	Preoperative blood RT-PCR-PSA not associated with characteristics or outcomes of prostate cancer HK2L correlation with risk of metastases.
Thomas 2002 [30]	PSA & PSMA RT-PCR	136 (pT1–pT2)	73 (54%)	59 months	Not significant	RT-PCR status did not predict pathologic stage or biochemical failure.
Helo 2009 [34]	PSA & PSMA RT-PCR	87 (cT1–cT2)	6 (7%)	28 months	Not significant	No association between KLK mRNA status and unfavorable localized disease features.

ND = Not described; cT1–cT2 = clinical stages T1–T2; pT1–pT2 = pathological stages T1–T2.

The first conclusion of our literature review is that very few studies investigated the clinical impact of CTC selectively in patients with localized prostate cancer. Overall, there is a definite trend toward a value of CTC in correlation with the pathological stage and toward a prognostic and predictive impact of CTC detection in early-stage prostate cancer since several studies have reported significant correlations of CTC numbers with the survival of patients and/or recurrence of the disease after treatment [19,20,24,28,40]. However, those studies used PSA RT-PCR to detect CTC, a method which has yielded contradictory results in other studies [25,27,34], thereby calling for further validation of those results in large cohorts of localized PCa patients. The diagnostic value of CTC detection in early-stage prostate cancer has been less extensively investigated than in metastatic patients. In this regard, the most interesting results come from Giesing et al. with a CTC test demonstrating an 86%sensitivity and an 82% specificity with a 69% positive predictive value and a 92% negative predictive value for PCa diagnosis [37]. However, the latter study reported on both early-stage (T1–T2) and locally advanced (T3–T4) PCa. Puche-Sanz et al. have also investigated the possibility of a diagnostic CTC test for PCa [54]. Yet, with a 14.2% sensitivity and a 78.4% specificity, their test would not further improve on the PSA screening test. The lack of further investigation of a potential diagnostic CTC test could, in fact, stem from the substantial difficulty of detecting CTC in a consistent and specific manner in the context of localized PCa.

The presence of circulating prostatic cells in benign prostatic hyperplasia and prostatitis impacts the specificity of certain CTC isolation techniques such as PSA RT-PCR [24,29]. Furthermore, the phenotypic heterogeneity of CTC has been established in the context of metastatic PCa [65]. Additionally, the occurrence of a phenotypic transition (EMT) in CTC from early-stage PCa patients, evidenced by Renier et al. [56], supports the notion of phenotypic heterogeneity among CTC from localized PCa patients as well. The heterogeneity of CTC is relevant to the potential theranostic interest of various CTC tests. Particularly, the expression of the androgen receptor (AR) is of substantial importance for therapy strategy decision in PCa. In fact, a recent review evaluating clinical trials in the context of metastatic PCa determined that the expression of a specific variant of the androgen receptor (AR-V7) was significantly correlated to the limited efficacy of abiraterone and enzalutamide treatments compared with taxane therapy [66]. Whether such an association still holds true in the context of early-stage PCa remains to be demonstrated. Miyamoto et al. also reported a considerable heterogeneity among prostate CTC, including heterogeneous patterns of AR splice variant expression, following microfluidic enrichment and single-cell RNA-seq analyses [67]. So far, only two studies have reported correlations of CTC detection in localized PCa patients with AR expression [54,57]. Puche-Sanz et al. observed a direct association of the expression of AR in the prostatic tissue and the presence of CTC in blood [54]. Russo et al. determined that the expression of AR and TYMS on CTC are frequent events but the implications of such results for a personalized treatment strategy in PCa remain to be elucidated [57]. Further studies are needed to evaluate the potential theranostic utility of CTC in the context of localized PCa.

5. Perspectives and Future Directions

The major issue concerning localized prostate cancer is the lack of a suitable marker which could identify benign cases from aggressive prostate cancers. The present study shows a trend toward a possible clinical impact of CTC detection in patients with localized prostate cancer. Despite this trend, the study raises key issues in particular about the technical approaches used, the need for CTC counting and characterization, the sensitivity and specificity controls and the timing of blood sampling. Overall, our analysis encourages the development of a CTC cell-based specific test able to identify and count CTC in a highly sensitive and specific manner in patients with localized cancers and single-cell

CTC analyses in patients with localized prostate cancer to specifically study the CTC heterogeneity. It also stimulates the use of a standardized approach to be employed in large clinical studies.

References

1. Pernar, C.H.; Ebot, E.M.; Wilson, K.M.; Mucci, L.A. The Epidemiology of Prostate Cancer. *Cold Spring Harb. Perspect. Med.* **2018**, *8*. [CrossRef] [PubMed]
2. Siegel, R.L.; Miller, K.D.; Jemal, A. Cancer statistics, 2018. *CA Cancer J. Clin.* **2018**, *68*, 7–30. [CrossRef] [PubMed]
3. Daniyal, M.; Siddiqui, Z.A.; Akram, M.; Asif, H.M.; Sultana, S.; Khan, A. Epidemiology, etiology, diagnosis and treatment of prostate cancer. *Asian Pac. J. Cancer Prev.* **2014**, *15*, 9575–9578. [CrossRef] [PubMed]
4. Cancer of the Prostate—Cancer Stat Facts. Available online: https://seer.cancer.gov/statfacts/html/prost.html (accessed on 31 May 2019).
5. Jahn, J.L.; Giovannucci, E.L.; Stampfer, M.J. The high prevalence of undiagnosed prostate cancer at autopsy: implications for epidemiology and treatment of prostate cancer in the Prostate-specific Antigen-era. *Int. J. Cancer* **2015**, *137*, 2795–2802. [CrossRef] [PubMed]
6. Baca, S.C.; Prandi, D.; Lawrence, M.S.; Mosquera, J.M.; Romanel, A.; Drier, Y.; Park, K.; Kitabayashi, N.; MacDonald, T.Y.; Ghandi, M.; et al. Punctuated evolution of prostate cancer genomes. *Cell* **2013**, *153*, 666–677. [CrossRef] [PubMed]
7. Heidenreich, A.; Bastian, P.J.; Bellmunt, J.; Bolla, M.; Joniau, S.; van der Kwast, T.; Mason, M.; Matveev, V.; Wiegel, T.; Zattoni, F.; et al. EAU guidelines on prostate cancer. part 1: Screening, diagnosis, and local treatment with curative intent-update 2013. *Eur. Urol.* **2014**, *65*, 124–137. [CrossRef] [PubMed]
8. Holmberg, L.; Bill-Axelson, A.; Helgesen, F.; Salo, J.O.; Folmerz, P.; Häggman, M.; Andersson, S.-O.; Spångberg, A.; Busch, C.; Nordling, S.; et al. A randomized trial comparing radical prostatectomy with watchful waiting in early prostate cancer. *N. Engl. J. Med.* **2002**, *347*, 781–789. [CrossRef] [PubMed]
9. Barry, M.J.; Simmons, L.H. Prevention of Prostate Cancer Morbidity and Mortality: Primary Prevention and Early Detection. *Med. Clin. N. Am.* **2017**, *101*, 787–806. [CrossRef]
10. Akizhanova, M.; Iskakova, E.E.; Kim, V.; Wang, X.; Kogay, R.; Turebayeva, A.; Sun, Q.; Zheng, T.; Wu, S.; Miao, L.; et al. PSA and Prostate Health Index based prostate cancer screening in a hereditary migration complicated population: Implications in precision diagnosis. *J. Cancer* **2017**, *8*, 1223–1228. [CrossRef]
11. Galper, S.L.; Chen, M. H.; Catalona, W.J.; Roehl, K.A.; Richie, J.P.; D'Amico, A.V. Evidence to support a continued stage migration and decrease in prostate cancer specific mortality. *J. Urol.* **2006**, *175*, 907–912. [CrossRef]
12. Tsodikov, A.; Gulati, R.; Heijnsdijk, E.A.M.; Pinsky, P.F.; Moss, S.M.; Qiu, S.; de Carvalho, T.M.; Hugosson, J.; Berg, C.D.; Auvinen, A.; et al. Reconciling the Effects of Screening on Prostate Cancer Mortality in the ERSPC and PLCO Trials. *Ann. Intern. Med.* **2017**, *167*, 449–455. [CrossRef] [PubMed]
13. Pound, C.R.; Partin, A.W.; Eisenberger, M.A.; Chan, D.W.; Pearson, J.D.; Walsh, P.C. Natural history of progression after PSA elevation following radical prostatectomy. *JAMA* **1999**, *281*, 1591–1597. [CrossRef] [PubMed]
14. Doyen, J.; Alix-Panabières, C.; Hofman, P.; Parks, S.K.; Chamorey, E.; Naman, H.; Hannoun-Lévi, J.-M. Circulating tumor cells in prostate cancer: A potential surrogate marker of survival. *Crit. Rev. Oncol. Hematol.* **2012**, *81*, 241–256. [CrossRef] [PubMed]
15. Hegemann, M.; Stenzl, A.; Bedke, J.; Chi, K.N.; Black, P.C.; Todenhöfer, T. Liquid biopsy: Ready to guide therapy in advanced prostate cancer? *BJU Int.* **2016**, *118*, 855–863. [CrossRef] [PubMed]
16. Borley, N.; Feneley, M.R. Prostate cancer: Diagnosis and staging. *Asian J. Androl.* **2009**, *11*, 74–80. [CrossRef] [PubMed]
17. Moreno, J.G.; Croce, C.M.; Fischer, R.; Monne, M.; Vihko, P.; Mulholland, S.G.; Gomella, L.G. Detection of hematogenous micrometastasis in patients with prostate cancer. *Cancer Res.* **1992**, *52*, 6110–6112.
18. Israeli, R.S.; Miller, W.H.; Su, S.L.; Samadi, D.S.; Powell, C.T.; Heston, W.D.; Wise, G.J.; Fair, W.R. Sensitive detection of prostatic hematogenous tumor cell dissemination using prostate specific antigen and prostate specific membrane-derived primers in the polymerase chain reaction. *J. Urol.* **1995**, *153*, 573–577.

19. Olsson, C.A.; de Vries, G.M.; Raffo, A.J.; Benson, M.C.; O'Toole, K.; Cao, Y.; Buttyan, R.E.; Katz, A.E. Preoperative reverse transcriptase polymerase chain reaction for prostate specific antigen predicts treatment failure following radical prostatectomy. *J. Urol.* **1996**, *155*, 1557–1562. [CrossRef]
20. Ennis, R.D.; Katz, A.E.; de Vries, G.M.; Heitjan, D.F.; O'Toole, K.M.; Rubin, M.; Buttyan, R.; Benson, M.C.; Schiff, P.B. Detection of circulating prostate carcinoma cells via an enhanced reverse transcriptase-polymerase chain reaction assay in patients with early stage prostate carcinoma. Independence from other pretreatment characteristics. *Cancer* **1997**, *79*, 2402–2408. [CrossRef]
21. Oefelein, M.G.; Ignatoff, J.M.; Clemens, J.Q.; Watkin, W.; Kaul, K.L. Clinical and molecular followup after radical retropubic prostatectomy. *J. Urol.* **1999**, *162*, 307–310. [CrossRef]
22. Okegawa, T.; Nutahara, K.; Higashihara, E. Preoperative nested reverse transcription-polymerase chain reaction for prostate specific membrane antigen predicts biochemical recurrence after radical prostatectomy. *BJU Int.* **1999**, *84*, 112–117. [CrossRef] [PubMed]
23. Sabile, A.; Louha, M.; Bonte, E.; Poussin, K.; Vona, G.; Mejean, A.; Chretien, Y.; Bougas, L.; Lacour, B.; Capron, F.; et al. Efficiency of Ber-EP4 antibody for isolating circulating epithelial tumor cells before RT-PCR detection. *Am. J. Clin. Pathol.* **1999**, *112*, 171–178. [CrossRef] [PubMed]
24. Mejean, A.; Vona, G.; Nalpas, B.; Damotte, D.; Brousse, N.; Chretien, Y.; Dufour, B.; Lacour, B.; Bréchot, C.; Paterlini-Bréchot, P. Detection of circulating prostate derived cells in patients with prostate adenocarcinoma is an independent risk factor for tumor recurrence. *J. Urol.* **2000**, *163*, 2022–2029. [CrossRef]
25. Llanes, L.; Páez, A.; Ferruelo, A.; Luján, M.; Romero, I.; Berenguer, A. Detecting circulating prostate cells in patients with clinically localized prostate cancer: Clinical implications for molecular staging. *BJU Int.* **2000**, *86*, 1023–1027. [CrossRef] [PubMed]
26. Slawin, K.M.; Shariat, S.F.; Nguyen, C.; Leventis, A.K.; Song, W.; Kattan, M.W.; Young, C.Y.; Tindall, D.J.; Wheeler, T.M. Detection of metastatic prostate cancer using a splice variant-specific reverse transcriptase-polymerase chain reaction assay for human glandular kallikrein. *Cancer Res.* **2000**, *60*, 7142–7148. [PubMed]
27. Shariat, S.F.; Gottenger, E.; Nguyen, C.; Song, W.; Kattan, M.W.; Andenoro, J.; Wheeler, T.M.; Spencer, D.M.; Slawin, K.M. Preoperative blood reverse transcriptase-PCR assays for prostate-specific antigen and human glandular kallikrein for prediction of prostate cancer progression after radical prostatectomy. *Cancer Res.* **2002**, *62*, 5974–5979. [PubMed]
28. Bianco, F.J.; Powell, I.J.; Cher, M.L.; Wood, D.P. Presence of circulating prostate cancer cells in African American males adversely affects survival. *Urol. Oncol.* **2002**, *7*, 147–152. [CrossRef]
29. Hara, N.; Kasahara, T.; Kawasaki, T.; Bilim, V.; Obara, K.; Takahashi, K.; Tomita, Y. Reverse transcription-polymerase chain reaction detection of prostate-specific antigen, prostate-specific membrane antigen, and prostate stem cell antigen in one milliliter of peripheral blood: Value for the staging of prostate cancer. *Clin. Cancer Res.* **2002**, *8*, 1794–1799.
30. Thomas, J.; Gupta, M.; Grasso, Y.; Reddy, C.A.; Heston, W.D.; Zippe, C.; Dreicer, R.; Kupelian, P.A.; Brainard, J.; Levin, H.S.; et al. Preoperative combined nested reverse transcriptase polymerase chain reaction for prostate-specific antigen and prostate-specific membrane antigen does not correlate with pathologic stage or biochemical failure in patients with localized prostate cancer undergoing radical prostatectomy. *J. Clin. Oncol.* **2002**, *20*, 3213–3218.
31. Gewanter, R.M.; Katz, A.E.; Olsson, C.A.; Benson, M.C.; Singh, A.; Schiff, P.B.; Ennis, R.D. RT-PCR for PSA as a prognostic factor for patients with clinically localized prostate cancer treated with radiotherapy. *Urology* **2003**, *61*, 967–971. [CrossRef]
32. Fizazi, K.; Morat, L.; Chauveinc, L.; Prapotnich, D.; De Crevoisier, R.; Escudier, B.; Cathelineau, X.; Rozet, F.; Vallancien, G.; Sabatier, L.; et al. High detection rate of circulating tumor cells in blood of patients with prostate cancer using telomerase activity. *Ann. Oncol.* **2007**, *18*, 518–521. [CrossRef] [PubMed]
33. Davis, J.W.; Nakanishi, H.; Kumar, V.S.; Bhadkamkar, V.A.; McCormack, R.; Fritsche, H.A.; Handy, B.; Gornet, T.; Babaian, R.J. Circulating tumor cells in peripheral blood samples from patients with increased serum prostate specific antigen: Initial results in early prostate cancer. *J. Urol.* **2008**, *179*, 2187–2191. [CrossRef] [PubMed]

34. Helo, P.; Cronin, A.M.; Danila, D.C.; Wenske, S.; Gonzalez-Espinoza, R.; Anand, A.; Koscuiszka, M.; Väänänen, R.-M.; Pettersson, K.; Chun, F.K.-H.; et al. Circulating prostate tumor cells detected by reverse transcription-PCR in men with localized or castration-refractory prostate cancer: Concordance with CellSearch assay and association with bone metastases and with survival. *Clin. Chem.* **2009**, *55*, 765–773. [CrossRef] [PubMed]
35. Maestro, L.M.; Sastre, J.; Rafael, S.B.; Veganzones, S.B.; Vidaurreta, M.; Martín, M.; Olivier, C.; de La Orden, V.B.; Garcia-Saenz, J.A.; Alfonso, R.; et al. Circulating tumor cells in solid tumor in metastatic and localized stages. *Anticancer Res.* **2009**, *29*, 4839–4843. [PubMed]
36. Eschwège, P.; Moutereau, S.; Droupy, S.; Douard, R.; Gala, J.-L.; Benoit, G.; Conti, M.; Manivet, P.; Loric, S. Prognostic value of prostate circulating cells detection in prostate cancer patients: A prospective study. *Br. J. Cancer* **2009**, *100*, 608–610. [CrossRef] [PubMed]
37. Giesing, M.; Suchy, B.; Driesel, G.; Molitor, D. Clinical utility of antioxidant gene expression levels in circulating cancer cell clusters for the detection of prostate cancer in patients with prostate-specific antigen levels of 4–10 ng/mL and disease prognostication after radical prostatectomy. *BJU Int.* **2010**, *105*, 1000–1010. [CrossRef] [PubMed]
38. Stott, S.L.; Lee, R.J.; Nagrath, S.; Yu, M.; Miyamoto, D.T.; Ulkus, L.; Inserra, E.J.; Ulman, M.; Springer, S.; Nakamura, Z.; et al. Isolation and characterization of circulating tumor cells from patients with localized and metastatic prostate cancer. *Sci. Transl. Med.* **2010**, *2*, 25ra23. [CrossRef] [PubMed]
39. Joung, J.Y.; Cho, K.S.; Kim, J.E.; Seo, H.K.; Chung, J.; Park, W.S.; Choi, M.K.; Lee, K.H. Prostate stem cell antigen mRNA in peripheral blood as a potential predictor of biochemical recurrence in high-risk prostate cancer. *J. Surg. Oncol.* **2010**, *101*, 145–148. [CrossRef]
40. Yates, D.R.; Rouprêt, M.; Drouin, S.J.; Comperat, E.; Ricci, S.; Lacave, R.; Sèbe, P.; Cancel-Tassin, G.; Bitker, M.-O.; Cussenot, O. Quantitative RT-PCR analysis of PSA and prostate-specific membrane antigen mRNA to detect circulating tumor cells improves recurrence-free survival nomogram prediction after radical prostatectomy. *Prostate* **2012**, *72*, 1382–1388. [CrossRef]
41. Lowes, L.E.; Lock, M.; Rodrigues, G.; D'Souza, D.; Bauman, G.; Ahmad, B.; Venkatesan, V.; Allan, A.L.; Sexton, T. Circulating tumour cells in prostate cancer patients receiving salvage radiotherapy. *Clin. Transl. Oncol.* **2012**, *14*, 150–156. [CrossRef]
42. Khurana, K.K.; Grane, R.; Borden, E.C.; Klein, E.A. Prevalence of circulating tumor cells in localized prostate cancer. *Curr. Urol.* **2013**, *7*, 65–69. [CrossRef] [PubMed]
43. Thalgott, M.; Rack, B.; Maurer, T.; Souvatzoglou, M.; Eiber, M.; Kreß, V.; Heck, M.M.; Andergassen, U.; Nawroth, R.; Gschwend, J.E.; et al. Detection of circulating tumor cells in different stages of prostate cancer. *J. Cancer Res. Clin. Oncol.* **2013**, *139*, 755–763. [CrossRef] [PubMed]
44. Loh, J.; Jovanovic, L.; Lehman, M.; Capp, A.; Pryor, D.; Harris, M.; Nelson, C.; Martin, J. Circulating tumor cell detection in high-risk non-metastatic prostate cancer. *J. Cancer Res. Clin. Oncol.* **2014**, *140*, 2157–2162. [CrossRef] [PubMed]
45. Kolostova, K.; Broul, M.; Schraml, J.; Cegan, M.; Matkowski, R.; Fiutowski, M.; Bobek, V. Circulating tumor cells in localized prostate cancer: Isolation, cultivation in vitro and relationship to T-stage and Gleason score. *Anticancer Res.* **2014**, *34*, 3641–3646. [PubMed]
46. Shao, C.; Liao, C.-P.; Hu, P.; Chu, C.-Y.; Zhang, L.; Bui, M.H.T.; Ng, C.S.; Josephson, D.Y.; Knudsen, B.; Tighiouart, M.; et al. Detection of live circulating tumor cells by a class of near-infrared heptamethine carbocyanine dyes in patients with localized and metastatic prostate cancer. *PLoS ONE* **2014**, *9*, e88967. [CrossRef] [PubMed]
47. Pal, S.K.; He, M.; Wilson, T.; Liu, X.; Zhang, K.; Carmichael, C.; Torres, A.; Hernandez, S.; Lau, C.; Agarwal, N.; et al. Detection and phenotyping of circulating tumor cells in high-risk localized prostate cancer. *Clin. Genitourin. Cancer* **2015**, *13*, 130–136. [CrossRef] [PubMed]
48. Thalgott, M.; Rack, B.; Horn, T.; Heck, M.M.; Eiber, M.; Kübler, H.; Retz, M.; Gschwend, J.E.; Andergassen, U.; Nawroth, R. Detection of Circulating Tumor Cells in Locally Advanced High-risk Prostate Cancer During Neoadjuvant Chemotherapy and Radical Prostatectomy. *Anticancer Res.* **2015**, *35*, 5679–5685.

49. Meyer, C.P.; Pantel, K.; Tennstedt, P.; Stroelin, P.; Schlomm, T.; Heinzer, H.; Riethdorf, S.; Steuber, T. Limited prognostic value of preoperative circulating tumor cells for early biochemical recurrence in patients with localized prostate cancer. *Urol. Oncol.* **2016**, *34*, 235.e11–235.e16. [CrossRef]
50. Todenhöfer, T.; Park, E.S.; Duffy, S.; Deng, X.; Jin, C.; Abdi, H.; Ma, H.; Black, P.C. Microfluidic enrichment of circulating tumor cells in patients with clinically localized prostate cancer. *Urol. Oncol.* **2016**, *34*, 483.e9–483.e16. [CrossRef]
51. Kuske, A.; Gorges, T.M.; Tennstedt, P.; Tiebel, A.-K.; Pompe, R.; Preißer, F.; Prues, S.; Mazel, M.; Markou, A.; Lianidou, E.; et al. Improved detection of circulating tumor cells in non-metastatic high-risk prostate cancer patients. *Sci. Rep.* **2016**, *6*, 39736. [CrossRef]
52. Tsumura, H.; Satoh, T.; Ishiyama, H.; Tabata, K.-I.; Takenaka, K.; Sekiguchi, A.; Nakamura, M.; Kitano, M.; Hayakawa, K.; Iwamura, M. Perioperative Search for Circulating Tumor Cells in Patients Undergoing Prostate Brachytherapy for Clinically Nonmetastatic Prostate Cancer. *Int. J. Mol. Sci.* **2017**, *18*. [CrossRef] [PubMed]
53. García, J.L.; Lozano, R.; Misiewicz-Krzeminska, I.; Fernández-Mateos, J.; Krzeminski, P.; Alfonso, S.; Marcos, R.A.; García, R.; Gómez-Veiga, F.; Virseda, Á.; et al. A novel capillary nano-immunoassay for assessing androgen receptor splice variant 7 in plasma. Correlation with CD133 antigen expression in circulating tumor cells. A pilot study in prostate cancer patients. *Clin. Transl. Oncol.* **2017**, *19*, 1350–1357. [CrossRef] [PubMed]
54. Puche-Sanz, I.; Alvarez-Cubero, M.J.; Pascual-Geler, M.; Rodríguez-Martínez, A.; Delgado-Rodríguez, M.; García-Puche, J.L.; Expósito, J.; Robles-Fernández, I.; Entrala-Bernal, C.; Lorente, J.A.; et al. A comprehensive study of circulating tumour cells at the moment of prostate cancer diagnosis: Biological and clinical implications of EGFR, AR and SNPs. *Oncotarget* **2017**, *8*, 70472–70480. [CrossRef] [PubMed]
55. Awe, J.A.; Saranchuk, J.; Drachenberg, D.; Mai, S. Filtration-based enrichment of circulating tumor cells from all prostate cancer risk groups. *Urol. Oncol.* **2017**, *35*, 300–309. [CrossRef] [PubMed]
56. Renier, C.; Pao, E.; Che, J.; Liu, H.E.; Lemaire, C.A.; Matsumoto, M.; Triboulet, M.; Srivinas, S.; Jeffrey, S.S.; Rettig, M.; et al. Label-free isolation of prostate circulating tumor cells using Vortex microfluidic technology. *NPJ Precis. Oncol.* **2017**, *1*, 15. [CrossRef]
57. Russo, G.I.; Bier, S.; Hennenlotter, J.; Beger, G.; Pavlenco, L.; van de Flierdt, J.; Hauch, S.; Maas, M.; Walz, S.; Rausch, S.; et al. Expression of tumour progression-associated genes in circulating tumour cells of patients at different stages of prostate cancer. *BJU Int.* **2018**, *122*, 152–159. [CrossRef] [PubMed]
58. Miyamoto, D.T.; Lee, R.J.; Kalinich, M.; LiCausi, J.A.; Zheng, Y.; Chen, T.; Milner, J.D.; Emmons, E.; Ho, U.; Broderick, K.; et al. An RNA-Based Digital Circulating Tumor Cell Signature Is Predictive of Drug Response and Early Dissemination in Prostate Cancer. *Cancer Discov.* **2018**, *8*, 288–303. [CrossRef]
59. Murray, N.P.; Aedo, S.; Fuentealba, C.; Reyes, E.; Minzer, S.; Salazar, A. The presence of secondary circulating prostate tumour cells determines the risk of biochemical relapse for patients with low- and intermediate-risk prostate cancer who are treated only with external radiotherapy. *Ecancermedicalscience* **2018**, *12*, 844. [CrossRef]
60. Bachman, J. Reverse-transcription PCR (RT-PCR). *Methods Enzymol.* **2013**, *530*, 67–74.
61. Pantel, K.; Denève, E.; Nocca, D.; Coffy, A.; Vendrell, J.-P.; Maudelonde, T.; Riethdorf, S.; Alix-Panabières, C. Circulating epithelial cells in patients with benign colon diseases. *Clin. Chem.* **2012**, *58*, 936–940. [CrossRef]
62. Poruk, K.E.; Valero, V.; He, J.; Ahuja, N.; Cameron, J.L.; Weiss, M.J.; Lennon, A.M.; Goggins, M.; Wood, L.D.; Wolfgang, C.L. Circulating Epithelial Cells in Intraductal Papillary Mucinous Neoplasms and Cystic Pancreatic Lesions. *Pancreas* **2017**, *46*, 943–947. [CrossRef] [PubMed]
63. NCI Dictionary of Cancer Terms. Available online: https://www.cancer.gov/publications/dictionaries/cancer-terms (accessed on 31 May 2019).
64. FDA-NIH Biomarker Working Group. *Understanding Prognostic versus Predictive Biomarkers*; Food and Drug Administration (US): Silver Spring, MD, USA, 2016.
65. Scher, H.I.; Graf, R.P.; Schreiber, N.A.; McLaughlin, B.; Jendrisak, A.; Wang, Y.; Lee, J.; Greene, S.; Krupa, R.; Lu, D.; et al. Phenotypic Heterogeneity of Circulating Tumor Cells Informs Clinical Decisions between AR Signaling Inhibitors and Taxanes in Metastatic Prostate Cancer. *Cancer Res.* **2017**, *77*, 5687–5698. [CrossRef] [PubMed]

66. Sciarra, A.; Gentilucci, A.; Silvestri, I.; Salciccia, S.; Cattarino, S.; Scarpa, S.; Gatto, A.; Frantellizzi, V.; Von Heland, M.; Ricciuti, G.P.; et al. Androgen receptor variant 7 (AR-V7) in sequencing therapeutic agents for castratrion resistant prostate cancer: A critical review. *Medicine (Baltim.)* **2019**, *98*, e15608. [CrossRef] [PubMed]
67. Miyamoto, D.T.; Ting, D.T.; Toner, M.; Maheswaran, S.; Haber, D.A. Single-Cell Analysis of Circulating Tumor Cells as a Window into Tumor Heterogeneity. *Cold Spring Harb. Symp. Quant. Biol.* **2016**, *81*, 269–274. [CrossRef] [PubMed]

15

Fibronectin Regulation of Integrin B1 and SLUG in Circulating Tumor Cells

Jeannette Huaman [1,2], Michelle Naidoo [1,2], Xingxing Zang [3] and Olorunseun O. Ogunwobi [1,2,4,*]

1 Department of Biological Sciences, Hunter College of The City University of New York, New York, NY 10065, USA; JHUAMAN@genectr.hunter.cuny.edu (J.H.); michelle.naidoo86@myhunter.cuny.edu (M.N.)
2 Department of Biology, The Graduate Center of The City University of New York, New York, NY 10016, USA
3 Departments of Microbiology and Immunology, and Medicine (Oncology), Albert Einstein College of Medicine, Bronx, NY 10461, USA; xingxing.zang@einstein.yu.edu
4 Department of Medicine, Weill Cornell Medicine, Cornell University, New York, NY 10065, USA
* Correspondence: ogunwobi@genectr.hunter.cuny.edu

Abstract: Metastasis is the leading cause of cancer death worldwide. Circulating tumor cells (CTCs) are a critical step in the metastatic cascade and a good tool to study this process. We isolated CTCs from a syngeneic mouse model of hepatocellular carcinoma (HCC) and a human xenograft mouse model of castration-resistant prostate cancer (CRPC). From these models, novel primary tumor and CTC cell lines were established. CTCs exhibited greater migration than primary tumor-derived cells, as well as epithelial-to-mesenchymal transition (EMT), as observed from decreased E-cadherin and increased SLUG and fibronectin expression. Additionally, when fibronectin was knocked down in CTCs, integrin B1 and SLUG were decreased, indicating regulation of these molecules by fibronectin. Investigation of cell surface molecules and secreted cytokines conferring immunomodulatory advantage to CTCs revealed decreased major histocompatibility complex class I (MHCI) expression and decreased endostatin, C-X-C motif chemokine 5 (CXCL5), and proliferin secretion by CTCs. Taken together, these findings indicate that CTCs exhibit distinct characteristics from primary tumor-derived cells. Furthermore, CTCs demonstrate enhanced migration in part through fibronectin regulation of integrin B1 and SLUG. Further study of CTC biology will likely uncover additional important mechanisms of cancer metastasis.

Keywords: metastasis; circulating tumor cells (CTCs); hepatocellular carcinoma (HCC); castration resistant prostate cancer (CRPC); epithelial-to-mesenchymal transition (EMT); fibronectin; integrin B1; SLUG; major histocompatibility complex class I (MHCI); immunomodulation

---

1. Introduction

Metastasis is associated with advanced stages of cancer. Resulting in 90% of cancer deaths worldwide [1], metastasis occurs in a series of steps. These steps include the dissociation of cells from the primary tumor, migration through surrounding tissue, intravasation, circulation through blood, followed by extravasation and re-colonization of distant sites throughout the body. At advanced stages of most cancers, there are limited treatment options [1–3]. As such, efforts are increasingly being focused on identification of novel metastasis-related molecular targets.

One way to potentially avoid the need for invasive tissue biopsies when studying cancer metastasis is through the use of circulating tumor cells (CTCs). CTCs are cells which have dissociated from the primary tumor and are found traveling in the blood [4–8]. Some CTCs will eventually form metastatic, secondary lesions. Because CTCs can be obtained from liquid biopsies (from blood), they enable the molecular profiling of potentially unresectable tumors in patients [9] and identification of molecular changes important for progression to advanced cancers [10]. However, there is a challenge with low

CTC numbers frequently found in the blood [11–15]. To address this potential obstacle to studying CTC biology, in this study, we established novel CTC cell lines and primary tumor-derived cells for molecular biological studies.

The two different cancer models used in this study were a syngeneic mouse model of hepatocellular carcinoma (HCC) and a xenograft mouse model of castration-resistant prostate cancer (CRPC). HCC is the most common form of liver cancer and is frequently diagnosed at very late stages. Consequently, it is one of the leading causes of cancer deaths worldwide [16–19]. Moreover, sorafenib, which is the main FDA approved drug to treat advanced HCC, extends life by only six months [20]. As such, better treatment options are needed. Similarly, CRPC is a form of prostate cancer (PCa) that is resistant to both medical and surgical castration [21,22]. However, androgen deprivation therapy (ADT) is the main standard of treatment for localized PCa [23,24]. This makes CRPC particularly challenging to treat. Over one third of CRPC patients will develop bone metastasis for which there is no cure [21,25]. Therefore, finding alternative treatments is critical for this cancer as well.

To this end, we propagated cell lines originating from primary tumors and CTCs. Our aim was to discover differences between these two cell types representing earlier and more advanced stages of cancers. Both HCC and CRPC CTCs demonstrate increased migration and evidence of epithelial-to-mesenchymal transition (EMT). Moreover, we discovered that in CTCs, fibronectin regulates integrin B1 and SLUG, which are known regulators of cell migration. Finally, we identified differences in CTC cell surface marker and cytokine secretion profiles that could have immunomodulatory implications. HCC CTCs had significantly reduced major histocompatibility complex class I (MHCI) expression, as well as significantly decreased secretion of endostatin, CXCL5, and proliferin as compared to primary tumor-derived cells. These findings may have implications for the function of metastatic cells and how they evade the immune system.

2. Materials and Methods

2.1. Cell Lines and Cell Culture

The BNL 1ME A 7R.1 cell line (purchased from ATCC), as well as the newly established primary tumor cell lines (TBOH1 and TBOH9) and circulating tumor cell lines (CBOH4 and CBOH9), were maintained in Dulbecco's Modified Eagle Media (DMEM) media supplemented with 10% fetal bovine serum, 1% penicillin/streptomycin, and L-glutamine. Trypsinization of cells occurred using 0.25% trypsin when 75–80% confluent.

The 22Rv1 cell line (purchased from ATCC), as well as the newly established primary tumor cell line T22OH and circulating tumor cell line C22OH, were maintained in Roswell Park Memorial Institute (RPMI) media supplemented with 10% fetal bovine serum and 1% penicillin/streptomycin. Trypsinization of cells occurred using 0.05% trypsin when 75–80% confluent. All cells lines were cultured in a 5% CO_2, 37 °C atmosphere.

2.2. Mouse Tumor Studies

Figure 1 shows the experimental mouse models used for this study and the subsequent establishment of novel cell lines. For the syngeneic mouse model of HCC, 7-week old, male, BALB/c mice were obtained from Taconic Biosciences Inc. Mice were implanted with 2.5×10^6 BNL 1ME A 7R.1 murine HCC cells. For the NOD scid gamma (NSG) human xenograft mouse model, 7-week old, male NSG mice were obtained from Jackson Laboratory. Mice were implanted with 2.5×10^6 22Rv1 human CRPC cells. Tumors were allowed to grow until reaching a max tumor volume of 2000 cm^3, at which point mice were euthanized, tumors removed, and samples of blood processed for CTCs as previously described by our lab [26]. Briefly, up to 1 mL of blood was obtained from intracardiac blood withdrawal from mice. The blood was spun down and the plasma removed. The rest of the blood sample, most importantly the buffy coat layer, which is where we expect our CTCs to be, was treated with red blood cell lysis buffer. After a series of spins and washes, samples were placed in media. Experiments on CTC cell lines were carried out for as long as forty-five passages (well over 6 months). Features remained consistent among different passages. While efficiency of cell line establishment was moderate to low, once established, the CTC cell lines exhibited high cell viability. In terms of our primary tumor cell lines, tumors were mechanically dissociated in media and given the chance to adhere to the plate to give rise to primary cell culture cell lines. H&E staining was performed to confirm metastasis to lungs using the core facilities at Albert Einstein School of Medicine. All mouse experiments were performed in compliance with Institutional Animal Care and Use Committee (IACUC)–approved protocols at Weill Cornell Medicine.

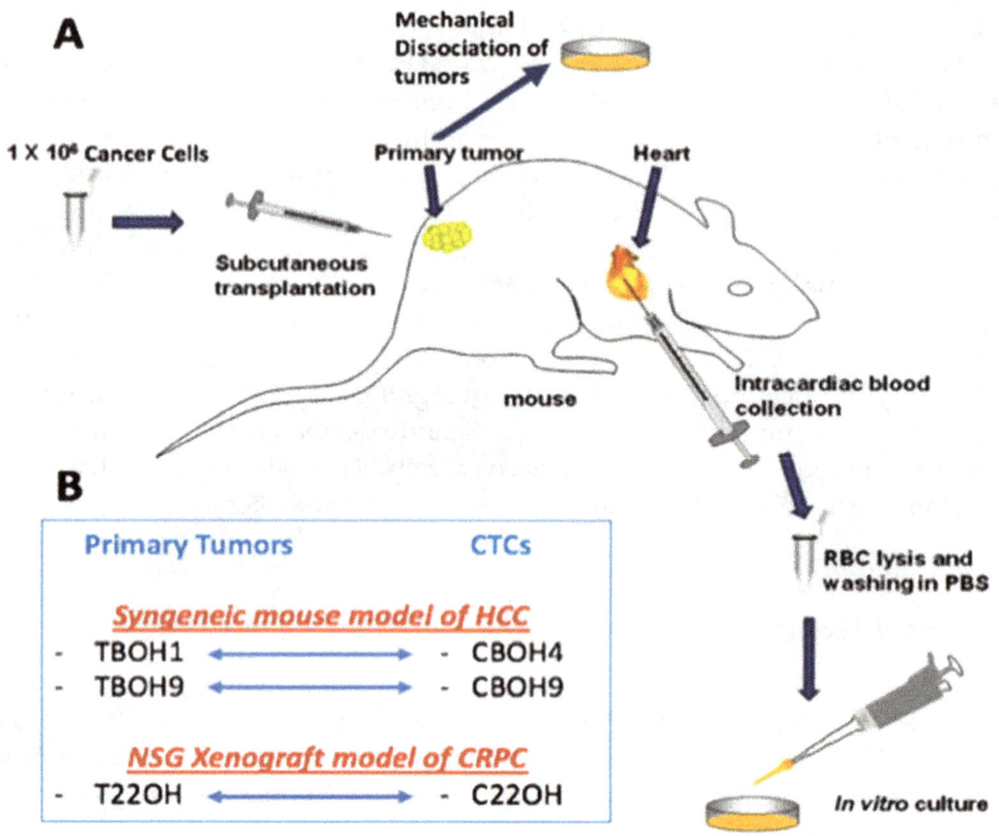

Figure 1. Establishment of novel primary tumor-derived cell lines and circulating tumor cell lines. (**A**) Schematic diagram summarizing how novel cell lines were established. This figure includes modifications to our previously described work [27]. For the syngeneic hepatocellular carcinoma (HCC) mouse model, BALB/c mice were subcutaneously implanted with the murine hepatocellular carcinoma cell line BNL 1ME A.7R.1. For the xenograft castration-resistant prostate cancer (CRPC) mouse model, NSG mice were subcutaneously implanted with the 22Rv1 human CRPC cell line. Mice were allowed to develop tumors over a period of 3–4 weeks, humanely sacrificed, primary tumors resected and mechanically dissociated, and then put in culture to establish the TBOH series of primary tumor-derived cell lines. The CBOH series of circulating tumor cells (CTCs) was established from cancer cells isolated from the bloodstream of the same mice implanted with either BNL 1ME A.7R.1 or 22Rv1 cells. (**B**) The newly established cell lines: the TBOH1 and CBOH4 pair and TBOH9 and CBOH9 pair were established from the HCC model. The T22OH and C22OH pair were established from the CRPC model.

2.3. *Immunofluorescence Staining*

Cells were incubated with coverslips and grown for 48 hr. The coverslips were collected, fixed with 4% paraformaldehyde for 15 min at room temperature, permeabilized with 0.5% Triton X-100 in 1X PBS/1% FBS for 10 min at room temperature. Cells were stained with CREB3L3 antibody (sc-377331; 1:500; Santa Cruz Biotechnology, TX, USA), or endostatin antibody (PA1-601; 1:200; Thermo Fisher

Scientific, Rockford, IL, USA) for 2 h, followed by incubation with Alexa Fluor 635 anti-rabbit secondary antibody (A31577; Thermo Fisher Scientific, Rockford, IL, USA) for 1 h. DAPI was used to counterstain nuclei, and slides were imaged using a Nikon A1 confocal microscope at a 60× magnification.

2.4. Migration Assays

For wound healing migration assays, 5×10^4 murine BNL 1ME A.7R.1 cells were grown in a 6-well tissue culture plate and permitted to reach 90–100% confluency. Using a plastic tip (1 mm thick), wounds were administered to monolayers of cells in each well. Wounded monolayers were washed with 1X PBS and incubated with media. Cells were observed, and images were taken using the Motic AE30 Inverted Microscope.

For transwell migration assays, 1×10^5 22Rv1 CRPC cells were seeded on the top of chambers containing 8 μm pores (Greiner Bio-one, Austria, cat #: 662 638) in serum-free media. The bottom chambers were filled with regular media to serve as a chemoattractant. After 48 hr, the top chambers were rinsed with 1X PBS, fixed with paraformaldehyde, treated with methanol, and stained with trypan blue staining. The chambers were placed on a slide and viewed using the Motic AE30 Inverted Microscope. All migration assays were carried out at least 3 times.

2.5. Protein Extraction and Western Blotting

Whole cell extracts were obtained by treating cells with radioimmunoprecipitation assay (RIPA) lysis buffer (Amresco, Ohio, USA, cat#: N653), supplemented with 10× protease inhibitors (Thermo Fisher Scientific, Rockford, IL, USA, cat#: 88665) and 100 mM PMSF (Amresco, OH, USA, cat#: M145). Protein concentration was calculated via the Bradford Assay using the Bio-Rad Protein Assay Dye Reagent Concentrate. For Western blot analysis, 30 μg of protein was run on precast SDS-PAGE gels and subsequently transferred onto nitrocellulose membranes. Membranes were blocked in 5% BSA in TBS-T for 1 h at room temperature, incubated with primary antibodies overnight at 4 °C, washed with 1× TBS-T, incubated with secondary antibodies for 1 h, washed, and imaged using the LI-COR Odyssey CLx imager with infrared fluorescence. The primary antibodies used were directed against fibronectin (ab2413; 1:1000; Abcam, Cambridge, UK), E-cadherin (3195S; 1:200; Cell Signaling, MA, USA), CREB3L3 (sc-377331; 1:500; Santa Cruz Biotechnology, TX, USA), AR-V7 (ab198394; 1:1000; Abcam, Cambridge, UK), PSA (sc-7316; 1:200; Santa Cruz Biotechnology, TX, USA), integrin B1 (4706S; 1:500; Cell Signaling, MA, USA), GAPDH (5174S; 1:1000; Cell Signaling, MA, USA), alpha-tubulin (sc-32293; 1:500; Santa Cruz Biotechnology, TX, USA), and beta-actin (A5441; 1:5000; Sigma, St.Louis, MO, USA). The secondary antibodies used were anti-mouse (925-32210; 1:15,000; LI-COR, Lincoln, NE, USA) and anti-rabbit (925-32211; 1:15,000; LI-COR, Lincoln, NE, USA). Analysis and quantification of western blots were performed using ImageJ software.

2.6. RNA Extraction and qPCR Analysis

The RNeasy Mini Kit (QIAGEN, Hilden Germany, cat#: 74104) was used to isolate total RNA from each of the cell lines used in this study, according to the protocol specified by the manufacturer. RNA concentration was measured using the spectrophotometer NanoDropTM 2000 (Thermo Fisher Scientific, Inc.). cDNA was obtained using 1 μg of RNA and the QuantiTect Reverse Transcription kit (QIAGEN, Hilden, Germany, cat#: 205311).

Expression of SLUG was measured by quantitative real time qPCR using SYBR-Green Master mix (Life Technologies, CA, USA, cat#: 4309155). Primers for SLUG and GAPDH were created with the OligoPerfect Designer program (ThermoFisher Scientific Inc; Wilmington, DE, USA). The following oligonucleotide sequences were used for primers: murine SLUG-F, 5′-CCTTTCTCTTGCCCTCACTG-3′, and murine SLUG-R, 5′-ACAGCAGCCAGACTCCTCAT-3′; murine GAPDH-F, 5′-TGATGGGTGTGAACCACGAG-3′, and GAPDH-R, 5′-AGTGATGGCATGGACTGTGG-3′; human SLUG-F, 5′-CTTTTTCTTGCCCTCACTGC-3′, and human SLUG-R, 5′-GCTTCGGAGTGAAGAAATGC-3′; human GAPDH-F, 5′-GAGTCAACGGATTTGGTCGT-3′, and human GAPDH-R, 5′-TTGATTTTGGAGGGATCTCG-3′. For each sample, SLUG was normalized with GAPDH expression. The comparative cycle threshold (Ct) method was used to quantify relative target gene expression. The instrument used was the Quantifect Studio System (Applied Biosystems).

2.7. Transfection of siRNAs

Cells were grown in 6-well plates. When a confluency of 60–70% was reached, the cells were transfected according to the manufacturer's instructions. Briefly, the cells were transfected with 10 nM of either fibronectin siRNA (Santa Cruz Biotechnology, TX, USA, cat#: sc-29315) or a non-targeting scramble control (Sigma, St.Louis, MO, USA) using Lipofectamine RNAiMAX (Thermo Fisher Scientific Inc.; Wilmington, DE, USA) diluted in Opti-MEM (ThermoFisher Scientific Inc.; Wilmington, DE, USA). The cells were then incubated for 24 h at 37 °C after which cells were harvested.

2.8. Flow Cytometry

Cells were incubated with trypsin for 3 min in order to harvest them. The pellets were washed, spun at 1200 rpm for 5 min, and re-suspended in 1% BSA in 1X PBS incubation buffer. Cells were spun and treated with mouse Fc block (BD Biosciences, NJ, USA, cat#: 553142) for 20 min. Either FITC anti-mouse MHCI antibody (BD Biosciences, NJ, USA, cat#: 553565) or its corresponding FITC Mouse Isotype control (BD Biosciences, NJ, USA, cat#: 553456) was added to cells and allowed to incubate for 1 h. The samples were fixed with 2% paraformaldehyde for 30 min at 4 °C, washed, and re-suspended in 1X PBS. Analysis was done using the BD FACs instrumentation and software version 7.0 (BD Biosciences, NJ, USA).

2.9. Cytokine Array

TBOH and CBOH cell lines were screened simultaneously for 111 murine cytokines using the Mouse XL Cytokine Array (R&D Biosystems, MN, USA, cat#: ARY028). Duplicate experiments were performed according to the manufacturer's instructions. Cytokine signals were quantified with densitometry using Image J software.

2.10. Statistical Analyses

Data from at least 3 different independent experiments were collected and presented as mean ± standard error of the mean (SEM). Statistical significance was evaluated using Student's *t*-test. p values of >0.05 were deemed significant.

3. Results

3.1. CTCs Obtained from Blood Express Tissue-Specific Markers

After the establishment of the novel cell lines (as shown in Figure 1), we confirmed tissue specificity of the established CTC lines. To confirm that the CTCs obtained from the bloodstream of the HCC syngeneic mouse models were of liver origin and not other cells potentially isolated from the blood, we performed immunofluorescence staining for CREB3L3, a validated liver specific marker [28,29]. As expected, the primary tumor cell lines TBOH1 and TBOH9 demonstrated strong CREB3L3 expression as shown by the red pigmentation in cells in Figure 2A,C. Similarly, the CTCs CBOH4 and CBOH9 also showed distinct CREB3L3 expression, confirming their liver origin, as well as their derivation from TBOH1 and TBOH9, respectively. No signal was observed in the negative controls in which cells were incubated with only the secondary antibody. Western blotting for CREB3L3 also revealed positive signals for both CBOH4 and CBOH9 as seen in Figure 2B,D.

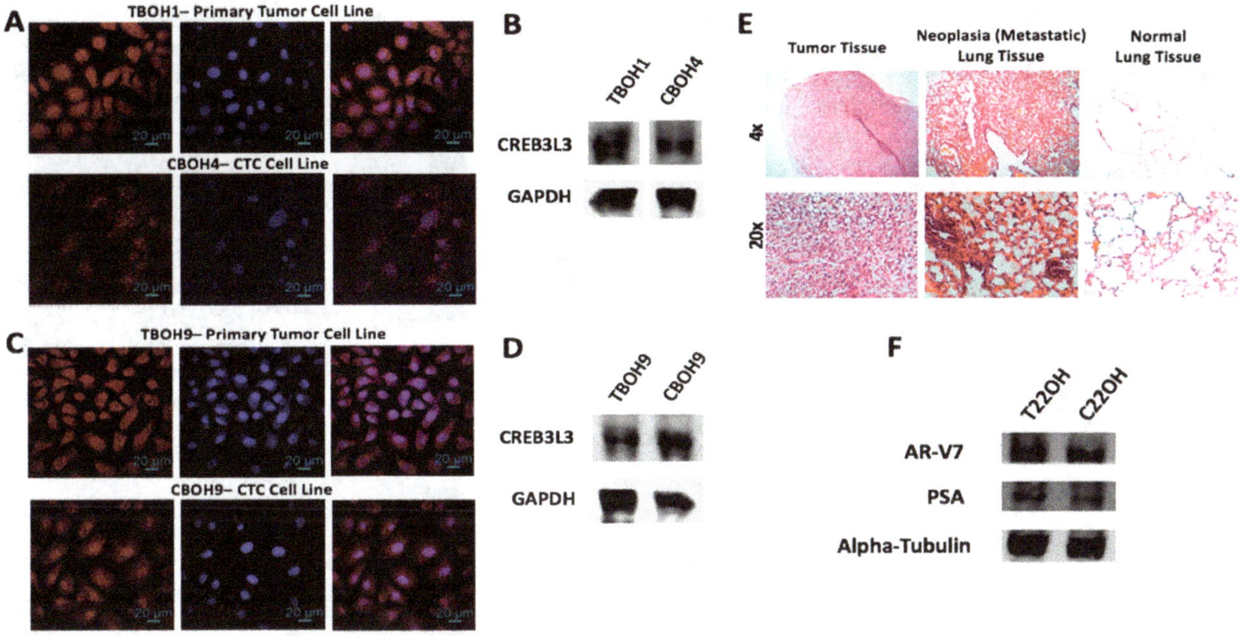

Figure 2. CTCs express tissue-specific markers and were obtained from mice that developed cancer metastasis. (**A**,**C**) Immunofluorescence staining for CREB3L3, a liver-specific marker; and (**B**,**D**) western blotting for CREB3L3. DAPI nuclear staining is shown in blue; CREB3L3 cytoplasmic staining is shown in red. (**E**) H&E staining of tumor and lung tissues from implantation of HCC cell line into BALB/c mice demonstrates evidence of cancer metastasis to lungs of cancer-bearing mice. (**F**) Western blotting was carried out to demonstrate prostate cancer origin for both CRPC primary tumor-derived and CTC lines. Experiments were carried out at least three times.

In addition to showing that the CTCs are hepatic-specific, we also performed H&E histological staining on lung tissues obtained from the mice implanted with the HCC cell line. This was to confirm the occurrence of HCC metastasis to the lungs, as shown in Figure 2E by the darker red pigmentation in the lungs of the HCC-implanted mice in comparison to the lungs of the non-cancer bearing mice.

Similarly, to confirm that the cells obtained from the blood of the NSG mice implanted with CRPC cells are of prostate origin, we performed western blotting for prostate specific antigen (PSA) and the AR-V7 variant androgen receptor protein known to be expressed by 22Rv1. These are markers characteristic of prostate cancer cells [23,30,31]. Both the primary tumor-derived cell line and the CTC cell line derived from NSG mice implanted with the 22Rv1 CRPC cell line demonstrated distinct AR-V7 and PSA expression as shown in Figure 2F, confirming their origin from the prostate and the derivation of C22OH from T22OH. Furthermore, the occurrence of metastasis was confirmed by visually finding multiple macroscopic tumors in distant sites during necropsy of the mice.

Our results show that we were successfully able to isolate CTCs from the blood from the syngeneic HCC and NSG CRPC mouse models. Functional assays were subsequently carried out to determine differences between CTCs and their corresponding primary tumor-derived cells.

3.2. CTCs Have a Greater Migratory Capacity than Primary Tumor-Derived Cells

Cancer cell migration is required for cancer metastasis [32,33]. To assess the migratory capability of both primary tumor-derived and CTC HCC cells, we performed wound healing migration assays. The rate at which wounds closed determined the migratory capability of cells. As shown in Figure 3A,C, we observed that CTCs (CBOH4 and CBOH9) were more migratory than their corresponding primary tumor-derived cell lines (TBOH1 and TBOH9, respectively). CBOH4 displayed a 55% increase in migration in comparison with its corresponding primary tumor-derived TBOH1 (Figure 3B), and CBOH9 demonstrated ~30% increased migration in comparison to TBOH9 (Figure 3D). The increased migratory capacity of CTCs is statistically significant.

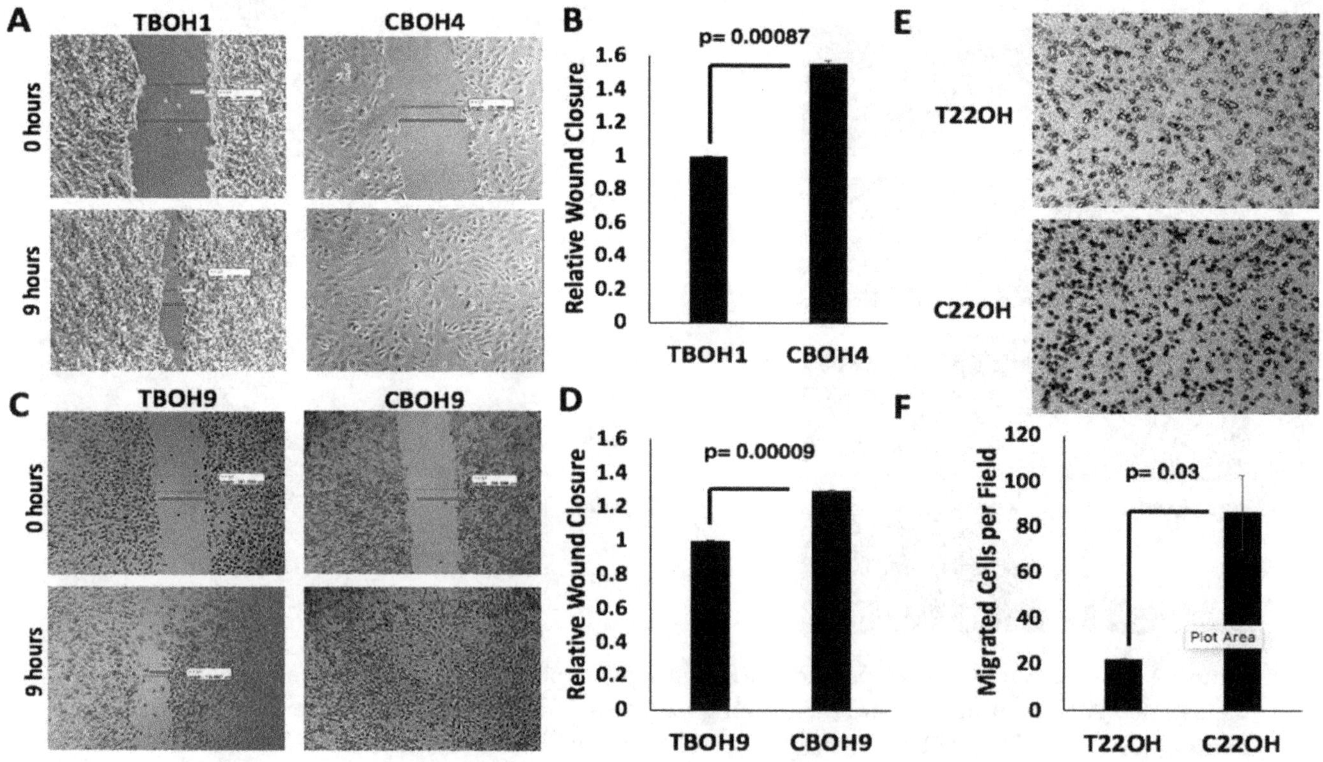

Figure 3. CTCs have greater migratory capacity than cancer cells from primary tumors. (**A**,**C**) Wound healing migration assays were performed on HCC cell lines. Cells were grown in 6-well plates. When confluent, wounds were made and measured at 0 and 9 h intervals. Wound closure by CTCs, but not by primary tumor-derived cell lines, was complete by 9 h. Images were taken at 10× magnification using Motic AE30 imaging software. (**B**,**D**) Migration (wound closure) was quantified; N = 3. (**E**) Transwell migration assays were performed on CRPC primary tumor-derived cell lines and CTC lines. Cells that migrated successfully through the pore membrane are represented by dark spots. Images shown were taken at 20× magnification. (**F**) Migration was quantified. Data provided on graphs are presented as mean ± standard error of the mean (SEM); N = 3.

To analyze the migratory capability of CRPC cells, we performed transwell migration assays. The number of cells passing through the chamber pores determined the migratory capability of cells. As shown in Figure 3E,F, the C220H CTC cell line demonstrated a four-fold increase in migration in comparison to the primary tumor-derived cell line T22OH.

3.3. CTCs Exhibit Epithelial to Mesenchymal Transition (EMT)

Having observed greater migration from CTCs in comparison to primary tumor-derived cells, we wanted to determine whether CTCs were undergoing EMT. This is a phenomenon frequently observed in cancer cells migrating and metastasizing [34–39]. Using western blotting, we examined protein expression of fibronectin, a well-known marker of migratory and mesenchymal cells [40–44]. As observed in Figure 4A–C, CTCs had greater fibronectin protein expression in all three pairs of cell lines. More specifically, CBOH4 showed a 4.81-fold increase in expression of fibronectin when compared to TBOH1; CBOH9 exhibited a 3.95-fold increase in fibronectin expression when compared with TBOH9, and C22OH had a 3.52-fold increase in fibronectin expression in comparison with T22OH. Another marker examined to assess EMT was E-cadherin, a well-known cell adhesion protein characteristic of epithelial cells [45,46]. E-cadherin expression was decreased 11.1-fold in CBOH4 in comparison to TBOH1; decreased 5.5-fold in CBOH9 in comparison with TBOH9, and decreased 2.1-fold in C22OH in comparison with T22OH.

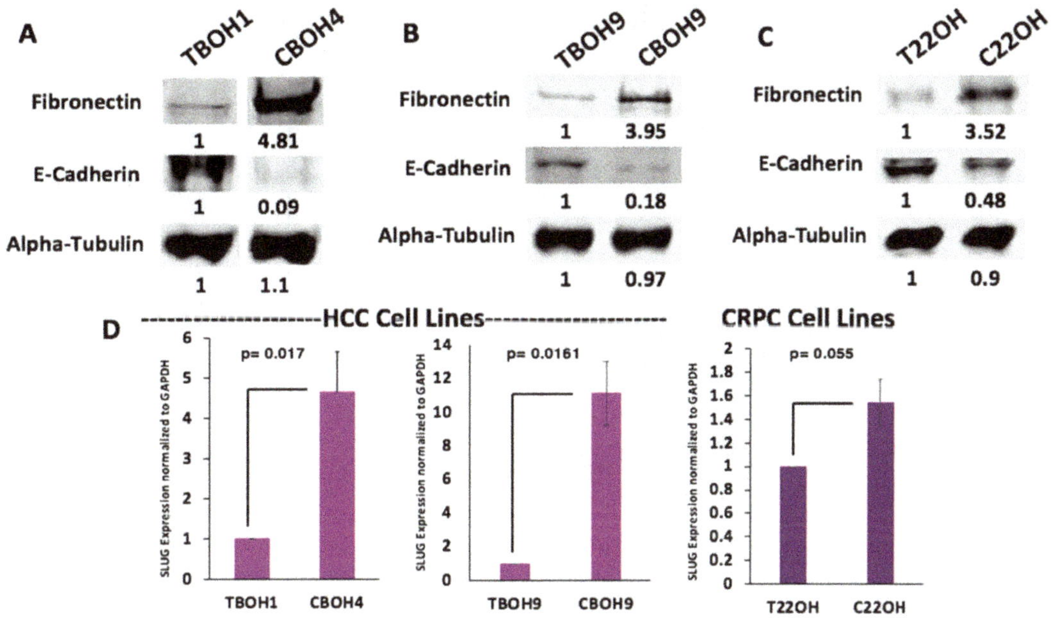

Figure 4. CTCs undergo epithelial-to-mesenchymal transition (EMT) as observed by increased fibronectin, decreased E-cadherin, and increased SLUG expression. (**A**,**B**) Fibronectin and E-cadherin protein expression by HCC primary tumor-derived cell lines and CTC lines. (**C**) Fibronectin and E-cadherin protein expression by CRPC primary tumor-derived cell line and CTC line. (**D**) SLUG expression was assessed using qPCR. Expression was normalized against GAPDH. Data provided on graphs are presented as mean ± standard error of the mean (SEM); N = 3.

EMT is also made possible by several transcription factors that initiate and maintain it [47]. One such transcription factor observed to be overexpressed in all three CTC cell lines was SLUG. The mRNA expression of this EMT transcription factor was assessed using qPCR. As shown in Figure 4D, CBOH4 exhibited a 4.6-fold increase in SLUG mRNA expression in comparison with TBOH1. CBOH9 demonstrated an 11.1-fold increase in SLUG expression in comparison to the corresponding primary tumor derived cell line TBOH9. Finally, C22OH exhibited a 1.5-fold increase in SLUG mRNA expression in comparison to the T22OH cell line. These findings demonstrate that EMT is occurring in CTC lines and may be the reason they are more migratory than primary tumor-derived cell lines.

3.4. Fibronectin Expression Regulates Integrin B1 and SLUG Expression in CTCs

Both HCC and CRPC CTCs expressed significantly more fibronectin than primary tumor-derived cells. We therefore investigated the molecular mechanisms of action of fibronectin in CTCs by knocking down fibronectin expression in CTCs. CBOH4 and C22OH were transfected for 24 h with either 25 pmol of fibronectin-specific siRNA or scramble siRNA. In comparison with primary tumor-derived cells, CTCs had higher expression of integrin B1 as shown in Figure 5A,B. Integrins are heterodimeric, transmembrane cell surface receptors that have been linked with metastasis and tumor migration [48–50]. The integrin B1 subunit specifically has been frequently upregulated in tumors [51,52]. Interestingly, knockdown of fibronectin in CTCs resulted in decreased integrin B1 expression (19.4% decrease in fibronectin expression in CBOH4 with knockdown of fibronectin; and 20.3% decrease in C22OH with knockdown of fibronectin).

Like integrin B1, SLUG is also a molecule that has been associated with greater migratory capacity by cancer cells [53–55]. To determine the effect of fibronectin knockdown on SLUG expression in CTCs, we performed qPCR analysis. As shown in Figure 5C, CBOH4 transfected with fibronectin-specific siRNA demonstrated a 40% decrease in SLUG expression in comparison with CBOH4 transfected with scramble siRNA. Similarly, as shown in Figure 5D, C22OH transfected with the fibronectin-specific siRNA demonstrated a 30% decrease in SLUG expression in comparison with C22OH transfected with scramble siRNA. Therefore, fibronectin has significant regulatory effects on integrin B1 and SLUG expression in CTCs.

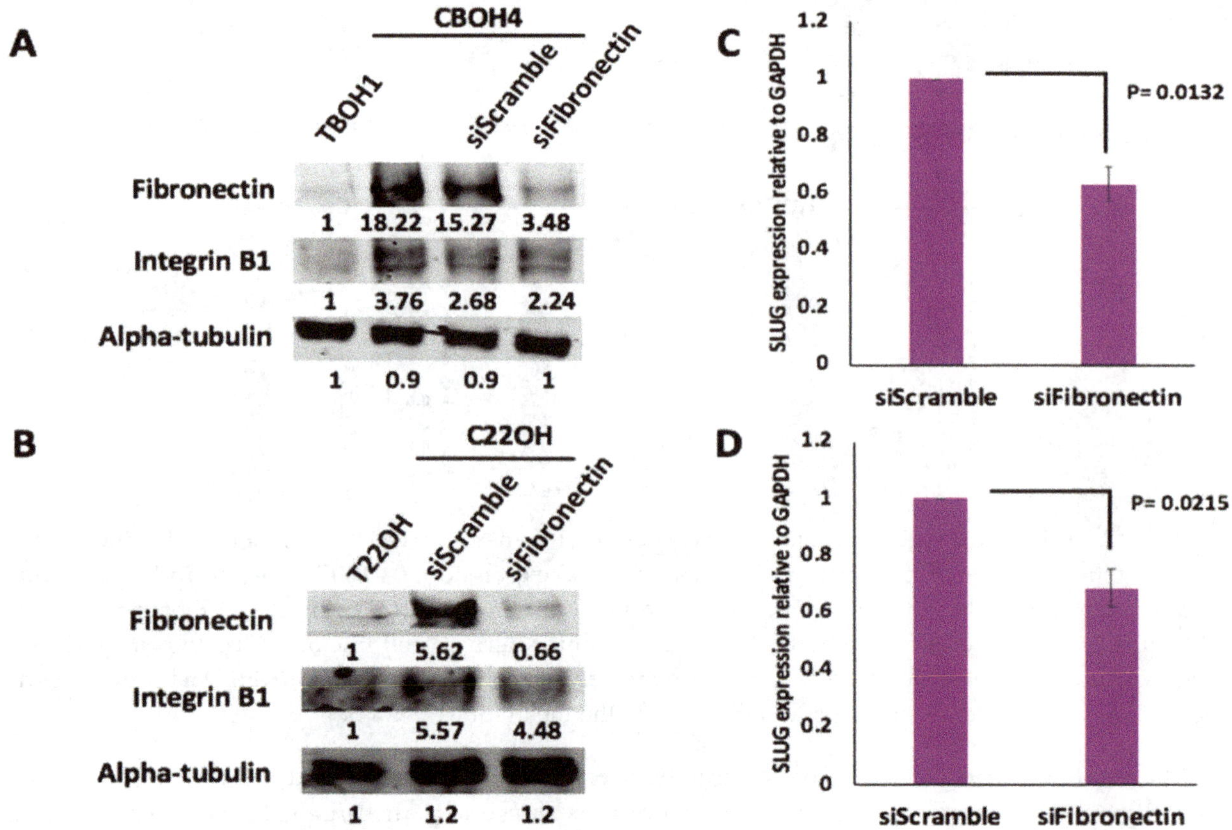

Figure 5. Fibronectin knockdown in CTCs caused decreased integrin B1 and SLUG expression. (**A**,**B**) After a 24 h transfection of CBOH4 and C22OH with fibronectin-specific siRNAs, western blotting was performed to determine the effects on integrin B1 expression. Western blotting revealed that fibronectin knockdown in CTCs was successful. The effect of fibronectin knockdown on integrin B1 expression was compared in control untransfected cells, CTCs transfected with scramble siRNAs, and CTCs transfected with fibronectin-specific siRNAs. Western blotting experiments were performed three separate times. (**C**,**D**) Effect of Fibronectin knockdown on SLUG expression in CBOH4 and C22OH CTCs was assessed by qPCR. Expression was normalized against GAPDH. Data provided on graphs are presented as mean ± standard error of the mean (SEM); N = 3.

3.5. HCC CTCs have Decreased MHCI Cell Surface Expression

To investigate potential immunomodulatory properties of CTCs, we used the cell lines derived from the syngeneic HCC mouse model. Using flow cytometry, we assayed for MHCI, a well-established cell surface molecule involved in self-recognition and identification of harmful entities for destruction by T-cells [56,57]. As shown in Figure 6A,B, we observed a 1.8-fold (45%) decrease in MHCI expression in the CBOH4 CTC line in comparison to the primary-tumor derived TBOH1 cell line. Similarly, as shown in Figure 6C,D, we observed a 1.5-fold (35%) decrease in MHCI expression in the CBOH9 CTC line in comparison to the TBOH9 primary tumor-derived cell line.

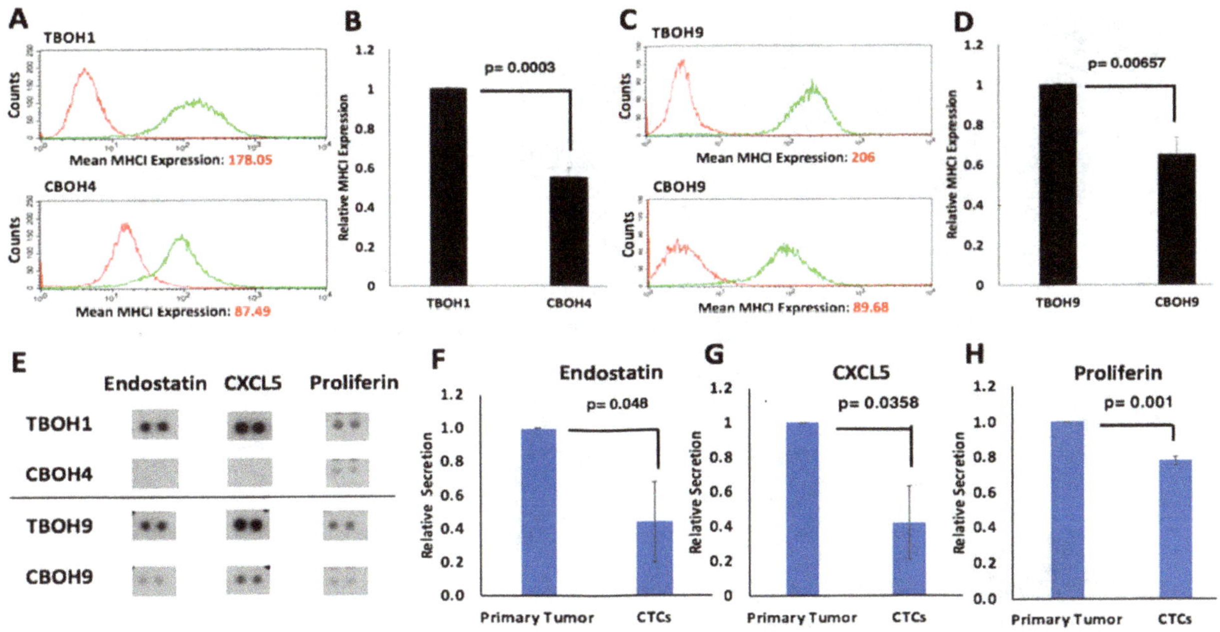

Figure 6. Immunomodulatory mechanisms of CTCs derived from a syngeneic mouse model of HCC. (**A–D**) Major histocompatibility complex class I (MHCI) expression was assessed using flow cytometry. CTCs have decreased MHCI cell surface protein expression in comparison to primary tumor-derived cell lines. Both isotype and MHCI antibodies were conjugated to fluorescein isothiocyanate (FITC). Isotype measuring background is shown in red. MHCI signal is shown in green; N = 5. (**E**) Analysis of 111 different cytokines secreted into cell media reveals consistent and significantly decreased secretion of endostatin, CXCL5, and proliferin in CTCs in comparison to primary tumor-derived cells; N = 2. (**F–H**) Endostatin, CXCL5, and proliferin signals from cytokine array were quantified. Data are presented as mean ± standard error of the mean (SEM).

3.6. HCC CTCs have Significantly Decreased Secretion of Endostatin, CXCL5, and Proliferin

Cytokines are secreted molecules that can activate signaling pathways and have effects on the immune response [58–60]. To determine if there are differences in the cytokine secretion profile between CTCs and primary tumor-derived cells, a cytokine array was used. Of the 111 cytokines assayed, three were consistently and significantly lower in CTCs (see Figure 6E). Endostatin, an anti-angiogenic molecule and inhibitor of tumor growth [61,62], was decreased 2.27-fold in CTCs in comparison to primary tumor-derived cells (Figure 6F). CXCL5, a molecule that plays a role in attracting leukocytes such as neutrophils [63,64], was also downregulated 2.38-fold in CTCs (Figure 6G). Finally, proliferin, a molecule with reported roles in cell growth regulation and differentiation [65–67], was decreased 1.28-fold in CTCs (Figure 6H). Statistical analyses found all three molecules to be significantly decreased with p values of <0.05.

3.7. CTCs, in Comparison to Primary Tumor-Derived Cells, Have Decreased Endostatin Expression

As a follow-up to the observation that CTCs secreted significantly reduced endostatin, we investigated intracellular endostatin expression in both of our HCC and CRPC CTC models using quantitative immunofluorescence. As shown in Figure 7A,B, CBOH4 had a 59% decrease in endostatin expression in comparison to TBOH1, and as shown in Figure 7C,D, CBOH9 had a 48% decrease in endostatin expression compared to TBOH9. Further, as shown in Figure 7E,F, C22OH had a 40% decrease in the expression of endostatin in comparison to T22OH. Thus, we can conclude that CTCs from both HCC and CRPC models examined expressed significantly less intracellular endostatin than their corresponding primary tumor-derived cells.

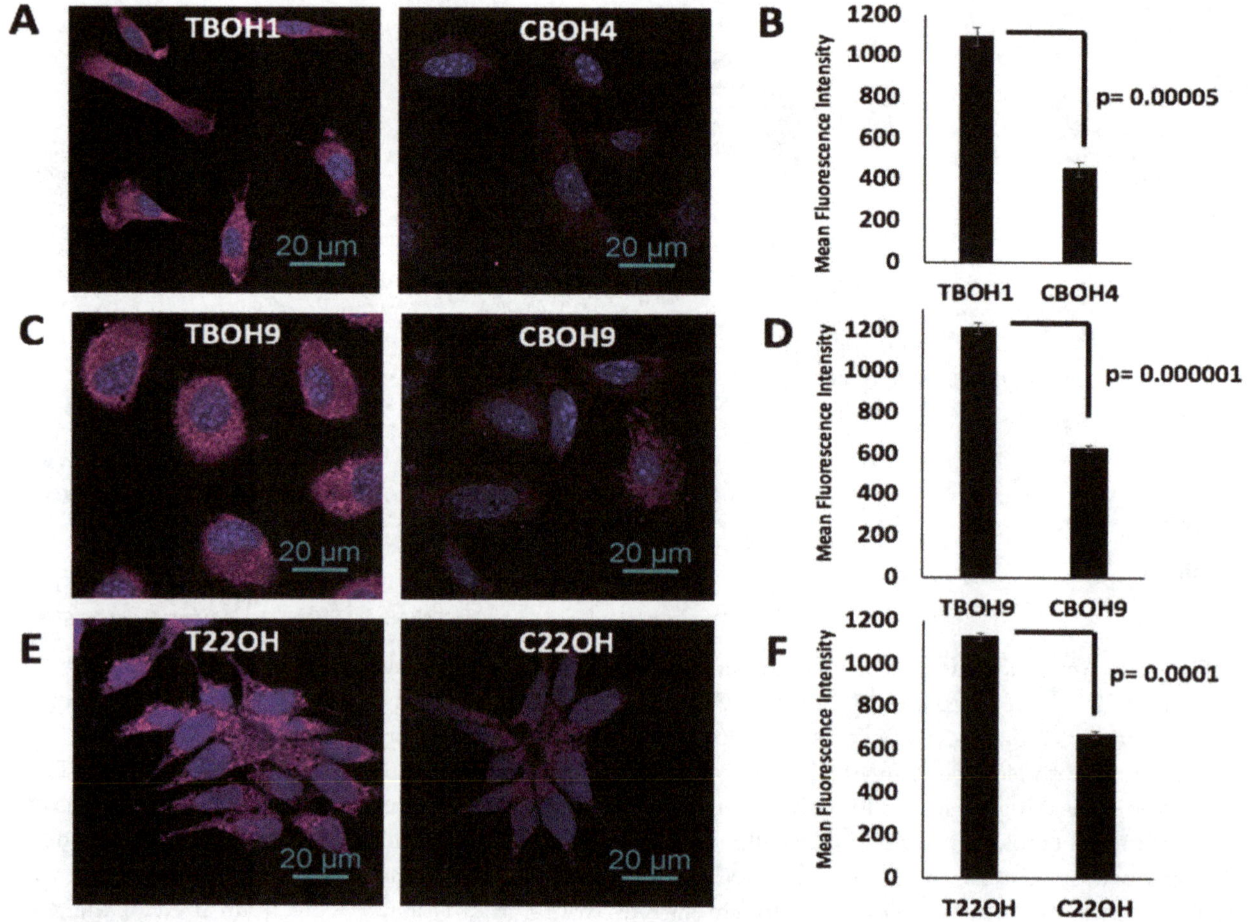

Figure 7. Decreased intracellular endostatin expression by CTCs. (**A**,**C**,**E**) Endostatin expression in TBOH1 and CBOH4, TBOH9 and CBOH9, and T22OH and C22OH was determined using immunofluorescence. (**B**,**D**,**F**) Mean fluorescence intensity was quantified using the NIS Elements software. Data are presented as mean ± standard error of the mean (SEM).

4. Discussion

Metastasis is the most common cause of death among cancer patients. Molecular mechanisms of cancer metastasis are not yet clear [1–3,33,68]. CTCs are an important step in the metastatic process. Consequently, in this study, we have focused on elucidating the molecular mechanisms of CTCs.

However, CTCs are not always abundant and are usually challenging to obtain in the human clinical setting [12–15]. To address these limitations, we have established an effective method for isolating CTCs, establishing long-term cultures, and propagating them into cell lines that are useful for studying CTC biology. The challenge involved in this task is underscored by the fact that only a few other groups have had success in doing this [69–72]. So far, it has been easier to establish short term CTC cultures, which when coupled with single-cell sequencing or short-term biochemical assays, have resulted in very useful information [73–82]. Further, these techniques can be physically intensive and expensive. Here, we have described an approach for propagating CTC cell lines and using them for functional characterization, identification of novel molecular pathways important to metastasis, as well as to gain insight into molecular receptors and secretions that might have immunomodulatory functions in CTCs.

In this study, we have focused on the use of HCC and CRPC models because they both have significant negative clinical outcomes. At present, there is no effective treatment for advanced HCC [16–18,20]. Similarly, metastatic CRPC currently has no cure [21–24,83].

In the present study, we have successfully propagated three pairs of CTC lines and their corresponding primary tumor-derived cell lines: two pairs of HCC origin and one pair of CRPC origin. Interestingly, all CTCs demonstrated significantly greater migratory capacity than their primary tumor-derived counterparts. Furthermore, we investigated the role of EMT in CTCs. EMT is a phenomenon that is frequently observed in cancer cells during cancer progression [34–36]. Accordingly, we observed evidence of EMT in CTCs as demonstrated by downregulation of E-cadherin and upregulation of fibronectin and SLUG expression. Therefore, our findings indicate that enhanced migratory capacity and EMT are characteristic of CTCs.

Because we observed fibronectin to be overexpressed by CTCs, we were also interested in identifying molecules regulated by fibronectin. Integrin receptors have long been associated with tumor migration [49]. While fibronectin binding and activation of these receptors is established [84,85], we report for the first time our observation that fibronectin has a regulatory role in the expression of the integrin B1 subunit. Fibronectin knockdown in CTCs resulted in decreased expression of integrin B1. Interestingly, another molecule significantly reduced by fibronectin knockdown in CTCs is SLUG. SLUG is a known regulator of EMT, and we observed SLUG overexpression in CTCs in comparison to primary tumor-derived cells. Upon fibronectin knockdown, SLUG expression was significantly decreased in CTCs. Thus, fibronectin expression regulates integrin B1 and SLUG expression in CTCs.

In this study, we also investigated major histocompatibility complex class I (MHCI) expression and secretion of cytokines by CTCs. The major histocompatibility complex class I (MHCI) is an important cell surface molecule that enables T cells to distinguish between "self" vs. "non-self". Upon recognition of harmful entities presented by MHCI, T cells will attack these MHCI-presenting cells [56,57]. We observed significant reduction of MHCI expression in CTCs arising from the syngeneic HCC mouse model. This has interesting implications on how CTCs may circumvent immunosurveillance. Additionally, we sought to identify molecules differentially secreted by CTCs in comparison to primary tumors. To this end, we assayed over 111 different cytokines using media secreted by cells. We found three molecules to be significantly decreased by CTCs: endostatin, CXCL5, and proliferin. We report for the first time that endostatin, CXCL5, and proliferin are significantly secreted less by CTCs than primary tumor derived cells. Further studies are imperative to clarify the implications of this differential secretion profile of CTCs to cancer metastasis.

Endostatin is known for its anti-angiogenic properties [61,62,86]. The fact that CTCs express less endostatin is intriguing and suggests that CTCs have acquired enhanced angiogenicity, which bodes well for the colonization of secondary sites. Moreover, it is conceivable that decreased endostatin secretion by CTCs may enhance their metastatic capability in other important ways [87–89]. CXCL5 is a chemokine that recruits and activates neutrophils. While some studies associate higher CXCL5 expression with a worse cancer prognosis [63,64,90,91], several reports indicate that a lower CXCL5 expression can also promote metastatic spread [91–94]. Proliferin is a placental growth hormone that has been observed to be pro-angiogenic [65–67]. The implications of our current finding that CTCs secrete reduced amounts of proliferin than primary tumor-derived cells are currently unclear and deserve further study. It is noteworthy, though, that previous studies have found that as tumors grow in volume, intra-tumoral cells become hypoxic due to lack of oxygen within the tumor mass. This, in turn, stimulates the production of pro-angiogenic factors, enabling new blood vessel formation inside the tumor to deliver oxygen [95]. This process may not be necessary for CTCs, and this may explain the reduced secretion of proliferin by CTCs observed.

Interestingly, endostatin expression was found to be downregulated in all our CTCs regardless of tissue of origin. This was even more compelling since fibronectin, which was upregulated in our CTCs, has been previously reported to bind to the same integrin $\alpha 5\beta 1$ receptor as does endostatin [85,96–100]. It is plausible that there could be a potential autocrine, competitive binding occurring between fibronectin and endostatin to the integrin $\alpha 5\beta 1$ receptor with resulting implications in a cancer cell's ability to migrate. Further studies are necessary to confirm or refute this.

It is important to also note that while our work highlights a novel methodology and the creation of new cell lines to uncover possible mechanisms of cancer cell action, our cell lines were obtained from murine models. Others have previously isolated CTCs from human patients using microfluidic devices to assess molecular phenotype and drug sensitivity [9,69–72]. We propose our methodology to contribute to the field by adding yet another way in which we can harness the information gained from studying CTCs. In doing so, we may refute or support novel ideas such as CTCs being hybrids between leukocytes and cancer cells [101,102]. Whether this explains the reason our CTCs demonstrate enhanced migration and higher levels of integrin B1, which was previously reported in leukocytes [103,104], remains to be proven.

In summary, we have described the establishment of novel primary tumor-derived cell lines and CTC lines from the same mouse models and have used them to elucidate novel molecular mechanisms of CTCs. This work has demonstrated that EMT, enhanced migration, and decreased endostatin, CXCL5, and proliferin secretion are consistently seen in CTCs. Additionally, we report a novel role for fibronectin in regulating integrin B1 subunit receptor expression in CTCs, and while fibronectin's ability to regulate SLUG activity in renal cell carcinoma cells to promote lung metastasis has been reported [105], we report for the first time an observation of this molecular phenomenon in CTCs of HCC and CRPC origin. Our findings suggest that fibronectin's regulation of SLUG in CTCs may contribute to their role in cancer metastasis. Taken together, our findings demonstrate that CTCs have unique and important molecular mechanisms with implications for cancer metastasis.

5. Conclusions

In conclusion, we have demonstrated the successful establishment of novel primary tumor-derived cell lines and CTC lines from a syngeneic HCC mouse model and a human xenograft CRPC mouse model. The CTCs in these models demonstrate enhanced migration and EMT when compared to primary tumor-derived cells. Here, we report our novel finding of fibronectin's regulation of integrin B1 and SLUG expression in CTCs. This may be a mechanism by which CTCs ensure greater migration and metastasis. Further, we report our observation of decreased MHCI expression and decreased secretion of endostatin, CXCL5, and proliferin by CTCs in comparison to primary tumor-derived cells. These molecular mechanisms of CTCs likely have important implications for cancer metastasis.

Author Contributions: Conceptualization, O.O.; analysis, J.H., M.N., and O.O.; resources, O.O. and X.Z.; funding acquisition, O.O.; methodology, J.H. and O.O; supervision, O.O.; writing—original draft preparation, J.H.;

Acknowledgments: We would like to thank Kim Ohaegbulam for assistance in facilitating work done using the core facilities at Albert Einstein School of Medicine.

References

1. Lee, W.C.; Kopetz, S.; Wistuba, I.I.; Zhang, J. Metastasis of cancer: When and how? *Ann. Oncol.* **2017**, *28*, 2045–2047. [CrossRef]
2. Chaffer, C.L.; Weinberg, R.A. A perspective on cancer cell metastasis. *Science* **2011**, *331*, 1559–1564. [CrossRef] [PubMed]
3. Gupta, G.P.; Massague, J. Cancer metastasis: Building a framework. *Cell* **2006**, *127*, 679–695. [CrossRef] [PubMed]
4. Van de Stolpe, A.; Pantel, K.; Sleijfer, S.; Terstappen, L.W.; Den Toonder, J.M. Circulating tumor cell isolation and diagnostics: Toward routine clinical use. *Cancer Res.* **2011**, *71*, 5955–5960. [CrossRef]
5. Gallerani, G.; Fici, P.; Fabbri, F. Circulating Tumor Cells: Back to the Future. *Front. Oncol.* **2016**, *6*, 275. [CrossRef] [PubMed]
6. Haber, D.A.; Velculescu, V.E. Blood-based analyses of cancer: Circulating tumor cells and circulating tumor DNA. *Cancer Discov.* **2014**, *4*, 650–661. [CrossRef] [PubMed]
7. Pantel, K.; Brakenhoff, R.H.; Brandt, B. Detection, clinical relevance and specific biological properties of disseminating tumour cells. *Nat. Rev. Cancer* **2008**, *8*, 329–340. [CrossRef]

8. Bailey, P.M.; Martin, S.S. Insights on CTC Biology and Clinical Impact Emerging from Advances in Capture Technology. *Cells* **2019**, *8*, 553. [CrossRef]
9. George, T.J., Jr.; Ogunwobi, O.O.; Sheng, W.; Fan, Z.H.; Liu, C. "Tissue is the issue": Circulating tumor cells in pancreatic cancer. *J. Gastrointest. Cancer* **2014**, *45*, 222–225. [CrossRef]
10. Micalizzi, D.S.; Maheswaran, S.; Haber, D.A. A conduit to metastasis: Circulating tumor cell biology. *Genes Dev.* **2017**, *31*, 1827–1840. [CrossRef]
11. Alix-Panabieres, C.; Pantel, K. Challenges in circulating tumour cell research. *Nat. Rev. Cancer* **2014**, *14*, 623–631. [CrossRef] [PubMed]
12. Alvarez Cubero, M.J.; Lorente, J.A.; Robles-Fernandez, I.; Rodriguez-Martinez, A.; Puche, J.L.; Serrano, M.J. Circulating Tumor Cells: Markers and Methodologies for Enrichment and Detection. *Methods Mol. Biol.* **2017**, *1634*, 283–303. [PubMed]
13. Cristofanilli, M.; Budd, G.T.; Ellis, M.J.; Stopeck, A.; Matera, J.; Miller, M.C.; Reuben, J.M.; Doyle, G.V.; Allard, W.J.; Terstappen, L.W.; et al. Circulating tumor cells, disease progression, and survival in metastatic breast cancer. *N. Engl. J. Med.* **2004**, *351*, 781–791. [CrossRef] [PubMed]
14. Miller, M.C.; Doyle, G.V.; Terstappen, L.W. Significance of Circulating Tumor Cells Detected by the CellSearch System in Patients with Metastatic Breast Colorectal and Prostate Cancer. *J. Oncol.* **2010**, *2010*, 617421. [CrossRef] [PubMed]
15. Zieglschmid, V.; Hollmann, C.; Bocher, O. Detection of disseminated tumor cells in peripheral blood. *Crit. Rev. Clin. Lab. Sci.* **2005**, *42*, 155–196. [CrossRef] [PubMed]
16. American Cancer Society. Cancer Facts and Figures 2018. Available online: https://www.cancer.org/content/dam/cancer-org/research/cancer-facts-and-statistics/annual-cancer-facts-and-figures/2018/cancer-facts-and-figures-2018.pdf (accessed on 19 June 2019).
17. Llovet, J.M.; Zucman-Rossi, J.; Pikarsky, E.; Sangro, B.; Schwartz, M.; Sherman, M.; Gores, G. Hepatocellular carcinoma. *Nat. Rev. Dis. Primers* **2016**, *2*, 16018. [CrossRef]
18. Llovet, J.M.; Montal, R.; Sia, D.; Finn, R.S. Molecular therapies and precision medicine for hepatocellular carcinoma. *Nat. Rev. Clin. Oncol.* **2018**, *15*, 599–616. [CrossRef]
19. Ogunwobi, O.H.; Harricharran, T.; Huaman, J.; Galuza, A.; Odumuwagun, O.; Tan, Y.; Ma, G.X.; Nguyen, M.T. Mechanisms of hepatocellular carcinoma progression. *World J. Gastroenterol.* **2019**, *25*, 2279–2293. [CrossRef]
20. Keating, G.M. Sorafenib: A Review in Hepatocellular Carcinoma. *Target Oncol.* **2017**, *12*, 243–253. [CrossRef]
21. Kirby, M.; Hirst, C.; Crawford, E.D. Characterising the castration-resistant prostate cancer population: A systematic review. *Int. J. Clin. Pract.* **2011**, *65*, 1180–1192. [CrossRef]
22. Suzman, D.L.; Antonarakis, E.S. Castration-resistant prostate cancer: Latest evidence and therapeutic implications. *Ther. Adv. Med. Oncol.* **2014**, *6*, 167–179. [CrossRef] [PubMed]
23. Pentyala, S.; Whyard, T.; Pentyala, S.; Muller, J.; Pfail, J.; Parmar, S.; Helguero, C.G.; Khan, S. Prostate cancer markers: An update. *Biomed. Rep.* **2016**, *4*, 263–268. [CrossRef] [PubMed]
24. Litwin, M.S.; Tan, H.J. The Diagnosis and Treatment of Prostate Cancer: A Review. *JAMA* **2017**, *317*, 2532–2542. [CrossRef] [PubMed]
25. Hong, J.H.; Kim, I.Y. Nonmetastatic castration-resistant prostate cancer. *Korean J. Urol.* **2014**, *55*, 153–160. [CrossRef] [PubMed]
26. Das, D.K.; Naidoo, M.K.; Ilboudo, A.; DuBois, P.; Durojaiye, V.; Liu, C.; Ogunwobi, O.O. Isolation and Propagation of Circulating Tumor Cells from a Mouse Cancer Model. *J. Vis. Exp.* **2015**. [CrossRef] [PubMed]
27. Ogunwobi, O.O.; Puszyk, W.; Dong, H.J.; Liu, C. Epigenetic upregulation of HGF and c-Met drives metastasis in hepatocellular carcinoma. *PLoS ONE* **2013**, *8*, e63765. [CrossRef] [PubMed]
28. Omori, Y.; Imai, J.; Watanabe, M.; Komatsu, T.; Suzuki, Y.; Kataoka, K.; Watanabe, S.; Tanigami, A.; Sugano, S. CREB-H: A novel mammalian transcription factor belonging to the CREB/ATF family and functioning via the box-B element with a liver-specific expression. *Nucleic Acids Res.* **2001**, *29*, 2154–2162. [CrossRef]
29. Barbosa, S.; Fasanella, G.; Carreira, S.; Llarena, M.; Fox, R.; Barreca, C.; Andrew, D.; O'Hare, P. An orchestrated program regulating secretory pathway genes and cargos by the transmembrane transcription factor CREB-H. *Traffic* **2013**, *14*, 382–398. [CrossRef]
30. Viswanathan, S.R.; Ha, G.; Hoff, A.M.; Wala, J.A.; Carrot-Zhang, J.; Whelan, C.W.; Haradhvala, N.J.; Freeman, S.S.; Reed, S.C.; Rhoades, J.; et al. Structural Alterations Driving Castration-Resistant Prostate Cancer Revealed by Linked-Read Genome Sequencing. *Cell* **2018**, *174*, 433–447. [CrossRef]

31. Wadosky, K.M.; Koochekpour, S. Androgen receptor splice variants and prostate cancer: From bench to bedside. *Oncotarget* **2017**, *8*, 18550–18576. [CrossRef]
32. Yilmaz, M.; Christofori, G. Mechanisms of motility in metastasizing cells. *Mol. Cancer Res.* **2010**, *8*, 629–642. [CrossRef] [PubMed]
33. Lambert, A.W.; Pattabiraman, D.R.; Weinberg, R.A. Emerging biological principles of metastasis. *Cell* **2017**, *168*, 670–691. [CrossRef] [PubMed]
34. Brabletz, T.; Kalluri, R.; Nieto, M.A.; Weinberg, R.A. EMT in cancer. *Nat. Rev. Cancer* **2018**, *18*, 128–134. [CrossRef] [PubMed]
35. Kalluri, R.; Weinberg, R.A. The basics of epithelial-mesenchymal transition. *J. Clin. Invest.* **2009**, *119*, 1420–1428. [CrossRef] [PubMed]
36. Lamouille, S.; Xu, J.; Derynck, R. Molecular mechanisms of epithelial-mesenchymal transition. *Nat. Rev. Mol. Cell Biol.* **2014**, *15*, 178–196. [CrossRef]
37. Tiwari, N.; Gheldof, A.; Tatari, M.; Christofori, G. EMT as the ultimate survival mechanism of cancer cells. *Semin. Cancer Biol.* **2012**, *22*, 194–207. [CrossRef] [PubMed]
38. Ye, X.; Weinberg, R.A. Epithelial-Mesenchymal Plasticity: A Central Regulator of Cancer Progression. *Trends Cell Biol.* **2015**, *25*, 675–686. [CrossRef]
39. Alix-Panabieres, C.; Mader, S.; Pantel, K. Epithelial-mesenchymal plasticity in circulating tumor cells. *J. Mol. Med.* **2017**, *95*, 133–142. [CrossRef]
40. Meng, X.N.; Jin, Y.; Yu, Y.; Bai, J.; Liu, G.Y.; Zhu, J.; Zhao, Y.Z.; Wang, Z.; Chen, F.; Lee, K.Y.; et al. Characterisation of fibronectin-mediated FAK signalling pathways in lung cancer cell migration and invasion. *Br. J. Cancer* **2009**, *101*, 327–334. [CrossRef]
41. Park, J.; Schwarzbauer, J.E. Mammary epithelial cell interactions with fibronectin stimulate epithelial-mesenchymal transition. *Oncogene* **2014**, *33*, 1649–1657. [CrossRef]
42. Yousif, N.G. Fibronectin promotes migration and invasion of ovarian cancer cells through up-regulation of FAK-PI3K/Akt pathway. *Cell. Biol. Int.* **2014**, *38*, 85–91. [CrossRef] [PubMed]
43. Ramos Gde, O.; Bernardi, L.; Lauxen, I.; Sant'Ana Filho, M.; Horwitz, A.R.; Lamers, M.L. Fibronectin Modulates Cell Adhesion and Signaling to Promote Single Cell Migration of Highly Invasive Oral Squamous Cell Carcinoma. *PLoS ONE* **2016**, *11*, e0151338. [CrossRef] [PubMed]
44. Wang, J.P.; Hielscher, A. Fibronectin: How its Aberrant Expression in Tumors may Improve Therapeutic Targeting. *J. Cancer* **2017**, *8*, 674–682. [CrossRef] [PubMed]
45. Jeanes, A.; Gottardi, C.J.; Yap, A.S. Cadherins and cancer: How does cadherin dysfunction promote tumor progression? *Oncogene* **2008**, *27*, 6920–6929. [CrossRef] [PubMed]
46. Wong, S.H.M.; Fang, C.M.; Chuah, L.H.; Leong, C.O.; Ngai, S.C. E-cadherin: Its dysregulation in carcinogenesis and clinical implications. *Crit. Rev. Oncol. Hematol.* **2018**, *121*, 11–22. [CrossRef] [PubMed]
47. Puisieux, A.; Brabletz, T.; Caramel, J. Oncogenic roles of EMT-inducing transcription factors. *Nat. Cell. Biol.* **2014**, *16*, 488–494. [CrossRef]
48. Desgrosellier, J.S.; Cheresh, D.A. Integrins in cancer: Biological implications and therapeutic opportunities. *Nat. Rev. Cancer* **2010**, *10*, 9–22. [CrossRef]
49. Hamidi, H.; Ivaska, J. Every step of the way: Integrins in cancer progression and metastasis. *Nat. Rev. Cancer* **2018**, *18*, 533–548. [CrossRef]
50. Seguin, L.; Desgrosellier, J.S.; Weis, S.M.; Cheresh, D.A. Integrins and cancer: Regulators of cancer stemness, metastasis, and drug resistance. *Trends Cell Biol.* **2015**, *25*, 234–240. [CrossRef]
51. Blandin, A.F.; Renner, G.; Lehmann, M.; Lelong-Rebel, I.; Martin, S.; Dontenwill, M. beta1 Integrins as Therapeutic Targets to Disrupt Hallmarks of Cancer. *Front. Pharmacol.* **2015**, *6*, 279. [CrossRef]
52. Howe, G.A.; Addison, C.L. beta1 integrin: An emerging player in the modulation of tumorigenesis and response to therapy. *Cell Adh. Migr.* **2012**, *6*, 71–77. [CrossRef] [PubMed]
53. Gu, A.; Jie, Y.; Yao, Q.; Zhang, Y.; Mingyan, E. Slug is associated with tumor metastasis and angiogenesis in ovarian cancer. *Reprod. Sci.* **2017**, *24*, 291–299. [CrossRef] [PubMed]
54. Uygur, B.; Wu, W.S. SLUG promotes prostate cancer cell migration and invasion via CXCR4/CXCL12 axis. *Mol. Cancer* **2011**, *10*, 139. [CrossRef] [PubMed]
55. Yu, M.; Chen, Y.; Li, X.; Yang, R.; Zhang, L.; Huangfu, L.; Zheng, N.; Zhao, X.; Lv, L.; Hong, Y.; et al. YAP1 contributes to NSCLC invasion and migration by promoting Slug transcription via the transcription co-factor TEAD. *Cell Death Dis.* **2018**, *9*, 464. [CrossRef] [PubMed]

56. Bubenik, J. MHC class I down-regulation: Tumour escape from immune surveillance? (review). *Int. J. Oncol.* **2004**, *25*, 487–491. [CrossRef] [PubMed]
57. Shastri, N.; Nagarajan, N.; Lind, K.C.; Kanaseki, T. Monitoring peptide processing for MHC class I molecules in the endoplasmic reticulum. *Curr. Opin. Immunol.* **2014**, *26*, 123–127. [CrossRef]
58. Jang, M.H.; Seoh, J.Y.; Miyasaka, M. Cytokines, chemokines, and their receptors: Targets for immunomodulation. Conference report: International Cytokine Society Conference 2005. *J. Leukoc. Biol.* **2006**, *80*, 217–219. [CrossRef]
59. Turner, M.D.; Nedjai, B.; Hurst, T.; Pennington, D.J. Cytokines and chemokines: At the crossroads of cell signalling and inflammatory disease. *Biochim. Biophys. Acta* **2014**, *1843*, 2563–2582. [CrossRef]
60. Lee, M.; Rhee, I. Cytokine Signaling in Tumor Progression. *Immune Netw.* **2017**, *17*, 214–227. [CrossRef]
61. Cao, Y.; Langer, R. A review of Judah Folkman's remarkable achievements in biomedicine. *Proc. Natl. Acad. Sci. USA* **2008**, *105*, 13203–13205. [CrossRef]
62. Dhanabal, M.; Ramchandran, R.; Volk, R.; Stillman, I.E.; Lombardo, M.; Iruela-Arispe, M.L.; Simons, M.; Sukhatme, V.P. Endostatin: Yeast production, mutants, and antitumor effect in renal cell carcinoma. *Cancer Res.* **1999**, *59*, 189–197. [PubMed]
63. Walz, A.; Burgener, R.; Car, B.; Baggiolini, M.; Kunkel, S.L.; Strieter, R.M. Structure and neutrophil-activating properties of a novel inflammatory peptide (ENA-78) with homology to interleukin 8. *J. Exp. Med.* **1991**, *174*, 1355–1362. [CrossRef] [PubMed]
64. Xia, J.; Xu, X.; Huang, P.; He, M.; Wang, X. The potential of CXCL5 as a target for liver cancer - what do we know so far? *Expert Opin. Ther. Targets* **2015**, *19*, 141–146. [CrossRef] [PubMed]
65. Linzer, D.I.; Lee, S.J.; Ogren, L.; Talamantes, F.; Nathans, D. Identification of proliferin mRNA and protein in mouse placenta. *Proc. Natl. Acad. Sci. USA* **1985**, *82*, 4356–4359. [CrossRef] [PubMed]
66. Jackson, D.; Volpert, O.V.; Bouck, N.; Linzer, D.I. Stimulation and inhibition of angiogenesis by placental proliferin and proliferin-related protein. *Science* **1994**, *266*, 1581–1584. [CrossRef] [PubMed]
67. Toft, D.J.; Rosenberg, S.B.; Bergers, G.; Volpert, O.; Linzer, D.I. Reactivation of proliferin gene expression is associated with increased angiogenesis in a cell culture model of fibrosarcoma tumor progression. *Proc. Natl. Acad. Sci. USA* **2001**, *98*, 13055–13059. [CrossRef] [PubMed]
68. Tabassum, D.P.; Polyak, K. Tumorigenesis: It takes a village. *Nat. Rev. Cancer* **2015**, *15*, 473–483. [CrossRef] [PubMed]
69. Cayrefourcq, L.; Mazard, T.; Joosse, S.; Solassol, J.; Ramos, J.; Assenat, E.; Schumacher, U.; Costes, V.; Maudelonde, T.; Pantel, K.; et al. Establishment and characterization of a cell line from human circulating colon cancer cells. *Cancer Res.* **2015**, *75*, 892–901. [CrossRef]
70. Gao, D.; Vela, I.; Sboner, A.; Iaquinta, P.J.; Karthaus, W.R.; Gopalan, A.; Dowling, C.; Wanjala, J.N.; Undvall, E.A.; Arora, V.K.; et al. Organoid cultures derived from patients with advanced prostate cancer. *Cell* **2014**, *159*, 176–187. [CrossRef]
71. Hamilton, G.; Burghuber, O.; Zeillinger, R. Circulating tumor cells in small cell lung cancer: Ex vivo expansion. *Lung* **2015**, *193*, 451–452. [CrossRef]
72. Yu, M.; Bardia, A.; Aceto, N.; Bersani, F.; Madden, M.W.; Donaldson, M.C.; Desai, R.; Zhu, H.; Comaills, V.; Zheng, Z.; et al. Cancer therapy. Ex vivo culture of circulating breast tumor cells for individualized testing of drug susceptibility. *Science* **2014**, *345*, 216–220. [CrossRef] [PubMed]
73. Bobek, V.; Kacprzak, G.; Rzechonek, A.; Kolostova, K. Detection and cultivation of circulating tumor cells in malignant pleural mesothelioma. *Anticancer Res.* **2014**, *34*, 2565–2569. [PubMed]
74. Bobek, V.; Matkowski, R.; Gurlich, R.; Grabowski, K.; Szelachowska, J.; Lischke, R.; Schutzner, J.; Harustiak, T.; Pazdro, A.; Rzechonek, A.; et al. Cultivation of circulating tumor cells in esophageal cancer. *Folia Histochem. Cytobiol.* **2014**, *52*, 171–177. [CrossRef] [PubMed]
75. Cegan, M.; Kolostova, K.; Matkowski, R.; Broul, M.; Schraml, J.; Fiutowski, M.; Bobek, V. In vitro culturing of viable circulating tumor cells of urinary bladder cancer. *Int. J. Clin. Exp. Pathol.* **2014**, *7*, 7164–7171. [PubMed]
76. Kang, J.H.; Krause, S.; Tobin, H.; Mammoto, A.; Kanapathipillai, M.; Ingber, D.E. A combined micromagnetic-microfluidic device for rapid capture and culture of rare circulating tumor cells. *Lab Chip* **2012**, *12*, 2175–2181. [CrossRef] [PubMed]
77. Kolostova, K.; Matkowski, R.; Gurlich, R.; Grabowski, K.; Soter, K.; Lischke, R.; Schutzner, J.; Bobek, V. Detection and cultivation of circulating tumor cells in gastric cancer. *Cytotechnology* **2016**, *68*, 1095–1102. [CrossRef]

78. Kolostova, K.; Zhang, Y.; Hoffman, R.M.; Bobek, V. In vitro culture and characterization of human lung cancer circulating tumor cells isolated by size exclusion from an orthotopic nude-mouse model expressing fluorescent protein. *J. Fluoresc.* **2014**, *24*, 1531–1536. [CrossRef]
79. Kulasinghe, A.; Perry, C.; Warkiani, M.E.; Blick, T.; Davies, A.; O'Byrne, K.; Thompson, E.W.; Nelson, C.C.; Vela, I.; Punyadeera, C. Short term ex-vivo expansion of circulating head and neck tumour cells. *Oncotarget* **2016**, *7*, 60101–60109. [CrossRef]
80. Ting, D.T.; Wittner, B.S.; Ligorio, M.; Vincent Jordan, N.; Shah, A.M.; Miyamoto, D.T.; Aceto, N.; Bersani, F.; Brannigan, B.W.; Xega, K.; et al. Single-cell RNA sequencing identifies extracellular matrix gene expression by pancreatic circulating tumor cells. *Cell Rep.* **2014**, *8*, 1905–1918. [CrossRef]
81. Zhang, Z.; Shiratsuchi, H.; Lin, J.; Chen, G.; Reddy, R.M.; Azizi, E.; Fouladdel, S.; Chang, A.C.; Lin, L.; Jiang, H.; et al. Expansion of CTCs from early stage lung cancer patients using a microfluidic co-culture model. *Oncotarget* **2014**, *5*, 12383–12397. [CrossRef]
82. Zhang, Z.; Shiratsuchi, H.; Palanisamy, N.; Nagrath, S.; Ramnath, N. Expanded Circulating Tumor Cells from a Patient with ALK-Positive Lung Cancer Present with EML4-ALK Rearrangement Along with Resistance Mutation and Enable Drug Sensitivity Testing: A Case Study. *J. Thorac. Oncol.* **2017**, *12*, 397–402. [CrossRef] [PubMed]
83. Nuhn, P.; De Bono, J.S.; Fizazi, K.; Freedland, S.J.; Grilli, M.; Kantoff, P.W.; Sonpavde, G.; Sternberg, C.N.; Yegnasubramanian, S.; Antonarakis, E.S. Update on Systemic Prostate Cancer Therapies: Management of Metastatic Castration-resistant Prostate Cancer in the Era of Precision Oncology. *Eur. Urol.* **2019**, *75*, 88–99. [CrossRef] [PubMed]
84. Wang, L.; Pan, D.; Yan, Q.; Song, Y. Activation mechanisms of alphaVbeta3 integrin by binding to fibronectin: A computational study. *Protein Sci.* **2017**, *26*, 1124–1137. [CrossRef] [PubMed]
85. Schaffner, F.; Ray, A.M.; Dontenwill, M. Integrin alpha5beta1, the Fibronectin Receptor, as a Pertinent Therapeutic Target in Solid Tumors. *Cancers* **2013**, *5*, 27–47. [CrossRef]
86. Folkman, J. Antiangiogenesis in cancer therapy—Endostatin and its mechanisms of action. *Exp. Cell Res.* **2006**, *312*, 594–607. [CrossRef]
87. Wang, S.; Lu, X.-A.; Liu, P.; Fu, Y.; Jia, L.; Zhan, S.; Luo, Y. Endostatin Has ATPase Activity, Which Mediates Its Antiangiogenic and Antitumor Activities. *Mol. Cancer Ther.* **2015**, *14*, 1192. [CrossRef]
88. Sato, Y. Endostatin as a Biomarker of Basement Membrane Degradation. *J. Atheroscler. Thromb.* **2017**, *24*, 1014–1015. [CrossRef]
89. Walia, A.; Yang, J.F.; Huang, Y.-H.; Rosenblatt, M.I.; Chang, J.-H.; Azar, D.T. Endostatin's Emerging Roles in Angiogenesis, Lymphangiogenesis, Disease, and Clinical Applications. *Biochim. Biophys. Acta* **2015**, *1850*, 2422–2438. [CrossRef]
90. Ocana, A.; Nieto-Jimenez, C.; Pandiella, A.; Templeton, A.J. Neutrophils in cancer: Prognostic role and therapeutic strategies. *Mol. Cancer* **2017**, *16*, 137. [CrossRef]
91. Coffelt, S.B.; Wellenstein, M.D.; De Visser, K.E. Neutrophils in cancer: Neutral no more. *Nat. Rev. Cancer* **2016**, *16*, 431–446. [CrossRef]
92. Granot, Z.; Henke, E.; Comen, E.A.; King, T.A.; Norton, L.; Benezra, R. Tumor entrained neutrophils inhibit seeding in the premetastatic lung. *Cancer Cell* **2011**, *20*, 300–314. [CrossRef] [PubMed]
93. Speetjens, F.M.; Kuppen, P.J.; Sandel, M.H.; Menon, A.G.; Burg, D.; Van de Velde, C.J.; Tollenaar, R.A.; De Bont, H.J.; Nagelkerke, J.F. Disrupted expression of CXCL5 in colorectal cancer is associated with rapid tumor formation in rats and poor prognosis in patients. *Clin. Cancer Res.* **2008**, *14*, 2276–2284. [CrossRef] [PubMed]
94. Blaisdell, A.; Crequer, A.; Columbus, D.; Daikoku, T.; Mittal, K.; Dey, S.K.; Erlebacher, A. Neutrophils Oppose Uterine Epithelial Carcinogenesis via Debridement of Hypoxic Tumor Cells. *Cancer Cell* **2015**, *28*, 785–799. [CrossRef] [PubMed]
95. Liao, D.; Johnson, R.S. Hypoxia: A key regulator of angiogenesis in cancer. *Cancer Metastasis Rev.* **2007**, *26*, 281–290. [CrossRef] [PubMed]
96. Faye, C.; Moreau, C.; Chautard, E.; Jetne, R.; Fukai, N.; Ruggiero, F.; Humphries, M.J.; Olsen, B.R.; Ricard-Blum, S. Molecular interplay between endostatin, integrins, and heparan sulfate. *J. Biol. Chem.* **2009**, *284*, 22029–22040. [CrossRef]

97. Rehn, M.; Veikkola, T.; Kukk-Valdre, E.; Nakamura, H.; Ilmonen, M.; Lombardo, C.; Pihlajaniemi, T.; Alitalo, K.; Vuori, K. Interaction of endostatin with integrins implicated in angiogenesis. *Proc. Natl. Acad. Sci. USA* **2001**, *98*, 1024–1029. [CrossRef]
98. Sudhakar, A.; Sugimoto, H.; Yang, C.; Lively, J.; Zeisberg, M.; Kalluri, R. Human tumstatin and human endostatin exhibit distinct antiangiogenic activities mediated by alpha v beta 3 and alpha 5 beta 1 integrins. *Proc. Natl. Acad. Sci. USA* **2003**, *100*, 4766–4771. [CrossRef]
99. Wickstrom, S.A.; Alitalo, K.; Keski-Oja, J. Endostatin associates with integrin alpha5beta1 and caveolin-1, and activates Src via a tyrosyl phosphatase-dependent pathway in human endothelial cells. *Cancer Res.* **2002**, *62*, 5580–5589.
100. Yokoyama, Y.; Ramakrishnan, S. Binding of endostatin to human ovarian cancer cells inhibits cell attachment. *Int. J. Cancer* **2007**, *121*, 2402–2409. [CrossRef]
101. Gast, C.E.; Silk, A.D.; Zarour, L.; Riegler, L.; Burkhart, J.G.; Gustafson, K.T.; Parappilly, M.S.; Roh-Johnson, M.; Goodman, J.R.; Olson, B.; et al. Cell fusion potentiates tumor heterogeneity and reveals circulating hybrid cells that correlate with stage and survival. *Sci. Adv.* **2018**, *4*, eaat7828. [CrossRef]
102. Laberge, G.S.; Duvall, E.; Haedicke, K.; Pawelek, J. Leukocyte-cancer cell fusion—Genesis of a deadly journey. *Cells* **2019**, *8*, 170. [CrossRef] [PubMed]
103. Sallusto, F.; Baggiolini, M. Chemokines and leukocyte traffic. *Nat. Immunol.* **2008**, *9*, 949–952. [CrossRef] [PubMed]
104. Chakraborty, A.K.; Funasaka, Y.; Ichihashi, M.; Pawelek, J.M. Upregulation of alpha and beta integrin subunits in metastatic macrophage-melanoma fusion hybrids. *Melanoma Res.* **2009**, *19*, 343–349. [CrossRef] [PubMed]
105. Knowles, L.M.; Gurski, L.A.; Engel, C.; Gnarra, J.R.; Maranchie, J.K.; Pilch, J. Integrin alphavbeta3 and fibronectin upregulate Slug in cancer cells to promote clot invasion and metastasis. *Cancer Res.* **2013**, *73*, 6175–6184. [CrossRef] [PubMed]

Permissions

All chapters in this book were first published by MDPI; hereby published with permission under the Creative Commons Attribution License or equivalent. Every chapter published in this book has been scrutinized by our experts. Their significance has been extensively debated. The topics covered herein carry significant findings which will fuel the growth of the discipline. They may even be implemented as practical applications or may be referred to as a beginning point for another development.

The contributors of this book come from diverse backgrounds, making this book a truly international effort. This book will bring forth new frontiers with its revolutionizing research information and detailed analysis of the nascent developments around the world.

We would like to thank all the contributing authors for lending their expertise to make the book truly unique. They have played a crucial role in the development of this book. Without their invaluable contributions this book wouldn't have been possible. They have made vital efforts to compile up to date information on the varied aspects of this subject to make this book a valuable addition to the collection of many professionals and students.

This book was conceptualized with the vision of imparting up-to-date information and advanced data in this field. To ensure the same, a matchless editorial board was set up. Every individual on the board went through rigorous rounds of assessment to prove their worth. After which they invested a large part of their time researching and compiling the most relevant data for our readers.

The editorial board has been involved in producing this book since its inception. They have spent rigorous hours researching and exploring the diverse topics which have resulted in the successful publishing of this book. They have passed on their knowledge of decades through this book. To expedite this challenging task, the publisher supported the team at every step. A small team of assistant editors was also appointed to further simplify the editing procedure and attain best results for the readers.

Apart from the editorial board, the designing team has also invested a significant amount of their time in understanding the subject and creating the most relevant covers. They scrutinized every image to scout for the most suitable representation of the subject and create an appropriate cover for the book.

The publishing team has been an ardent support to the editorial, designing and production team. Their endless efforts to recruit the best for this project, has resulted in the accomplishment of this book. They are a veteran in the field of academics and their pool of knowledge is as vast as their experience in printing. Their expertise and guidance has proved useful at every step. Their uncompromising quality standards have made this book an exceptional effort. Their encouragement from time to time has been an inspiration for everyone.

The publisher and the editorial board hope that this book will prove to be a valuable piece of knowledge for researchers, students, practitioners and scholars across the globe.

List of Contributors

Su Bin Lim
NUS Graduate School for Integrative Sciences & Engineering, National University of Singapore, Singapore 117456, Singapore
Department of Biomedical Engineering, National University of Singapore, Singapore 117583, Singapore

Chwee Teck Lim
NUS Graduate School for Integrative Sciences & Engineering, National University of Singapore, Singapore 117456, Singapore
Department of Biomedical Engineering, National University of Singapore, Singapore 117583, Singapore
Mechanobiology Institute, National University of Singapore, Singapore 117411, Singapore
Institute for Health Innovation and Technology (iHealthtech), National University of Singapore, Singapore 117599, Singapore

Wan-Teck Lim
Division of Medical Oncology, National Cancer Centre Singapore, Singapore 169610, Singapore
Office of Academic and Clinical Development, Duke-NUS Medical School, Singapore 169857, Singapore
IMCB NCC MPI Singapore Oncogenome Laboratory, Institute of Molecular and Cell Biology (IMCB), Agency for Science, Technology and Research (A*STAR), Singapore 138673, Singapore

Verena Martini and Birte Kulemann
Department of General and Visceral Surgery, Medical Center-University of Freiburg, Faculty of Medicine, University of Freiburg, 79106 Freiburg, Germany

Sylvia Timme-Bronsert
Institute for Surgical Pathology, Medical Center-University of Freiburg, Faculty of Medicine, University of Freiburg, 79106 Freiburg, Germany

Stefan Fichtner-Feigl and Jens Hoeppner
Department of General and Visceral Surgery, Medical Center-University of Freiburg, Faculty of Medicine, University of Freiburg, 79106 Freiburg, Germany
Comprehensive Cancer Center Freiburg-CCCF, Medical Center-University of Freiburg, 79106 Freiburg, Germany

Marta Popeda, Natalia Bednarz-Knoll, Anna Jurek, Aleksandra Markiewicz and Anna J. Zaczek
Laboratory of Translational Oncology, Intercollegiate Faculty of Biotechnology, Medical University of Gdansk, 80-211 Gdansk, Poland

Tomasz Stokowy
Department of Clinical Science, University of Bergen, 5021 Bergen, Norway

Magdalena Niemira, Agnieszka Bielska and Adam Kretowski
Clinical Research Centre, Medical University of Bialystok, 15-276 Bialystok, Poland

Leszek Kalinowski
Department of Medical Laboratory Diagnostics and Central Bank of Frozen Tissues & Genetic Specimens, Medical University of Gdansk, 80-211 Gdansk, Poland
Biobanking and Biomolecular Resources Research Infrastructure Poland (BBMRI.PL), 80-211 Gdansk, Poland

Jolanta Szade
Department of Pathomorphology, Medical University of Gdansk, 80-211 Gdansk, Poland

Karin Larsson, Jan-Erik Damber and Karin Welén
Department of Urology, Sahlgrenska Cancer Center, Institute of Clinical Sciences, Sahlgrenska Academy, University of Gothenburg, 405 30 Gothenburg, Sweden

Andreas Josefsson
Department of Urology, Sahlgrenska Cancer Center, Institute of Clinical Sciences, Sahlgrenska Academy, University of Gothenburg, 405 30 Gothenburg, Sweden
Department of Surgical and Perioperative Sciences, Urology & Andrology, Umeå University, 901 87 Umeå, Sweden
Wallenberg Center for Molecular Medicine, Umeå University, 901 87 Umeå, Sweden

Eva Freyhult
Department of Medical Sciences, National Bioinformatics Infrastructure Sweden, Science for Life Laboratory, Uppsala University, 751 85 Uppsala, Sweden

Afroditi Nanou, Leonie L. Zeune and Leon W.M.M. Terstappen
Department of Medical Cell BioPhysics, Faculty of Science and Technology, University of Twente, 7522 NH Enschede, The Netherlands

Bianca C. Troncarelli Flores, Emne Ali Abdallah and Alexcia Camila Braun
International Research Center, A.C. Camargo Cancer Center, São Paulo 01508-010, Brazil

Celso A.L. Mello and Virgilio Souza e Silva
Department of Medical Oncology, A.C. Camargo Cancer Center, São Paulo 01509-900, Brazil

Maria Letícia Gobo Silva
Department of Radiotherapy, A.C. Camargo Cancer Center, São Paulo 01509-900, Brazil

Gustavo Gomes Mendes
Department of Radiology, A.C. Camargo Cancer Center, São Paulo 01509-900, Brazil

Samuel Aguiar Junior
Department of Pelvic Surgery, A.C. Camargo Cancer Center, São Paulo 01509-900, Brazil
National Institute for Science and Technology in Oncogenomics and Therapeutic Innovation, São Paulo 01509-900, Brazil

Ludmilla Thomé Domingos Chinen
International Research Center, A.C. Camargo Cancer Center, São Paulo 01508-010, Brazil
National Institute for Science and Technology in Oncogenomics and Therapeutic Innovation, São Paulo 01509-900, Brazil

Elisabetta Rossi
Department of Surgery, Oncology and Gastroenterology, Oncology Section, University of Padova, 35124 Padova, Italy
Veneto Institute of Oncology IOV-IRCCS, 35128 Padua, Italy

Francesco Fabbri
Istituto Scientifico Romagnolo per lo Studio e la Cura dei Tumori (IRST) IRCCS, 47014 Meldola, Italy

Kumuditha M.Weerakoon-Ratnayake, Malgorzata A. Witek, Shalee Mog and Kavya Dathathreya
Department of Chemistry, The University of Kansas, Lawrence, KS 66047, USA
Center of BioModular Multiscale Systems for Precision Medicine, Lawrence, KS 66045, USA

Swarnagowri Vaidyanathan
Center of BioModular Multiscale Systems for Precision Medicine, Lawrence, KS 66045, USA
Bioengineering, The University of Kansas, Lawrence, KS 66045, USA

Nicholas Larkey and Mengjia Hu
Center of BioModular Multiscale Systems for Precision Medicine, Lawrence, KS 66045, USA
Department of Pathology & Laboratory Medicine, University of Kansas Medical Center, Kansas City, KS 66160, USA

Jilsha Jose
Center of BioModular Multiscale Systems for Precision Medicine, Lawrence, KS 66045, USA

Keith August
Children's Mercy Hospital, Kansas City, MO 64108, USA

Andrew K. Godwin
Department of Pathology & Laboratory Medicine, University of Kansas Medical Center, Kansas City, KS 66160, USA

Mateusz L. Hupert
Biofluidica Inc., BioFluidica Research Laboratory, Lawrence, KS 66047, USA

Steven A. Soper
Department of Chemistry, The University of Kansas, Lawrence, KS 66047, USA
Center of BioModular Multiscale Systems for Precision Medicine, Lawrence, KS 66045, USA
Department of Pathology & Laboratory Medicine, University of Kansas Medical Center, Kansas City, KS 66160, USA
Biofluidica Inc., BioFluidica Research Laboratory, Lawrence, KS 66047, USA
Department of Mechanical Engineering, The University of Kansas, Lawrence, KS 66045, USA

Laure Cayrefourcq and Françoise Garima
Laboratory of Rare Human Circulating Cells (LCCRH), University Medical Centre of Montpellier, UPRES EA2415, 34093 Montpellier, France

Catherine Alix-Panabières
Laboratory of Rare Human Circulating Cells (LCCRH), University Medical Centre of Montpellier, UPRES EA2415, 34093 Montpellier, France
Laboratory of Rare Human Circulating Cells (LCCRH), University Medical Centre of Montpellier–UM EA2415, 34295 Montpellier, France

Aurélie De Roeck, Caroline Garcia, Pierre-Emmanuel Stoebner, Fanny Fichel and Laurent Meunier
Department of Dermatology, Nîmes University Hospital, University of Montpellier, 30029 Nîmes, France

Françoise Perriard and Jean-Pierre Daures
UPRES EA2415, University Institute of Clinical Research (IURC), Montpellier University, 34093 Montpellier, France

Olga A. Sindeeva and Roman A. Verkhovskii
Laboratory of Biomedical Photoacoustics, Saratov State University, 83 Astrakhanskaya St, 410012 Saratov, Russia

Mustafa Sarimollaoglu
Arkansas Nanomedicine Center & Winthrop P. Rockefeller Cancer Institute, University of Arkansas for Medical Sciences, Little Rock, AR 72205, USA

Vladimir P. Zharov
Laboratory of Biomedical Photoacoustics, Saratov State University, 83 Astrakhanskaya St, 410012 Saratov, Russia
Arkansas Nanomedicine Center & Winthrop P. Rockefeller Cancer Institute, University of Arkansas for Medical Sciences, Little Rock, AR 72205, USA

Galina A. Afanaseva, Alexander S. Fedonnikov, Evgeny Yu. Osintsev and Elena N. Kurochkina
Laboratory of Biomedical Photoacoustics, Saratov State University, 83 Astrakhanskaya St, 410012 Saratov, Russia
Saratov State Medical University, 112 Bolshaya Kazachia St., 410012 Saratov, Russia

Dmitry A. Gorin
Laboratory of Biophotonics, Skolkovo Institute of Science and Technology, 3 Nobelya Str., 121205 Moscow, Russia

Sergey M. Deyev
Shemyakin-Ovchinnikov Institute of Bioorganic Chemistry, Russian Academy of Sciences, Miklukho-Maklaya St., 16/10, Moscow 117997, Russia

Ekaterina I. Galanzha
Laboratory of Biomedical Photoacoustics, Saratov State University, 83 Astrakhanskaya St, 410012 Saratov, Russia
Laboratory of Lymphatic Research, Diagnosis and Therapy (LDT), University of Arkansas for Medical Sciences, Little Rock, AR 72205, USA

Mohammed Nimir, Yafeng Ma, Thomas Opperman and Francis Young
Centre for Circulating Tumour Cell Diagnostics and Research, Ingham Institute for Applied Medical Research, 1 Campbell St, Liverpool, NSW 2170, Australia
South Western Clinical School, University of New South Wales, Goulburn St, Liverpool, NSW 2170, Australia

Tanzila Khan and Sarah A. Jeffreys
Centre for Circulating Tumour Cell Diagnostics and Research, Ingham Institute for Applied Medical Research, 1 Campbell St, Liverpool, NSW 2170, Australia
School of Medicine, Western Sydney University, Campbelltown, NSW 2560, Australia

Pei Ding and Adam Cooper
Centre for Circulating Tumour Cell Diagnostics and Research, Ingham Institute for Applied Medical Research, 1 Campbell St, Liverpool, NSW 2170, Australia
School of Medicine, Western Sydney University, Campbelltown, NSW 2560, Australia
Liverpool Hospital, Elizabeth St & Goulburn St, Liverpool, NSW 2170, Australia

Therese M. Becker
Centre for Circulating Tumour Cell Diagnostics and Research, Ingham Institute for Applied Medical Research, 1 Campbell St, Liverpool, NSW 2170, Australia
South Western Clinical School, University of New South Wales, Goulburn St, Liverpool, NSW 2170, Australia
School of Medicine, Western Sydney University, Campbelltown, NSW 2560, Australia

Wei Chua and Bavanthi Balakrishnar
Liverpool Hospital, Elizabeth St & Goulburn St, Liverpool, NSW 2170, Australia

Paul De Souza
Centre for Circulating Tumour Cell Diagnostics and Research, Ingham Institute for Applied Medical Research, 1 Campbell St, Liverpool, NSW 2170, Australia
School of Medicine, Western Sydney University, Campbelltown, NSW 2560, Australia
Liverpool Hospital, Elizabeth St & Goulburn St, Liverpool, NSW 2170, Australia
School of Medicine, University of Wollongong, Wollongong, NSW 2522, Australia

Fabienne Schochter, Thomas W. P. Friedl, Amelie deGregorio, Sabrina Krause, Jens Huober, Brigitte Rack and Wolfgang Janni
Department of Obstetrics and Gynecology, University Hospital Ulm, University of Ulm, 89075 Ulm, Germany

Claudia Hille, Tobias M. Gorges, Sabine Riethdorf and Klaus Pantel
Department of Tumor Biology, University Medical Center Hamburg-Eppendorf, 20246 Hamburg, Germany

Martine Mazel
Laboratory of Rare Human Circulating Cells (LCCRH), University Medical Centre of Montpellier–UM EA2415, 34295 Montpellier, France

Thomas Steuber
Martini Clinic, University Medical Center Hamburg-Eppendorf, 20246 Hamburg, Germany

Gunhild Von Amsberg
Department of Hematology and Oncology, University Medical Center Hamburg-Eppendorf, 20246 Hamburg, Germany

Frank König
ATURO, Urology Practice, 14197 Berlin, Germany

Sven Peine
Department of Transfusion Medicine, University Medical Center Hamburg-Eppendorf, 20246 Hamburg, Germany

Lucile Broncy
INSERM Unit 1151, Faculté de Médecine, Université Paris Descartes, 75014 Paris, France

Patrizia Paterlini-Bréchot
INSERM Unit 1151, Faculté de Médecine, Université Paris Descartes, 75014 Paris, France
Laboratoire de Biochimie A, Hôpital Necker-Enfants Malades, 75015 Paris, France

Jeannette Huaman and Michelle Naidoo
Department of Biological Sciences, Hunter College of The City University of New York, New York, NY 10065, USA
Department of Biology, The Graduate Center of The City University of New York, New York, NY 10016, USA

Xingxing Zang
Departments of Microbiology and Immunology, and Medicine (Oncology), Albert Einstein College of Medicine, Bronx, NY 10461, USA

Olorunseun O. Ogunwobi
Department of Biological Sciences, Hunter College of The City University of New York, New York, NY 10065, USA
Department of Biology, The Graduate Center of The City University of New York, New York, NY 10016, USA
Department of Medicine, Weill Cornell Medicine, Cornell University, New York, NY 10065, USA

Index

Printed in the USA
CPSIA information can be obtained
at www.ICGtesting.com
JSHW052339110923
48263JS00003BA/3